Understanding Photovoltaics

2024 Edition

Design & Installation of Residential Solar PV Systems

Jay Warmke

BRS Press

www.solarPVtraining.com

Published by BRS Press
ISBN: 978-1-957113-05-0
Copyright © 2024 Jay and Annie Warmke
9th Edition

Tables & Figures (unless otherwise noted) © Jay and Annie Warmke
Layout and design by BRS Press
Edited by Deborah Lauer
Cover Design by Ryan Evans
Thanks to Eduardo Sandoval, Elyse Perruchon and Laura Darlak, for being great students as well as serving as cover models.

Disclaimer: The information contained in this text is for general informational purposes only. The information is provided by Blue Rock Station LLC and while we endeavor to keep the information up to date and correct, we make no representations or warranties of any kind, express or implied, about the completeness, accuracy, reliability, suitability or availability with respect to the text or the information, products, services, or related graphics contained within for any purpose. Any reliance you place on such information is therefore strictly at your own risk. Blue Rock Station LLC assumes no responsibility for actual or consequential damages incurred as a result of any user's reliance on the information contained within this text.

This text is intended as a basic introduction to the concepts associated with the design and installation of solar photovoltaic systems. The content should not be used for construction, fabrication, or any other application without verification by an engineer or other appropriate design/construction professional that it meets any and all codes and engineering and other requirements applicable to such application.

Contact us at:
Blue Rock Station LLC
1190 Virginia Ridge Road
Philo, OH 43771
740-674-4300
Annie@bluerockstation.com
www.solarPVtraining.com

Table of Contents

Chapter 1: Introduction to Photovoltaics 1
    Recent Changes in the PV Industry 1
    Growth by Country 2
    Declining PV Module Prices 3
    Government Incentives 5
        State/Utility Cash Incentives 6
        Federal Tax Credits 6
        Renewable Energy Certificates (RECs) 6
        Tariffs on Imported Solar Panels 7
    Connecting to the Grid 8
        Interconnection 8
        Net-Metering 8
        Grid-Tied Growing Pains 9
        Technical Concerns of Utilities 10
    History of Photovoltaics 11
    PV Certification Programs 13
        North American Board of Certified Energy Practitioners (NABCEP) 13
        Electronics Technicians Association, International 14
    Electrical Codes and Standards 14
        Fire Codes 15
        Other Electrical Codes and Standards 16
        Canadian Electrical Code 16
        IEC 60364 17
        BS 7671 18
        IEEE 1547 18
    Approved and Tested Products 19
        UL Marks 19
        Electrical Testing Labs/Intertek (ETL) 19
        Canadian Standards Association (CSA) 19
        CE Mark 20
    Safety Regulations 20
        The Occupational Safety and Health Administration (OSHA) 20
        Canadian Centre for Occupational Health and Safety (CCOHS) 21
        European Agency for Safety and Health at Work (EU-OSHA) 21
    Chapter 1 Review Questions 22

Chapter 2: Solar Cells and Solar Modules 27
    How PV Cells Work 27
    Types of Solar Cells 29
        Number of Cells 31
        Split Cell Panels 31

- Emerging Technologies..32
  - Multi-Layer Photovoltaic Cells..32
  - Perovskite...32
  - PERC Solar Cells..33
  - Bifacial Modules..34
  - Busbar Changes...34
  - Bifacial Solar Cells...35
- Solar Cells, Solar Modules, Solar Panels and Solar Arrays.................35
- Flat Panels and Module Standards..36
- How Solar Modules are Rated...37
  - STC versus PTC versus NOCT versus NOMT.........................38
- Panel IV (current-volt) Curves..39
- Factors that Impact Solar Modules...40
  - Physical Size...40
  - Reflectivity..41
  - Size and Placement of the Electrical Contacts......................41
  - Temperature...41
  - Shading...42
  - Age..43
- Blocking Diodes..43
- Typical Panel Ratings..44
  - Resistance to Hail Damage...46
- Chapter 2 Review Questions...48
- Lab Exercises : Chapter 2..54

Chapter 3: Types of Photovoltaic Systems...55
- Stand-Alone or Off-Grid..55
- Grid-Tied Systems...56
- Grid Interactive with Battery Backup...58
  - Grid-Coupled Battery Sizing..60
  - DC-Coupled 61
  - AC-Coupled 62
- Integrating Multiple Generating Sources (Hybrid Systems)..............63
- PV System Voltage Limits...64
  - Battery Bank Voltage Limits..64
  - Advantages of Higher Voltage...65
- Determining the System Load..65
- Load Analysis Spreadsheet...66
  - Energy Use after Energy Efficiency Measures.......................67
- Maximum Power Draw versus Daily Load..67
  - Calculating Unusual Loads..69
- Chapter 3 Review Questions...72
- Lab Exercises : Chapter 3..76

Chapter 4: Basic Electrical Concepts..77
- What is Electricity?...77
- How Electricity is Measured?..78
- Electrical Safety..79

The Difference between Watts, Watt-Hours and Watts/Hour.................80
What is the Grid?.................80
    Residential Voltages.................82
    Single Phase Power.................82
    Three Phase Power.................84
Who Owns the Grid?.................85
    State Regulation.................86
Electrical Circuits.................86
    Connecting in Series and Parallel.................87
Direct and Alternating Current.................89
Resistance.................90
    Minimizing the Impact of Resistance on a PV System.................91
    Voltage Drop Equation.................92
    Measuring Resistance as a Continuity Test.................94
Voltage Variations.................95
Bonding and Grounding.................96
Chapter 4 Review Questions.................98
Lab Exercises : Chapter 4 103

Chapter 5: Parts of the PV System.................105

Solar Panels.................105
Panel Junction Box.................107
Solar Panel Connectors.................108
Combiner Box.................109
Junction Box 110
Lightning & Surge Protection.................110
Wiring the System.................111
    Temperature Ratings of Conductors and Terminations.................112
    Effect of Temperature on Wire 113
    Wires in Conduit.................113
    Types of Wire.................115
    Color Coding of Wires.................116
DC Disconnect.................118
    Service Disconnect.................119
Overcurrent Protection.................120
Charge Controller.................121
    MPPT Charge Controllers.................122
    Battery Management Systems.................124
    Charging Lithium Ion Batteries.................124
Inverters.................125
    Stand-Alone Inverters.................126
    Inverter Waveforms.................127
    Grid-Tied Inverters.................128
    Rapid Shutdown Initiator.................130
    Inverter Standards.................131
    Smart Inverters.................131
    Selecting a Grid-Tied Inverter.................132
    Bimodal Inverters.................133

- Module Level Power Electronics (MLPE)..133
  - Micro Inverters..134
  - Power Optimizers..135
- Solar Power Generation Monitoring..136
- AC Disconnect...137
  - Current Transformers..137
- Electrical Service Panel..138
  - Smart Electrical Panels..138
- Bi-Directional Meter...139
- Some Tools of the Trade...140
  - Ammeter..140
  - Crimping Tool...140
  - Wire Strippers..141
  - Other Typical Installation Tools...141
- Chapter 5 Review Questions...142
- Lab Exercises : Chapter 5 150

Chapter 6: Conducting a Site Survey..151
- General Site Information & System Goals..151
  - Available Sunlight..151
  - Solar Panel Orientation...152
  - Angle of the Array..155
  - Measuring the Angle of the Roof...158
  - Sun Charts..159
- Locating the System on the Site...159
  - Mounting Options..160
- Shading Issues..164
  - Solar Pathfinder...165
  - Remote Site Assessments...166
  - Total Solar Resource..168
- Other Issues Impacting the Location of the Array..................................170
- Determine the Location of the Balance of Systems (BOS)....................170
  - Enclosures...172
  - Battery Location...172
- Existing Electrical Equipment...173
  - The Service Panel...174
  - Load Side Connection...174
  - Supply Side Connection..175
  - Conduit Pathways and Existing Grounds...176
  - Generators..176
  - Site Hazard Assessment..177
- Economics of Solar...178
- System Type Economic Comparison...178
  - Grid-Tied with String Inverter Configuration....................................179
  - Grid-Tied with Micro Inverters Configuration..................................180
  - Grid-Tied with Power Optimizer Configuration...............................180
  - DC Coupled Configuration with 10 kWh Battery.............................180
  - AC Coupled Configuration with 10 kWh Battery.............................181

Understanding Photovoltaics: Design & Installation of Residential Solar PV Systems

  Stand-Alone Configuration...181
  PV Systems Compared to Utility Rates..182
 Chapter 6 Review Questions 184
 Lab Exercises : Chapter 6 190

Chapter 7: Design/Install the Array and Inverter..............................191
 Determining the Size of the Array...191
  Future-Proofing Designs..192
  Array Size Example..193
  Selecting a Solar Module...196
 Calculating String Lengths...199
  Determine the Maximum Number of Panels in each String.............199
  Determine the Minimum Number of Panels in each String..............201
  Determine how many Strings are Required..................................203
  String Calculations for a Micro Inverter Systems.........................204
  String Calculations for Power Optimizer Systems.......................205
 Selecting the Inverter..206
  Matching Inverter Output with Panel Output.............................207
  Inverter Specifications..209
  Sizing an AC Coupled Inverter...211
  Grounded or Ungrounded Inverters..213
  Functionally Versus Solidly Grounded Systems...........................213
 Chapter 7 Review Questions 215
 Lab Exercises : Chapter 7 219

Chapter 8: Wiring the System..221
 The PV Wiring System..221
  Connectors..221
  Combiner Box...223
  Junction Box...225
  Array Circuits..225
 DC Wiring Circuits..227
  Sizing Overcurrent Protection..227
  Wiring the Combiner to the DC Disconnect................................229
  PV System Wiring Checklist...229
  DC Circuits in Metal Conduit...230
 Conduits and Raceways..231
  Installing Conduit..231
 DC Disconnect..238
 Wiring the Inverter Input Circuit..239
 Wiring the Inverter Output Circuit..239
  Grid-Tied Systems..239
  Micro Inverter Systems..239
 AC Disconnect..241
  Connecting to the Utility...242
  Supply Side Connections...245
 Grounding and Bonding System..248

    Bonding..248
    Ground Faults...251
  Chapter 8 Review Questions 253
  Lab Exercises : Chapter 8...257

Chapter 9: Energy Storage Systems...259

  Major Types of Storage Systems..259
    Lead Acid Deep Cycle..259
    Lithium Ion Batteries..260
  Battery Cycles..261
    Effect of Temperature on Batteries...263
    How Batteries are Rated...263
    Charging Batteries...265
    Self Discharge..267
  Battery Safety..267
    Battery Standards..269
    Battery Bank Location...269
    AC Coupled and High Voltage Battery Systems...............................271
  Electric Vehicle Bi-Directional Charging..273
    Electric Vehicle Charging Stations...274
  Chapter 9 Review Questions 276
  Lab Exercises : Chapter 9...280

Chapter 10: Mounting Systems..281

  Racking System..281
  Roof Mounted Systems..282
    Calculating Space Required for the Array..283
    Loading Issues..285
    Pull-Out and Shear Loads...286
    Setbacks..291
    Attaching Panels to Rails..293
    Torque..294
    Galvanic Corrosion..294
    Flat Roof Systems (Ballasted)...296
    Squirrel Guards..297
    Grounding and Bonding the Solar Array..297
  Ground Mounted Systems..298
    Foundation Types...298
    Site Preparation..301
  Online Design Tools for Racking Systems...303
  Chapter 10 Review Questions 304
  Lab Exercises : Chapter 10...306

Chapter 11: Designing a Stand-Alone System...307

  Designing a Stand-Alone System...307
    Stand Alone System Components...308
    Selecting a Stand-Alone Inverter...312
  Sizing a Stand-Alone System Array..314

    String Calculations for a Stand-Alone System..................................316
  Sizing a Stand-Alone Battery Bank 316
    Sample Design..316
    Installing and Testing Lead Acid Battery Banks..................................321
  Wiring a Stand-Alone System..323
    PV Source Circuit..323
    Charge Controller Output Circuit..323
    Inverter Input Circuit..325
  Chapter 11 Review Questions 328
  Lab Exercises : Chapter 11..331

Chapter 12: Job Site Safety..333
  Safety First...333
    Safety Begins Before Arriving at the Site..334
    Safety Policies and Procedures..335
    Personal Protective Equipment..335
    Lock-Out Tag-Out Procedures...342
    MSDS or SDS Documents..342
  Chapter 12 Review Questions 344
  Lab Exercises: Chapter 12...346

Chapter 13: Paperwork..349
  Project Documentation..349
    Permits...349
    Utility..350
    Authority Having Jurisdiction..351
    Application Package...351
    The Permitting Process..353
    Labels and Warning Signs..355
    Commissioning Forms..358
    Operation and Maintenance Documentation....................................358
  Chapter 13 Review Questions 360
  Lab Exercises: Chapter 13...362

Chapter 14: Commissioning the System...363
  Commissioning the System...363
    Final Installation Checkout...364
    Visual Inspection of the System...364
    Verification of Code Compliance...366
    Electrical Verification Tests..367
    String Inverter Startup Sequence..368
    System Function Tests..369
    Verify Array Power and Energy Production..369
  Chapter 14 Review Questions 373
  Lab Exercises : Chapter 14 374

Chapter 15: System Maintenance & Troubleshooting 377
    Monitoring Performance..377
    Troubleshooting 378
        Typical System Problems..378
    System Maintenance...381
        Battery Bank Maintenance..381
    Chapter 15 Review Questions 383

Formulas..385

Acronyms..387

Figures...393

Tables..399

Index..401

Chapter 1

Introduction to Photovoltaics

Chapter Objectives:

- Become familiar with the recent changes in the PV industry.
- Understand the scope of and reasons for the dramatic decline in PV system prices.
- Explore the various incentive programs used to promote the adoption of PV and how they are changing.
- Comprehend how the interconnection of PV systems affects the electrical grid.
- Review the history of photovoltaics.
- Become familiar with PV installer certification programs.
- Introduce national electric codes and standards.
- Identify product service marks.
- Be familiar with workplace safety rules and administrative organizations.

Recent Changes in the PV Industry

The solar electric (photovoltaic or PV) industry continues to undergo remarkable changes. These changes will rapidly transform the way society looks at energy. While the global PV capacity has experienced spectacular growth over the past decade, it is anticipated to continue to grow at a remarkable rate (more than 20% per year) for the coming years, as illustrated in Figure 1-1.

Photovoltaics (derived from the Greek word phōs meaning light and the word volt, named for Italian physicist Alessandro Volta (1745-1827)) is at the heart of the energy transition - moving away from a dependency on fossil fuels towards the age of renewables.

Worldwide, by the end of 2012, there were just over 100 gigawatts (100 billion watts) of installed photovoltaic generating capacity. To put 100

PHOTOVOLTAICS

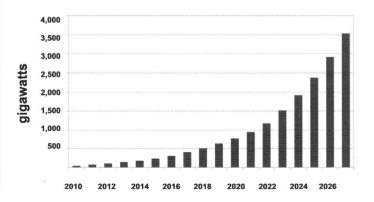

FIGURE 1-1: GROWTH OF SOLAR ELECTRIC (2010- 2027)

gigawatts into perspective, this is enough generating capacity to power approximately 10 million average American homes. Two years later (2014), this number had doubled. Two years on (2016), the world will have added another 100 gigawatts of capacity.

By 2020, every three months the world was adding more PV capacity than existed worldwide prior to 2012.

PV growth is especially remarkable considering that for all practical purposes, the commercial photovoltaic industry is only about 15 years old – just a teenager.

Growth by Country

The modern solar PV market grew out of technological innovations developed primarily in the United States. And while adoption of solar was quite modest by today's standards, the US dominated the global adoption of solar PV up until the mid 1990s. By 1996 the total US installed PV capacity amounted to about 77 megawatts.

In the late 1990s and early 2000s, a number of earthquakes in Japan pushed that nation to expand their solar capacity to restore power to a disrupted grid. Japan quickly became the major solar player, installing 1,132 megawatts by 2004.

For the next decade, aggressive renewable energy policies by the German government ensured that Germany became the focus of worldwide solar PV expansion. By 2016 the cumulative installed solar capacity in that nation exceeded 40 gigawatts (40,000 megawatts).

FIGURE 1-2: SOLAR PV CAPACITY BY COUNTRY (AS OF 2022)

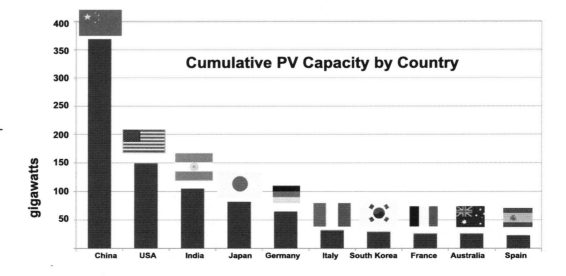

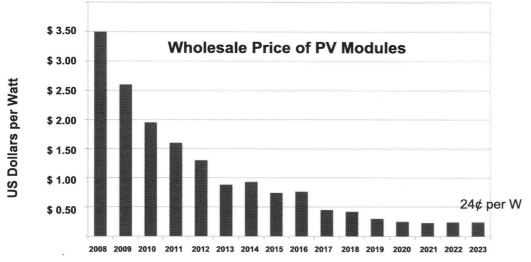

FIGURE 1-3: DECLINE IN THE WHOLESALE PRICE OF PV MODULES (2008 - 2023)

In 2015, China passed Germany as the nation that installed the most solar - and continues to this day as the dominant nation in installed PV capacity, as shown in Figure 1.2.

Declining PV Module Prices

Most decisions in industry (and life for that matter) are driven by economics.

During recent years there has been a dramatic decrease in the cost of **photovoltaic cells** and modules. From 2004 through 2008, the cost of **PV modules** remained fairly flat, averaging around $3.50 to $4.00 per watt of energy produced. This price stability has been attributed to the fact that the German and Spanish governments had embarked on an aggressive incentive program to encourage the installation of PV systems, keeping demand high for a very limited supply.

These two countries offered incentives that allowed developers to pay relatively high prices for PV modules and still make a profit. Also, due to escalating demand, there was a worldwide shortage of **polysilicon** (the main ingredient in PV panels) that constrained production.

But in September of 2008, the Spanish government abruptly abandoned their PV incentive program. At the same time, advances in the solar cell manufacturing process (the basic component used in the manufacture of solar modules), increases in the efficiency of low-cost modules, economic saving in increasing the scale of production, the moving of production facilities to low-wage locations (such as China), and an increased availability in polysilicon (production of this raw material increased 32% that year) all combined to lower the cost of solar panel production. By the end of 2009, PV manufacturers were able to drop prices from $4/W to $2/W (see Figure 1.3) and still make a profit.

(PHOTOVOLTAIC CELLS)

(PHOTOVOLTAIC MODULES)

(POLYSILICON)

These same factors have continued to force the price of PV modules down since 2009. By January, 2019 the average wholesale price of a polysilicon (Taiwan) solar module had dropped to around $0.22/W.

BALANCE OF SYSTEMS (BOS)

While the price of modules and inverters have declined rapidly, so too has the cost of the **balance of systems (BOS),** all equipment other than the panels and the inverter(s).

In a typical PV system (according to the Solar Energy Industries Association), the hardware used in a solar installation only accounts for about 37% of the total cost of installation. Of the total cost of installation (Figure 1-4), the costs can be further broken down as: 17% for the modules, 5% for the inverter, 5% for the racking, and 10% for other system components (such as wire, junction boxes, conduits, etc).

SOFT COSTS

The majority of the costs involved in the installation of a residential PV system are referred to as **soft costs.** These soft costs include (percentages according to the US Dept of Energy): permitting fees (2%), permitting and interconnection labor (2%), sales tax (5%), transaction costs (6%), installer/designer profit (9%), installer/designer overhead (9%), customer acquisition (9%), installation labor (11%), supply chain costs (10%).

NATIONAL RENEWABLE ENERGY LABORATORIES (NREL)

In addition to declining hardware costs, changes in installation procedures, innovations in racking systems, and decreasing profit margins have combined to lower the installed cost of PV systems. Actual installed costs in 2010 for residential-scale PV installations were about $7.24/W (according to **NREL - National Renewable Energy Laboratory**).

Residential systems (5 kW) installed costs fell dramatically to an average of $3.44/W by 2014 and continue to decline. By 2020 the retail fully installed cost of a residential system (before incentives) in the US averaged around $2.70/W. In 2020 with the outbreak of the worldwide pandemic, supply

FIGURE 1-4: COST BREAKDOWN OF INSTALLING RESIDENTIAL PV SYSTEM

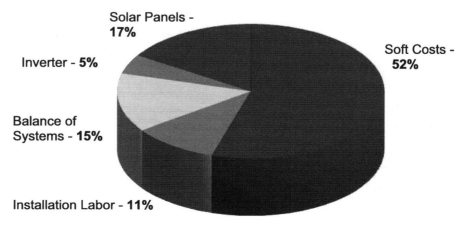

chain disruption has resulted in prices increasing slightly. As of 2023, the average price of a small residential system in the US was around $3 per watt.

The point where the cost of installing a photovoltaic system is less than or equal to the cost of the electricity (if purchased from the local electric company) produced over the life of that system is known as **grid parity**. Industry experts estimate the PV industry in the US hit grid parity in 2012 or 2013.

GRID PARITY

How quickly a PV system will recover the cost of installation in energy savings (known as the **payback period**) varies widely, depending on the cost of installation (including incentives) and the cost of electricity at that location. But generally it is reasonable to expect the system to pay for itself in 7-10 years. The average life expectancy of a solar installation is between 25-30 years.

PAYBACK PERIOD

Government Incentives

For the past decade or so, it has been government policy (national, state and local) to attempt to "kick start" the development of photovoltaic systems by offering incentives to offset the relatively high cost of installing a system.

PV systems installed on residential and/or business buildings are often referred to as **behind-the-meter** (because they are privately owned systems, on the customer's side of the electric meter) - as opposed to **front-of-the-meter** which refers to systems owned by the utility or that feed the grid on the utility side of the meter.

BEHIND-THE-METER

FRONT-OF-THE-METER

These incentives have largely been in the form of:
- a combination of cash incentives (provided through state and/or utility PV incentive programs),
- federal and/or state investment tax credits (ITC),
- federal grants in lieu of tax credits,
- quota obligations with tradable renewable energy certificates (REC),
- feed-in tariffs (FIT),
- feed-in premiums (FIP),
- net metering policies,
- loan guarantees,
- soft loans,
- bridge loans,
- investment grants,
- and accelerated depreciation of the capital investments made on the installation of solar energy systems.

Feed-in tariffs (FIT) guarantee the retail price to renewable energy generators for a period of time, funding the added cost through tax revenues or by allowing utilities to pass along the costs to consumers.

FEED-IN TARIFF (FIT)

FEED-IN PREMIUMS (FIP)

In a **feed-in premium (FIP)** scheme, renewable energy generators have to market their generated electricity directly within the electricity market and then receive an additional payment on top of the market price - generally paid by the government as an incentive or subsidy.

State/Utility Cash Incentives

Most of the state or utility incentive programs were designed to provide up-front cash to help reduce the final cost of an installed PV system. These are based either on system capacity, a percentage of installed cost, or a projection of the annual energy production of the system.

A study by the Lawrence Berkeley National Labs found that these cash incentives peaked in 2002 at around $5/W for installed systems. Since that time, the programs have declined significantly as prices for PV systems have fallen. By 2010, the average pre-tax cash incentive was $1.6/W for residential systems and $1.8/W for commercial systems. The decline continued, with cash incentive programs essentially disappearing from the solar installation pricing model.

Federal Tax Credits

INVESTMENT TAX CREDIT (ITC)

Beginning on January 1, 2006, the U.S. federal **investment tax credit (ITC)** for commercial PV systems rose from 10% to 30% of project costs, and a 30% ITC (capped at $2,000) was established for residential PV systems. A tax credit is a reduction of federal taxes owed, as opposed to a tax deduction, which reduces the portion of income upon which tax liability is calculated.

In 2009, the $2,000 cap was lifted on residential systems. The 30% income tax credit remained in effect through 2019, then declined to 26% in 2020 and further declined to 22% in 2021.

INFLATION REDUCTION ACT (IRA)

In 2022 the US Congress passed the **Inflation Reduction Act (IRA)**, which is arguably the most sweeping renewable energy bill ever passed. The IRA extended the investment tax credits for all solar installations (residential, commercial and utility-scale) at the 30% level through 2032. In 2033 it steps down to 26%, and steps down to 22% in 2034.

ENERGY COMMUNITY

The IRA provides an additional 10% tax credit for non-residential systems if the system includes a predetermined percentage of product manufactured within the United States (an attempt to jump-start domestic manufacturing). It also provides a further 10% tax credit if the system is installed in an **"energy community"** - a region that has previously relied on fossil fuels for a significant portion of the area's employment base.

Renewable Energy Certificates (RECs)

RENEWABLE ENERGY CERTIFICATE (REC)

Another form of incentive available (in some countries and states) to owners of installed PV systems can be realized by selling **renewable energy**

certificates, or RECs. Many states, in order to encourage the development of renewable energy systems, have established goals that require a certain percentage of all the energy used within the state to be generated through renewable sources. These are referred to as **renewable portfolio standards (RPS)**. Often these standards require that a portion of this renewable energy must be generated using photovoltaic systems, a provision known as a **solar set-aside or SREC**.

If utilities do not meet these goals, there are fines that are levied against them by the state. As a result, a market has developed where utilities can purchase renewable energy credits from system owners to demonstrate they have met a portion of the mandated goal.

As might be expected, in states with aggressive goals and larger fines, the price of a REC is higher than in states where there are modest or no goals in place.

In 2010, the average price of SRECs in states with an RPS in place ranged from a high of $600/MWh in New Jersey to a low of about $60/MWh in New Hampshire -with the average ranging between $300-$400/MWh.

Amortized over the expected life of a PV system (say 20 years), the net effect of an SREC incentive at this price is to lower the installed cost by $3-$4/W for the average system.

Changes in RPS legislation in the few states that still offer RECs have resulted in wide variations in prices. As of Jan 2024, SRECs were selling for as much as $425.00/MWh for solar power generated in Washington DC, but as little as $4.00/MWh for similar systems sited in Ohio.

Tariffs on Imported Solar Panels

Until 2011, the US and Germany were the major exporters of solar panels. But as China began to dominate the manufacturing of solar cells and panels, companies operating in the US and Germany found they could no longer compete with the low-cost labor and government subsidies of China. By 2018 China accounted for nearly 80% of the global solar panel production.

Beginning in 2012, the US government began imposing trade tariffs on panels manufactured in China - citing unfair trade practices. In 2018, the **tariff** program was expanded. These tariffs began at 30% and dropped by 5% each year through 2021 (2018 - 30%, 2019 - 25%, 2020 - 20%, 2021 - 15%, 2022 - 0%). In 2022, tariffs of between 14 percent and 15 percent on imported crystalline silicon solar panels were re-imposed for a period of four years.

In 2022, about 88% of US solar panel were imported, primarily from Asia.

Chapter 1: Introduction to Photovoltaics   Page 7

Connecting to the Grid

In the early days of PV systems, most were stand-alone, relying on battery banks charged by the PV panels to provide energy in remote locations where no electrical service was available or practical. In more recent times, the vast majority of PV systems in developed countries are grid-tied, allowing customers to use the solar power from their systems when it is available, and to use power from the grid when there is not enough sunlight available to generate the electricity required to meet their immediate needs.

> DISTRIBUTED ENERGY SYSTEM

How seamlessly this new **distributed energy system** is linked into a utility's grid varies dramatically from utility to utility. Some utility systems that are in need of additional generating capacity embrace these PV systems, viewing them as additional power plants that the utility has access to without having to make any investment in construction. Other utilities that are having trouble selling all the power they can currently generate often attempt to block these "competitors" from their system.

Interconnection

> INTERCONNECTION

With this in mind, most states have adopted rules that regulate the **interconnection** (the attaching of distributed energy systems to the grid) of PV systems. These regulations typically address issues such as: who qualifies, size of systems allowed, disconnection systems required, insurance required, dispute resolution, billing agreements, interconnection and engineering charges, etc.

The interconnection process for residential solar systems can take anywhere from two to four weeks on average, depending on the utility company and system size. For larger utility-scale systems, the time required to obtain interconnection approval can be four to five years (in some cases, even longer).

Net-Metering

> NET-METERING

In most states, legislatures have put in place net-metering requirements for investor-owned utilities (typically municipal systems and energy co-ops are excluded from these rules). **Net-metering** essentially requires that utilities must purchase excess energy sold back to them by the owner of the PV system.

For example, on a sunny day when the owner of a home with a PV system is at work and energy consumption at the home is low, the system may be generating more energy than is being used. This excess energy is "sold" to the utility company. In effect, the PV system becomes a small power plant, feeding energy onto the grid. As it does so, the electric meter runs backward, taking kilowatt hours off the "clock" - so to speak.

In the evening, when the homeowner returns and starts turning on lights and appliances, the amount of energy sold back to the utility will slow. As the sun sets, the energy produced by the system will drop to zero, and the homeowner will receive all their power from the grid.

In effect, the homeowner is using the utility grid as a large battery backup system. At the end of the stated period of time (a month, a year), the homeowner will pay only the net effect of their energy use (how much they used less how much they generated back to the utility).

In this case, the homeowner is effectively being paid full retail for the electricity sold to the utility. But increasingly, utilities are changing net-metering policies to pay owners of PV systems wholesale rates, but charge retail.

These net-metering agreements can get fairly complicated, addressing issues such as: who qualifies, system capacity limits, overall program capacity limits, restrictions on "rollover" credits from one period to another, who pays for the electric meter and how many are required, who owns the RECs, what insurance is required, what other fees can be assessed (access fees, distribution fees, etc), what price will be paid for the excess energy, and many more.

Grid-Tied Growing Pains

The electrical grid has little storage capacity. It operates by matching supply and demand on a second-by-second basis. When more power is needed, the utilities ramp up production.

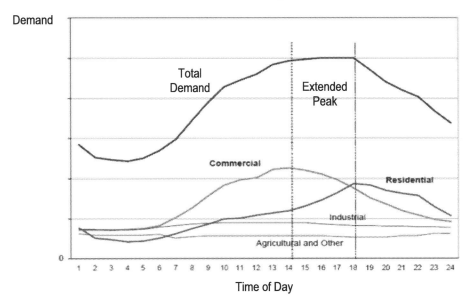

Figure 1-5: Traditional electric utility demand curve

Chapter 1: Introduction to Photovoltaics   Page 9

However, as more rooftop solar units come on line, the production of electricity is increasingly taken away from the utilities and put in the hands of the consumer.

At first blush it might appear that solar is an excellent match with electrical demand. As indicated in Figure 1-5, peak demand for electricity typically falls in the late afternoon, the time of day when solar generation is at its highest.

However, a comprehensive study by the California Independent Systems Operator (CAISO, California's bulk electric system operator) anticipates a growing mismatch as solar comes to dominate the grid. Figure 1-6, often referred to as the **duck curve**, anticipates two peak daily demand points.

DUCK CURVE

The first will occur in the early morning as people prepare for work and the sun is not yet up. Then the availability of solar will reduce the demand from the grid during the day. In the evening (as the sun sets), demand will once again spike. The grid of today is unprepared to deal with such sudden and dramatic spikes.

Technical Concerns of Utilities

The grid was designed to send power from a central generating facility, through a number of transformers, to the customer. When additional generating facilities are added to the existing system - generation sources not controlled by the utility - unanticipated problems may occur.

These problems include:
- **voltage violations,**
- increased **demand ramping,**
- system instability,
- simultaneous tripping and re-closing.

VOLTAGE VIOLATION

DEMAND RAMPING

FIGURE 1-6: DUCK SHAPED ELECTRIC UTILITY DEMAND CURVE AS ROOFTOP SOLAR ADDED
(GRAPH FROM CAISO)

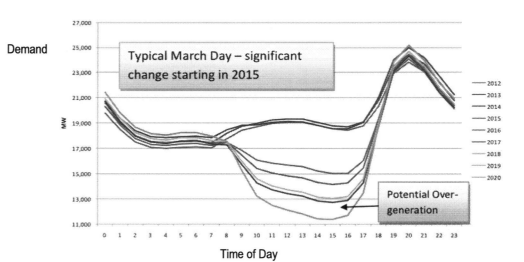

The grid is designed to provide power when needed. This is a very complex task, requiring constant adjustment to ensure there is enough power flowing to meet demand, but not too much power - which will cause an increase in voltage within the system.

Utilities can control their own generation, but have little or no control over power pushed onto the grid by owners of distributed energy systems. Power flowing onto the grid from a PV system may cause the voltage on the grid to rise at the point where it connects, or the **point of common coupling (PCC)**.

POINT OF COMMON COUPLING (PCC)

This rise is due to **impedance** on the grid, as there is no load demand ready to accept the power generated by the PV system.

IMPEDANCE

This voltage fluctuation can be made even more profound due to the variable nature of solar; output rising and falling unpredictably as the sun shines brighter or is covered by clouds.

If this voltage rise is extreme enough, the inverter within the PV system may shut off, disconnecting power flow to the grid. It then must sit idle for a period of time before it can reconnect to the grid. This unwanted tripping not only stresses the PV system, but results in the loss of potential power as the PV system sits idle.

A second problem PV systems cause utilities is an increase in demand ramping. As illustrated by the duck curve, peaks and valleys in demand will result in either the utility maintaining unused generating capacity for long periods of time, or require them to purchase high-cost power from other providers during **peak load demand**.

PEAK LOAD DEMAND

History of Photovoltaics

Although the PV industry is for all practical purposes only about a decade or so old, it had its start much earlier. In fact, much earlier than one might expect.

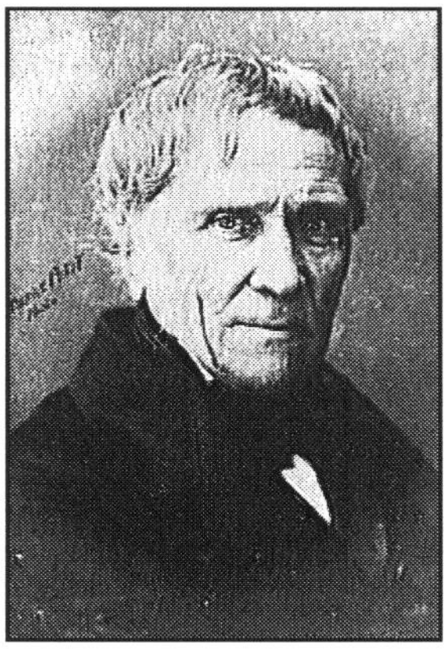

FIGURE 1-7: ANTOINE-CÉSAR BECQUEREL (1788- 1878)

(PHOTO FROM WIKI MEDIA COMMONS)

In 1839, a French physicist named Antoine-César Becquerel (pictured in Figure 1-7) discovered that an electrode he had immersed in a conductive liquid would conduct a small amount of electricity when exposed to light (the **photovoltaic effect**).

PHOTOVOLTAIC EFFECT

During the 1860's, a British electrician named Willoughby Smith began testing underwater telegraph lines for faults. He accidentally discovered that selenium, when exposed to light, was an excellent conductor of electricity. But when no light was present, the material would not conduct electricity at all. A decade later, two

American scientists (William Adams and Richard Day) expanded upon this discovery and found that it was indeed the sun's energy that created the electrical flow through **selenium**.

For the next decade, scientists worked to understand the relationship between selenium and electricity. In 1883, American inventor Charles Fritts is credited with inventing the very first PV cell by placing a layer of selenium on a metal plate and coating it with gold leaf. When placed in sunlight, the solar cell generated a very weak electrical current, estimated to be less than one percent efficient in converting the energy contained in the sunlight into electricity. But it was a start.

In 1904, a young patent clerk named Albert Einstein (Figure 1-8) set about to explain how light could create electricity. He argued that light was made up of tiny packets of energy that moved like a wave. He called these packets of energy "lichtquant" - or light quantum. This term later became known as a **photon**.

FIGURE 1-8: ALBERT EINSTEIN (1879- 1955)

(PHOTO FROM WIKI MEDIA COMMONS)

He speculated that these particles of energy could knock loose electrons from some materials such as selenium and silicon (later known as semi-conductors). These free electrons moving through a conductive wire created an electrical flow of energy. Einstein was later (in 1921) awarded the Nobel Prize in Physics for his work on photovoltaics (not for, as you might assume, the Theory of Relativity).

Scientists continued working with selenium, attempting to generate electricity through thin wafers – but without much luck. Then in 1953, three scientists from **Bell Labs** (Gerald Pearson, Daryl Chapin and Calvin Fuller) found that silicon, rather than selenium might prove a better material to use in photovoltaic cells. In fact, it turned out that it was five times more efficient than selenium.

Within a year of experimentation, the team had developed cells that were 50 times as efficient as cells created with selenium in transmitting electrons along connected conducting wires.

The space race and **NASA** provided the motivation and the funding to promote the early development of PV systems to provide power for satellites and spacecraft. But prices remained stubbornly high. In 1971 the price of electricity generated through PV was still over 200 times that of power generated through traditional fossil fuel methods.

PV Certification Programs

As any industry matures, standards and industry best-practices begin to emerge, elevating the quality of work and products within that industry. Certification programs emerge so consumers will have an unbiased way to assess the qualifications of people they may wish to employ.

Since the year 2000, two primary PV certification programs have emerged within the United States. These are offered by: **NABCEP (North American Board of Certified Energy Practitioners)**, and **ETA-I (Electronic Technicians Association International)**.

North American Board of Certified Energy Practitioners (NABCEP)

NABCEP was founded in 2000 as the nation's first non-profit organization to offer credentials to professionals working in the renewable energy industry (specifically solar and wind power).

FIGURE 1-9: NORTH AMERICAN BOARD OF CERTIFIED ENERGY PRACTITIONERS

WWW.NABCEP.ORG

The NABCEP Solar PV Installation Professional Certification is designed for individuals already working within the PV industry.

Candidates must:
- have experience in the field acting as the person responsible for installing PV systems. Depending on the category selected, all candidates for the Solar PV Installation Professional exam must provide documentation for three (3) or five (5) installations where they have acted in the role of contractor, lead installer, foreman, supervisor, or journeyman,
- a minimum of 58 hours of advanced PV training and
- an OSHA 10-hour construction industry card or equivalent.

NABCEP does offer an entry-level examination (the Associate Program), however this exam does not lead to any certification offered by the organization.

NABCEP also offers a number of certifications that are more narrowly targeted for individuals working within the PV industry.

These include:
- PV Technical Sales Professional (PVTS)
- PV Commissioning & Maintenance Specialist (PVCMS)
- PV Design Specialist (PVDS)
- PV Installer Specialist (PVIS)
- PV System Inspector (PVSI)

FIGURE 1-10:
ELECTRONICS TECHNICIANS ASSOCIATION

WWW.ETA-I.ORG

Electronics Technicians Association

The Electronics Technicians Association (ETA) was founded in 1978 and has certified over 150,000 individuals around the world in over 80 technical programs.

In 2009, the ETA expanded their certification programs to include the renewable energy industry. Addressing the need for an industry entry-level certification program in PV, ETA created a PV certification program. Requirements include:

PV Level I Certification
- Attend an ETA-approved school (40-hours) and pass Level 1 written exam as well as a hands-on assessment

PV Level II Certification
In 2019 ETA instituted a Level II PV certification program designed for those who have:

- been working in the industry and document (signed off by customers and/or supervisors) on-the-job installation experience checklist.
- a minimum of 60 hours of PV training and,
- an OSHA 10-hour construction industry card or equivalent.
- obtain Customer Service Specialist certification.

Renewal Process
The ETA certification is valid for a period of four years. It may be renewed under the following conditions:

- if working in the industry that pertains to the certification, then submit the Employer Verification Form and show proof of 24 hours of continuing education during the certification period. Or retest to renew the certification.
- if not working in the industry that pertains to the certification the candidate must attend an ETA approved hands-on refresher course and show proof of 24 hours continuing education during the certification period. Or retest to renew the certification.

FIGURE 1-11:
NATIONAL ELECTRICAL CODE

WWW.NFPA.ORG

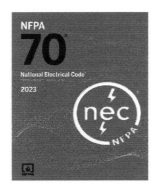

Electrical Codes and Standards

In the US, the **National Electrical Code (NFPA 70 or NEC)**, Figure 1-11, is published every three years by the **National Fire Protection Association** and seeks to create a safe standard for the installation of electrical wiring and equipment.

The NEC is not a federal law, but must be adopted by local or state governing bodies to have the effect of law. Often these bodies are slow to adopt the most recent version of the code (most recent update released in 2023), so note that older versions of the code may be the required guide in many parts of the country (as indicated in Figure 1-12).

NATIONAL ELECTRICAL CODE (NFPA 70)

NATIONAL FIRE PROTECTION ASSOC.

When a roof is repaired or major components within an existing system are replaced, local authorities may require that an installed system be upgraded to comply with the currently adopted NEC for that location. Check with local building authorities if there is any question as to which NEC version applies.

PV systems are electrical in nature, so much of the code is important in designing and installing a complete system, **Article 690** of the code specifically addresses photovoltaics. Article 705 addresses the specific requirements of interconnected systems, and Article 250 focuses on bonding and grounding of electrical systems.

ARTICLE 690

Other sections of specific interest to PV installers include:

- Article 691 - Large Scale Photovoltaic Electric Power Production Facility
- Article 706 - Energy Storage Systems
- Article 710 - Stand Alone Systems
- Article 712 - DC Microgrids

FIGURE 1-12: STATE-BY-STATE ADOPTION OF NEC

(IMAGE FROM ELECTRICAL SAFETY FOUNDATION INTERNATIONAL)

In addition to the United States, the NEC has been adopted within Mexico, Costa Rica, Venezuela and Colombia.

Fire Codes

The two main fire codes in use today within the United States are the IFC and the NFPA 1.

The **International Fire Code (IFC)** provides the primary foundation for fire safety regulation in 42 states (all but FL, WV, MD, VT, MA, RI, ME, HI), Guam, Puerto Rico, the U.S. Virgin Islands and the District of Columbia.

INTERNATIONAL FIRE CODE (IFC)

The IFC establishes minimum requirements necessary to protect people and property from fire and explosion hazards within buildings and structures. The IFC also provides comprehensive regulations for the storage of hazardous materials.

NFPA 1 is similar to the IFC in that it provides guidelines for building design and construction to minimize the risk of fire. It has been adopted by the states that have not adopted the IFC (FL, WV, MD, VT, MA, RI, ME, HI). This code is written and developed by the National Fire Protection Association.

Because different non-governmental organizations develop these two fire standards, they often overlap in the requirements they contain, and in some cases, they contradict each other. Also, many states and local jurisdictions often adopt different components of both the IFC and NFPA 1. So it is important that PV designers and installers know which code(s) are applicable in their areas.

Other Electrical Codes and Standards

The **International Energy Conservation Code (IECC)** contains specifications for onsite renewable energy generation in commercial systems.

The **IEC 60364** published by the International Electrotechnical Commission is used as a basis for electrical codes in many European countries.

The British Standard, BS 7671 is the set of regulations for electrical wiring used within the United Kingdom.

Australian/New Zealand Standard, **AS/NZS 3000:2007 Wiring Rules** is used in Australia and New Zealand.

Canadian Electrical Code

The **Canadian Electrical Code (CEC)** published by the CSA (Canadian Standards Association) is used within Canada. Like the NEC, the CEC is updated every three years (2018, 2021, 2024, etc).

While the two codes are largely similar, there are differences that must be addressed when installing a system. Varying wire gauge designations, differences in clearances and ampacity limitations are examples of requirements that must be checked and complied with.

In 2012 the CEC added *Section 64: Renewable Energy Systems* to the code to more fully address the rapidly evolving technologies. However, solar electric systems remained in Section 50. In the 2018 revision, solar technologies were moved to Section 64.

This section includes:

- 64-002 - Special terminology
- 64-058 - Overcurrent protection
- 64-060 - Disconnecting means
- 64-062 - Wiring methods
- 64-066 - Ungrounded renewable energy power systems
- 64-070 - Equipment bonding
- 64-110 - Unbalanced interconnections (inverters)
- 64-112 - Utility-interactive point of connection (inverters)
- 64-202 - Voltage of solar photovoltaic systems
- 64-210 - Wiring methods
- 64-212 - Insulated conductor marking or colour coding
- 64-214 - Overcurrent protection for apparatus and conductors
- 64-216 - Photovoltaic DC arc-fault protection
- 64-218 - Rapid shutdown
- 64-220 - Attachment plugs and similar wiring devices
- 64-222 - Photovoltaic module bonding

IEC 60364

The **International Electrotechnical Commission (IEC)** produces standards, including the *IEC 60364: Electrical Installations for Buildings*, for use in its 82 member countries. The IEC is world-wide in scope, influencing the way work takes place throughout all continents, but its influence is most felt in Europe.

Unlike the NEC, IEC 60364 is not intended to be used directly by designers, installers or enforcement/inspection officials. It is intended to be used only as a guide to those developing national wiring rules.

The portion of the standard that applies to the design of photovoltaics is IEC 60364-7-712 (2017).

IEC 62446-1:2016 defines the documentation required following the installation of a grid connected PV system. It also describes the commissioning tests, inspection criteria and documentation expected to verify the safe installation and correct operation of the system. It is for use by system designers and installers of grid connected solar PV systems as a template to provide effective documentation to a customer.

IEC 60904-2:2023 gives requirements for the classification, selection, packaging, marking, calibration and care of photovoltaic reference devices. This standard covers photovoltaic reference devices used to determine the electrical performance of photovoltaic cells, modules and arrays under natural and simulated sunlight. It does not cover photovoltaic reference devices for use under concentrated sunlight.

The **Central European Normalization Electrotechnique (CENELEC)** is the European committee for electrotechnical standardization. Designated as a

European Standards Organization by the European Commission, CENELEC is a non-profit technical organization that publishes the European standard for Electrical installations and protection, referred to as HD 60364.

BS 7671

British standard *BS 7671 "Requirements for Electrical Installations* is the national standard in the United Kingdom for electrical installation and the safety of electrical wiring in domestic, commercial, industrial, and other buildings. This standard is essentially the UK version of the HD 60364. Requirements that are UK specific and not derived from HD 60364 can be identified by the number 200, 201, 202 etc. For example, Regulation 412.1.201:

The current version BS 7671:2018+A2:2022 was published in March 2022 and came into effect on 28 March 2022 (the 18th Edition).

The part of the standard that addresses PV specifically is *Section 712, Solar photovoltaic (PV) power supply systems*.

IEEE 1547

The **IEEE Standard 1547** was created to establish a technical standard for interconnecting distributed energy systems (referred to in the standard as **distributed energy resources (DER)**) with electrical power systems (EPSs) - basically the grid. As technology became more sophisticated and more and more systems were connected to it, issues arose that needed to be addressed.

The latest revision, IEEE 1547-2018, changed the testing standards for distributed energy systems to create harmonized interconnection requirements and offer flexibility in performance requirements. Active IEEE standards must undergo a revision process at least every 10 years.

These standards are of significant importance to utility grid operators as well as equipment manufacturers in order to ensure compliance.

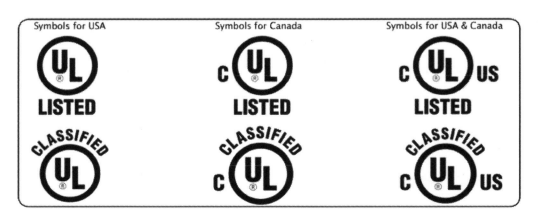

FIGURE 1-13: UL SERVICE MARKS

(IMAGE FROM UNDERWRITERS LABORATORIES)

Approved and Tested Products

The use of equipment that has been tested and approved for its specific application is required by most codes and standards (such as the NEC).

Manufacturers submit their products to facilities (referred to as a **national recognized testing laboratory**, or **NRTL**) that then test them under laboratory conditions to ensure that they conform with the appropriate industry standard.

Approved equipment is identified with a listing label indicating that it conforms with a specific standard.

UL Marks

Underwriters Labs (UL) is the most recognizable of the NRTLs within North America. They offer a number of service marks for products (illustrated in Figure 1-13) that have been tested and/or evaluated within their facilities.

A **UL Listing Mark**, indicates that the product has been tested and found to be in compliance with a specific standard. This mark is generally accepted by the code bodies as evidence that the product has been approved as conforming with the standard's requirements.

The **UL Classification Mark** indicates that the product has only been tested under limited circumstances.

The UL marks generally indicate if they have been tested for the United States, Canada or both - as indicated with the letter "C" for Canada and the letters "US" for the United States.

Electrical Testing Labs/Intertek (ETL)

Electrical Testing Labs/Intertek (ETL) is another widely recognized NRTL that also tests products for conformance with various standards.

FIGURE 1-14: ETL SERVICE MARK

(IMAGE FROM ETL/INTERTEK)

There is essentially no difference in the testing process or acceptance of the designations issued by UL or ETL (Figure 1-14).

Canadian Standards Association (CSA)

In addition to writing codes and standards, CSA also provides NRTL services, testing and listing products that conform to applicable industry standards. Their marks are recognized throughout North America and are considered interchangeable with UL and ETL.

FIGURE 1-15: CSA SERVICE MARK

(IMAGE FROM CSA)

A CSA mark that contains the indicator "US or "NRTL" demonstrates that the product is certified for the US market (Figure 1-15).

FIGURE 1-16:
CE SERVICE MARK

CE Mark

The **CE Mark** (an abbreviation of "Conformité Européenne" - French for "European Conformity") indicates that the product conforms with health, safety, and environmental protection standards for products sold within the European Economic Area (EEA). The marking (Figure 1-16) is also found on products sold outside the EEA that have been manufactured to EEA standards.

Unlike the UL, ETA or CSA marks, this designation is not subject to an independent third-party evaluation. The manufacturer of the product self-assesses whether the product meets the requirements outlined by the European Economic Area. As a result, it does not necessarily meet the code requirements as an approved product with relation to a specific standard.

Safety Regulations

Most nations have, over the years, developed a "safety bill of rights," that seeks to assure the safe and healthful conditions for working men and women. These are largely created to minimize the risk of injury and death caused by accidents on the job.

The Occupational Safety and Health Administration (OSHA)

FIGURE 1-17:
OCCUPATIONAL SAFETY AND HEALTH ADMINISTRATION

The **Occupational Safety and Health Administration (OSHA)** was established in the United States in 1971 as a result of the Occupational Safety and Health Act of 1970. It's mission was to ensure safe and healthful working conditions for working men and women by setting and enforcing standards and by providing training, outreach, education and assistance.

While no accurate statistics were maintained prior to 1970, it is estimated that in 1970 around 14,000 workers were killed on the job. That number fell to approximately 4,340 in 2009. Over that period of time, U.S. employment nearly doubled.

OSHA officials are empowered to conduct site inspections (without notice) and issue fines or citations when violations are found.

OSHA requires that all employers:
- Follow all relevant OSHA safety and health standards.
- Find and correct safety and health hazards.
- Inform employees about chemical hazards through training, labels, alarms, color-coded systems, chemical information sheets and other methods.

- Notify OSHA within 8 hours of a workplace fatality or within 24 hours of any work-related inpatient hospitalization, amputation or loss of an eye
- Provide required personal protective equipment at no cost to workers.
- Keep accurate records of work-related injuries and illnesses.
- Post OSHA citations, injury and illness data and OSHA information in the workplace where workers will see them.
- And not retaliate against any worker for insisting upon enforcing their safety rights under the law.

A number of organizations offer training in workplace safety practices. The OSHA Outreach Training Program provides workers with basic and advanced training about common safety and health hazards on the job. Students receive an OSHA 10-hour (commonly referred to as an **OSHA 10 certification**) or 30-hour course completion card (**OSHA 30 certification**) at the end of the training.

Canadian Centre for Occupational Health and Safety (CCOHS)

The **Centre for Occupational Health and Safety** is the Canadian equivalent of OSHA. The CCOHS was created in 1978 by an Act of Parliament - Canadian Centre for Occupational Health and Safety Act. The act in part stated that all Canadians had "…a fundamental right to a healthy and safe working environment.".

European Agency for Safety and Health at Work (EU-OSHA)

The **European Agency for Safety and Health at Work (EU-OSHA)** was founded in 1994, with a goal to collect, analyze, and distribute information to those involved in occupational safety and health.

The main difference between the US and the EU agencies is that the US OSHA has the power to create laws and levy fines to companies throughout the United States. In the EU, each individual country is responsible for conducting workplace inspections, establishing work safety laws, and enforcing the regulations.

Chapter 1 Review Questions

1. Converting the energy of the sun from light to electricity is known as _____.
   A) solar thermal
   B) photovoltaics
   C) polycrystalline
   D) megawatts

2. A point where the cost of electricity from a solar energy system is the same price as electricity purchased from the local electric company is known as _____.
   A) grid parity
   B) the feed-in tariff
   C) delightful
   D) the solar set-aside

3. From a very practical perspective, the photovoltaics industry is:
   A) only about 15 years old.
   B) reaching a point of market saturation.
   C) is an industry in decline.
   D) controlled by only one or two dominant manufacturers.

4. The 2022 law that extended the investment tax credits for all solar installations (residential, commercial and utility-scale) at the 30% level through 2032 was known as:
   A) the Inflation Reduction Act
   B) the Feed-In Tariff
   C) the Energy Community Rider
   D) Empowering America Law

5. When installing a residential PV system, the hardware (panels, inverters, switches, wire, etc) accounts for about:
   A) 20% of the total installed cost
   B) 40% of the total installed cost
   C) 60% of the total installed cost
   D) 80% of the total installed cost

6. A system that guarantees the retail price to renewable energy generators for a period of time is known as a:
   A) investment tax credit (ITC)
   B) renewable energy certificate (REC)
   C) payback period (PP)
   D) feed-in tariff (FIT)

7. Owners of solar energy systems can often make additional money by selling _____ to utilities, who must produce a certain amount of their power from renewable sources according to the state's renewable portfolio standard.
   A) renewable set-asides
   B) easement reduction credits
   C) utility portfolio standards certificates
   D) renewable energy certificates

8. When a state determines that a certain amount of energy must be generated through a renewable source, this is referred to as the state's:
   A) solar set-aside program (SSP)
   B) renewable energy manifesto (REM)
   C) renewable energy certificate (REC)
   D) renewable portfolio standard (RPS)

9. An SREC is a:
   A) special regional energy commission
   B) silicon resistant emitting crystal
   C) sustainable renewable energy concept
   D) solar renewable energy certificate

10. Net-metering laws require:
    A) that utilities provide meters free of charge to all distributed energy system owners.
    B) that utilities allow solar energy system owners to sell excess energy to the utility.
    C) that utilities provide a certain amount of energy (varies from state-to-state) from renewable sources such as wind, solar or hydro.
    D) that utilities have "first option" on all power generated through distributed energy systems owned by their customers, meaning that the owner must sell to the utility prior to making the energy available to any other potential customers.

11. A system that produces energy at or near the point where it will be used, and is generally owned by the customer (such as a solar array or a generator) is referred to as a:
    A) distributed energy system
    B) disbursed energy system
    C) interconnected energy system
    D) net metered energy system

12. Attaching a photovoltaic system to the grid is known as:
    A) commissioning
    B) net-metering
    C) interconnecting
    D) a solar set-aside

13. Which of the following is **NOT** a technical concern when adding distributed energy sources to the electrical grid?
    A) voltage violations
    B) renewable portfolio standard
    C) system instability
    D) increased demand ramping

14. A graphic representation of the daily electrical demand curve that indicates adding more and more solar resources to the grid may result in an over supply of power during the afternoon, is often referred to as:
    A) the duck curve
    B) lichtquant
    C) feed-in premium (FIP)
    D) feed-in tariff (FIT)

15. The photovoltaic effect:
    A) refers to changes our society has experienced as more and more people adopt solar electric systems.
    B) occurs when heat is released as light photons strike a solid surface, such as the surface of a PV panel.
    C) refers to the unusual characteristic of photons, where they can behave as a particle as well as a wave.
    D) occurs when the light from the sun is converted into electricity.

16. Albert Einstein argued that light was made up of tiny packets of energy that moved like a wave. He called these packets of energy "lichtquant" - or light quantum. This term later became known as:
    A) a photon
    B) photovoltaics
    C) a semi-conductor
    D) an electron

17. In 1953, three scientists from _____ determined that silicon performed well as a component within photovoltaic cells.
    A) Germany
    B) Bell Labs
    C) NASA
    D) California

18. Who is largely credited with first discovering the photovoltaic effect, which later led to the development of photovoltaic cells?
    A) Albert Einstein
    B) Alexander Graham Bell
    C) Antoine-César Becquerel
    D) Charles Fritts

19. The National Electrical Code is published every three years by the:
    A) National Electrical Commission
    B) National Renewable Energy Laboratories
    C) National Fire Protection Association
    D) North American Board of Certified Energy Practitioners

20. The portion of the NEC that specifically addresses photovoltaics is:
    A) Article 690
    B) Chapter 250
    C) Section 8.5
    D) NEC 2014

21. The NEC document is also referred to as:
    A) industry best practices
    B) the law of the land
    C) Ugly's Reference Guide
    D) NFPA 70

22. Prior to selling products in the marketplace, reputable manufacturers must submit them to _____ to ensure that they conform with the appropriate industry standard.
    A) NABCEP
    B) NFPA
    C) IEEE
    D) a national recognized testing laboratory (NRTL)

23. Which of the following is **NOT** a common listing or service mark found on products that indicates they comply with the appropriate industry standard?
    A) OSHA
    B) ETL
    C) CSA
    D) UL

24. In the US, which agency is tasked with the responsibility to ensure safe and healthful working conditions for working men and women?
    A) OSHA
    B) ETL
    C) CSA
    D) UL

25. The document that details industry best practices from a **SAFETY** perspective in the U.S. is:
    A) the American National Standards Institute (ANSI).
    B) Underwriters Laboratories (UL).
    C) the National Electrical Code (NEC).
    D) the Electronic Technicians Association International (ETA-I).

Chapter 2
Solar Cells and Solar Modules

Chapter Objectives:

- Become familiar with how solar cells convert sunlight into electricity
- Evaluate the advantages and disadvantages of the various types of solar modules available today.
- Understand the conditions under which the performance characteristics of solar modules are evaluated.
- Explore how amps and volts are measured and produced by PV panels.
- Comprehend the various environmental factors that will affect the performance of a solar panel.
- Understand the terms used to describe the various performance levels measured when evaluating solar panels.

How PV Cells Work

A typical solar module is made up of a number of solar collectors called **solar cells**. These cells are constructed from a fairly common mineral called **silicon** (sand, quartz and even glass is made up largely of silicon).

While some materials make good **conductors** (like copper, which allows electricity to easily flow through it), other materials are good **insulators** (such as plastic) that prevent electricity from moving through it (which is why the copper wire is wrapped in a plastic insulating coating).

Silicon is known as a **semiconductor**, which really just means that it has properties that fall somewhere between those of a conductor and those of an insulator.

It might be helpful to think of a solar cell as a silicon sandwich (made up of two wafers with a gap in the middle). Impurities are introduced into each silicon wafer in a process known as **doping** that change the electrical characteristics of both layers - but in very different ways.

A small number of phosphorus or arsenic atoms are introduced into the top layer. When combined with the silicon, this creates a compound with a single free electron in its outer layer, which is relatively free to move around, because it is not "locked" within a layer of the atom's structure. Since electrons have a

SOLAR CELLS

SILICON

CONDUCTOR

INSULATOR

SEMICONDUCTOR

DOPING

negative charge, the top wafer is referred to as being composed of an **N-Type** material (negatively charged), since it has one too many electrons than would normally be present in the atom.

At the same time, the bottom wafer is doped with boron or gallium. The resulting compound has one too few electrons in its outer atomic layer, often referred to as an **electron hole**. This missing electron results in a positively charged, or **P-Type** material.

If the wafers were made of a material that had good conductive properties, when connected, the negative and positive charges would quickly combine and reach a neutral (or balanced) state.

But silicon is a semi-conductor. When the two layers are pressed together (as illustrated in Figure 2-1), a very thin barrier is formed (which is chemical in nature) between the bottom and the top wafers. This is referred to as the **P-N junction**. This barrier layer behaves like an insulator and does not allow the electrons (the electrical energy) to flow from one layer to the other.

But when a photon from the sunlight strikes the PV cell, it passes through the layers and strikes electrons within the cell, dislodging them from the silicon atom. If the photon has enough energy, that electron can move through the depletion zone (the **band gap**) between the two layers.

Silicon has a relatively low **band gap energy level** (1.1 electron volts – or 1.1 eV). So any photon with an energy level higher than 1.1 eV can cause an electron to flow from the bottom level of the solar cell to the top layer.

But after crossing the band gap, there will not be enough energy present for that electron to cross back to the bottom layer. In this way, the band gap acts as a simple **diode** – allowing energy to flow only in one direction.

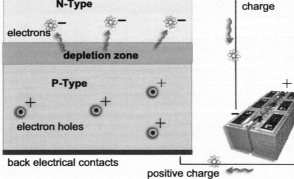

Figure 2-1: Cross section of a PV cell

In order for electricity (the electrons) to flow, it is then necessary to create a **circuit**, connecting the top layer (now negatively charged due to the dislodged electrons) to the bottom layer (positively charged). To accomplish this, small wires are placed on top of the panel (referred to as **busbars**), connecting all the cells together. The bottom of the panel is typically made of metal, creating a good conductive surface.

When the wires from the top of the panel are connected to the metal backing plate at the bottom of the panel, an electric circuit is created and the electrons can flow from the negatively charged top layer of silicon to the positively charged bottom layer of silicon, once again equalizing the state of charge.

It is important to note that no electrons will flow across the band gap unless there is a load to service (similar to how water in a pipe cannot flow unless the tap is opened). A panel sitting in the sun will not generate any power unless connected to a load. Both voltage and current are required for electricity to flow. There is voltage (i.e., potential) present in the panel, but no current when the load is disconnected.

Types of Solar Cells

Wafer-based silicon solar cells are the basis for most of the solar modules (or panels) on the market today. They account for about 90% of the total PV module market share.

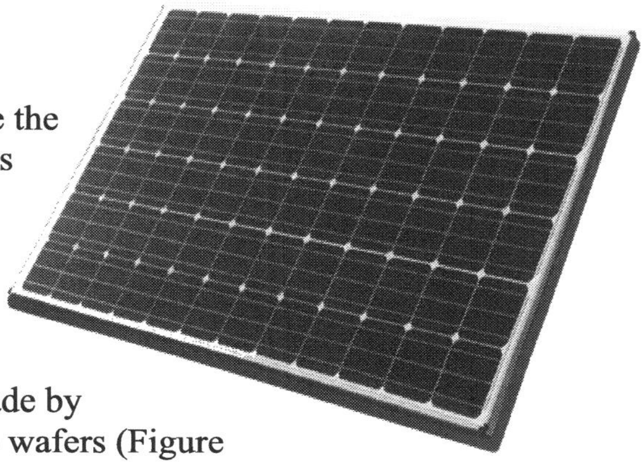

FIGURE 2-2: MONOCRYSTALLINE PV MODULE

MONOCRYSTALLINE PV MODULE

Monocrystalline modules are made by slicing a single silicon crystal into wafers (Figure 2-2). They are generally more expensive, but are also more efficient in converting the energy of the sun into electricity than other panel options. Typical commercially-available monocrystalline panels available in 2023 had an average efficiency range of between 19% - 23%.

POLYCRYSTALLINE PV MODULE

Polycrystalline modules (Figure 2-3) are not constructed by slicing wafers from a single pure silicon crystal, but rather by pouring molten silicon into bars, and then slicing wafers from the hardened material. This type of PV cell is less expensive to produce, but is also a bit less efficient, with efficiency ranges between 16% - 20%.

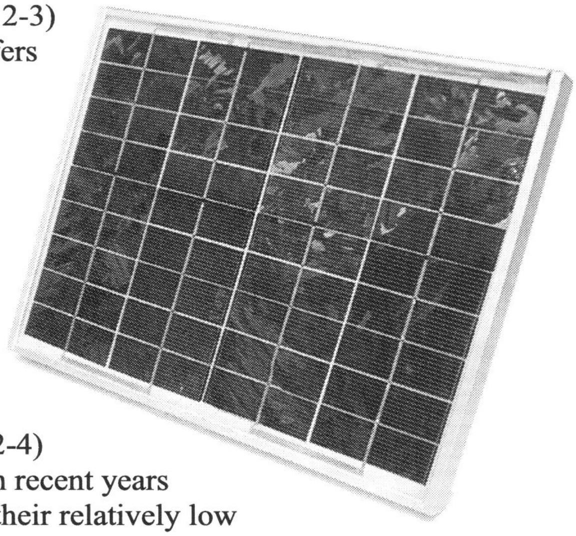

FIGURE 2-3: POLYCRYSTALLINE PV MODULE

THIN-FILM PV MODULE

Thin-Film solar panels (Figure 2-4) have been gaining in popularity in recent years for utility-scale applications, but their relatively low efficiency has restricted their use for residential rooftop solar.

FIGURE 2-4: THIN-FILM PV MODULE

These panels are typically less efficient than crystalline modules and the efficiency varies depending on the type of material used in the cells. But in general they tend to have efficiencies between 7% to 18%.

Thin-Film technology has a higher theoretical efficiency than silicon and are expected to grow at a compound annual rate of 23% from 2020-2025.

The three main types of solar thin-film cells include:
- Amorphous Silicon (a-Si): This type of cell is made from amorphous silicon (a-Si), which is a non-crystalline silicon, making them much easier and less expensive to produce than mono or polycrystalline solar cells.
- Cadmium Telluride (CdTe): This is the second most common solar cell type in the world after crystalline cells. They are made from a chemical compound called Cadmium Telluride, which is very good at capturing sunlight and converting it to energy in low light levels.
- Copper Indium Gallium Selenide (CIGS): These cells are made by placing layers of Copper, Indium, Gallium, and Selenide on top of each other to create a powerful semiconductor.

Globally, the PV marketshare of thin-film technologies remains between 5%-10% as of 2023, thin-film technology has become increasingly more popular in the United States. CdTe cells alone accounted for nearly 30% of all new utility-scale installations in 2022.

Reasons for this include:
- CdTe panels are made in the US (First Solar) and are not subject to import tariffs.
- They operate at higher voltages than silicon - which can be a problem with residential installations but is generally not a concern when designing utility-scale systems.
- Thin-film panels work better in hot climates as compared to crystalline panels. This makes them the panel of choice for regions with consistently high temperatures, such as deserts or tropical areas.
- Thin-film panels tend to perform better in low-light conditions compared to crystalline silicon panels. While they will not produce as much power during peak sunlight, they will produce more power during low-level light (mornings, evenings, cloudy conditions) than will silicon panels.

(AMORPHOUS)

Another advantage of thin-film technology (also referred to as **amorphous**, meaning "without shape") is that they are flexible and can be more easily integrated into the building design – such as solar shingles, for example.

Number of Cells

The number of cells in a solar panel can vary from 36 cells to 144 cells. The two most common solar panel options on the market today are **60-cell** and **72-cell.** 60-cell solar panels are much more common for residential solar installations, while 72-cell solar panels are more commonly used for commercial or other large-scale projects.

> 60-CELL PANEL
>
> 72-CELL PANEL

The most obvious difference is that 72-cell solar panels have more photovoltaic cells, therefore they are larger than 60-cell panels. 60-cell panels are usually built six cells wide and ten cells tall. 72-cell panels are also six cells wide but have an additional two rows of cells that make them a bit taller (about 12 inches or 305 mm).

But the increase in cells also results in higher voltages from the panel. Since each cell generates about ½ volt, then a 60-cell panel will generate about 30 volts, while a 72-cell panel will generate about 36 volts.

Increasingly, high voltage panels (96-cell) are becoming more common in the marketplace. Be aware that not all equipment (particularly inverters) are compatible with all solar panel cell configurations.

Split Cell Panels

Split modules and **half cut cells** are another innovation in module production designed to improve efficiency. Panels are created by cells that are half the traditional size. Half-cell modules have solar cells that are cut in half, which improves the module's performance and durability. Traditional 60- and 72-cell panels will have 120 and 144 half-cut cells, respectively.

> SPLIT MODULES
>
> HALF CUT CELLS

FIGURE 2-5: SPLIT-CELL PV MODULE

The panel is configured into two distinct modules (each with the same voltage but with half the current of a traditional module) as seen in Figure 2-5. These modules are combined in series within the junction box (doubling the voltage) located at the center back (rather than top) of the panel.

By reducing the current within the panel, there are lower resistive losses through the busbars, allowing for smaller conductors and less loss due to shading. The lower current also translates to lower cell temperatures, which reduces the potential formation and severity of **hot spots** (areas of high temperature that affect only one zone of the solar panel and result in a localized decrease in efficiency) that can result from shading, dirt or cell damage.

> HOT SPOTS

Experts expect the market share of half-cut cells to grow from 5% in 2018 to nearly 20% in 2020 to 60% by 2030.

Emerging Technologies

As is the way of things, there is a constant push to develop ever more efficient systems. It is no different with solar PV technology.

Multi-Layer Photovoltaic Cells

There is a limit to just how efficient a single-junction solar cell can be. This theoretical maximum efficiency, known as the **Shockley–Queisser limit**, for a single-junction silicon solar cell is about 32%.

> SHOCKLEY-QUEISSER LIMIT

One way to gain efficiency from a solar cell is to stack multiple layers of semiconductor material on top of each other - each capable of generating electricity from different ranges of wavelengths within the electro-magnetic spectrum.

> HETEROJUNCTION CELLS

Recently, multiple layer or **heterojunction cells** have been tested to efficiencies as high as 47%. Silicon heterojunction cells (SHJs) contain a base layer of crystalline silicon coated in thin layers of amorphous silicon. The 2022 efficiency record (47.1%) was achieved by a cell that incorporated six layers of conductors. The efficiency record for a three-layer cell (in 2022) was 39.5%.

> TANDEM CELLS

Multi-layer cells that incorporate only two layers of semiconductor material are often called **tandem cells**.

Perovskite

In order to make tandem solar cells, solar researchers need to look for new materials (other than silicon) that exhibit the necessary properties. One promising technology includes **perovskites**.

> PEROVSKITE

Perovskites are a class of materials first discovered in Russia in 1839. They are not a single element (like silicon), but rather can be made from a number of elements and form a unique crystal structure that offers properties desirable in solar cells.

Advantages of perovskites include:
- They can be made without special high-heat processes, making them much cheaper than traditional silicon.
- They can be deposited onto surfaces via liquid or vapor (essentially "painted" on any surface).
- They capture energy at a slightly different bandwidth than silicon, allowing them to be used in tandem with silicon to expand cell efficiency.
- Perovskite cells can be transparent or many-colored, making them ideal for solar windows or other building integrated applications.

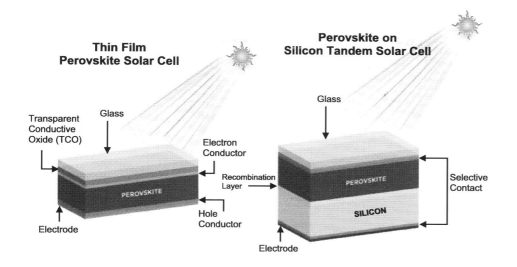

Figure 2-6: Thin Film Perovskite Cells and Silicon-Perovskite tandem cells

Experiments with this type of material in solar cells began at Brown University in 2009, achieving cell efficiencies of around 3%. Efficiencies of cells made exclusively of perovskites have grown dramatically. In 2023, researchers achieved efficiency results in excess of 28%.

The first commercial uses for perovskites will be in heterojunction and tandem solar cells (as shown in Figure 2-6). One company, Oxford PV, announced the commercial release of silicon-perovskite tandem solar cells in late 2023 - promising efficiency of 27% at a 15% lower cost. Other companies have announced plans to make tandem perovskite cells available starting in 2025.

PERC Solar Cells

PERC (Passivated Emitter and Rear Cell or Passivated Emitter and Rear Contact) solar cells are a relatively new innovation in solar cell technology that is gaining wide acceptance. It is estimated that in 2023 PERC cells comprised over 75% of the total worldwide market share.

PERC Cells

Most traditional crystalline solar cells contain at least three elements.
- Absorber – typically a semi-conductor such as silicon which absorbs incoming photons;
- Membrane – usually a PN-junction which prevents the excited electron from recombining to its original layer; and,
- Contacts – that collect the electrons and connect the cells together and to a load.

A PERC cell adds an additional layer (a dielectric passivation layer) to the bottom of the cell that:
- makes the flow of electrons more steady and consistent, thereby producing more electric current;

- Increases the cell's ability to capture light – unabsorbed light is reflected back up to the solar cell for a second absorption attempt, producing additional energy;
- Reflecting specific solar wavelengths that normally generate heat - keeping the cell cooler and therefore more efficient.

PERC cells are approximately 6-12% more efficient than standard PV cells, without adding much additional cost.

Bifacial Modules

Bifacial modules produce solar power from both sides of the panel. Both the front and back of the panel contain exposed solar cells, as shown in Figure 2-7.

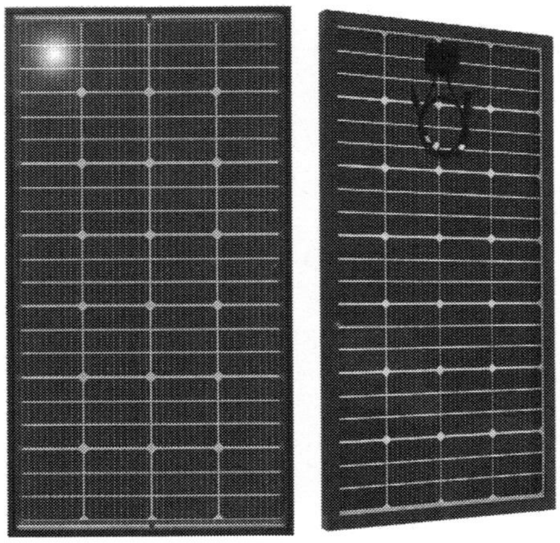

FIGURE 2-7: BIFACIAL PV MODULE
(PHOTO FROM CANADIAN SOLAR)

When bifacial modules are installed on a highly reflective surface (like a white commercial roof or on the ground covered with snow or sand), bifacial module manufacturers claim up to a 30% increase in production from the extra power generated from the **albedo** energy (the light energy reflected up from the surface below the panel) collected by the cells located on the rear of the panel.

Several field studies have found, however, that actual increases in production are substantially less (3-10%).

Bifacial panels typically cost 6-10% more than equivalent **monofacial** modules. But given that modules are only a fraction of the overall installed cost, a fairly modest increase in energy output may more than justify the additional cost of bifacial modules.

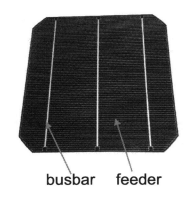
busbar feeder

FIGURE 2-8: BUSBAR AND FEEDER CONDUCTORS ON A SILICON SOLAR CELL

Busbar Changes

Changing the busbars of solar panels is also emerging as a way to increase a panel's efficiency. Busbars are the ribbon conductors on the surface of the cell. The busbars are connected to tiny **feeder conductors** (as shown in Figure 2-8), and together they transport current from the cell to the junction box.

As PV cells have become more efficient, they generate more current. As a result, most manufacturers have moved from 3 busbars to 5 or 6 busbars.

Some have even begun incorporating up to 12 very thin round wires rather than flat busbars.

More busbars shade a greater portion of the surface of the cell, reducing conversion efficiency. However, multiple busbars provide lower resistance and a shorter path for the electrons to travel, resulting in higher performance.

Bifacial Solar Cells

Not to be confused with bi-facial modules, which have solar cells on both the front and back of the panel, **bifacial solar cells** is a technology that has been around for a while, but is becoming increasingly popular as significant reductions in the cost of solar glass have made this technology an affordable option.

Bifacial solar cells are designed to allow light to enter from both sides; the front and the back of the panel. They typically use a front surface design similar to that used in industry-standard panels, with the major difference being the structure of the rear surface contact (back of the panel). Rather than cover the entire back surface with a reflective aluminum contact, a surface is used in its place that allows reflected and diffused sunlight to enter the panel through the rear.

Solar Cells, Solar Modules, Solar Panels and Solar Arrays

Proper terminology can be helpful as well as, at times, confusing. The National Electric Code (NEC) offers a number of definitions that can prove both helpful as well as confusing.

A solar cell is defined as *"the basic photovoltaic device that generates electricity when exposed to light."*

When a number of these cells are connected together, they form a **solar module**. The NEC defines this as *"a complete, environmentally protected unit consisting of solar cells, optics, and other components, exclusive of tracker, designed to generate dc power when exposed to sunlight."* In other words, what is normally thought of as a solar panel.

The NEC code makes a distinction between solar modules and **solar panels**. In today's marketplace, this is largely a distinction without a difference, and the terms are used interchangeably (as they will be in this text). The NEC defines the term solar panel as *"a collection of modules mechanically fastened together, wired, and designed to provide a field-installable unit."*

When solar panels (or solar modules) are electrically connected together, this grouping is referred to as a **solar array.** The NEC states that a solar array is *"a mechanically integrated assembly of modules or panels with a support*

Figure 2-9:
Relationship between solar cell, solar module, solar panel, and solar array

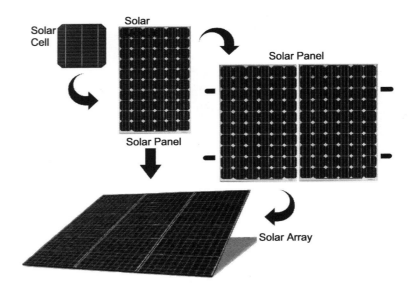

structure and foundation, tracker, and other components, as required, to form a dc or ac power-producing unit."

The relationship between cells, modules, panels and an array can be seen in Figure 2-9.

Flat Panels and Module Standards

Markets across the world require a safety standard for module construction and operation. For the United States, this safety standard for solar panels has traditionally been **ANSI/UL1703**. For Europe and Asia, the standard is **IEC 61730**. Both standards are similar and ensure that the module designs have the basic requirements for safety of operation and construction.

- ANSI/UL 1703
- IEC 61730
- UL 61730

UL 61730, a more recent addition to solar panel testing and certifications, combines the testing procedures and standards of UL 1703 with IEC 61730, allowing for complete international approval. It has become increasingly common to see this certification listed on panel specifications as opposed to either the UL or IEC standard.

IEC 61215 is one of the core testing standards for residential solar panels. If a solar panel module successfully meets IEC 61215 standards, it completed several stress tests and performs well regarding quality, performance, and safety.

IEC 61215 standards apply to monocrystalline and polycrystalline PV modules, the most common types of solar panels. In North America, UL test to the IEC 61215 standard and panels are often listed as **UL 61215** certified.

- UL 61215

The IEC sets different testing standards for other solar electric technologies, such as thin-film solar products (IEC 61646).

There are a number of other standards that apply to solar panels in very specific locations and/or conditions.

IEC 62716: *Ammonia corrosion testing of photovoltaic (PV) modules.* Installing PV modules close to farms and livestock can expose the panels to high levels of ammonia. Ammonia corrosion can accelerate degradation in a panel, leading to lower overall electricity production.

IEC 61701: *Salt mist corrosion testing.* Installing PV modules near the ocean or on a boat can expose them to high concentrations of salt mist, leading to corrosion or delamination. Panels that successfully pass IEC 61701 tests are a suitable choice for beach-front systems or those located near roads experiencing high levels of salting in the winter.

IEC 60068-2-68: *Blowing sand resistance testing.* Systems installed in sandy desert environments are frequently exposed to abrasive sand that can lead to physical or mechanical defects over time.

How Solar Modules are Rated

Environmental conditions (such as temperature, light intensity, and even altitude) impact how a solar panel works. With this in mind, the industry has come up with some **standard test conditions (STC)** under which to rate solar panels (comparing apples to apples, if you will).

These test conditions are concerned with three main environmental factors:

1) The amount of sunlight hitting the panel (known as **irradiance**). On a cloudless day at noon at sea level, the energy contained within the sun is about 1,000 watts per square meter (W/m²). The STC assumes this amount of energy will be present.

2) The thickness of the air mass the sunlight must pass through. In space (at the edge of the Earth's atmosphere), sunlight contains about 1,300 W/m². Known as the **Solar Constant**, the official measurement of the Sun's energy at the outside edge of the Earth's atmosphere is 1,366 W/m². However since the Earth travels in an elliptical orbit around the Sun, even the solar constant varies by +/- 3% throughout the year.

As it passes through the atmosphere, some of that energy is lost (scattered, absorbed, or deflected back into space). The more atmosphere it must pass through, the less energy available when it hits a panel. So a panel located on top of a mountain will often receive more energy than a panel located at sea level. This is NOT because the panel on the mountain is closer to the sun – but because the light must pass through more of the atmosphere to reach the panel on the seashore.

At the equator at sea level, if the sun was directly overhead, the air mass would be one. But north of the Tropic of Cancer and south of the Tropic of Capricorn, the sun is never directly overhead (it will pass through more atmosphere as the angle increases at higher latitudes). So the STC assumes an air mass of 1.5 (a reading that is found at sea level in the more temperate latitudes).

3) Air temperature also affects the performance of a solar panel. The warmer the temperature, the LESS well the panel performs. In colder temperatures, a solar panel will actually produce more energy than in warmer temperatures. The STC used when rating PV panels assumes cell temperature of 25 degrees C (or 77 degrees F).

STC versus PTC versus NOCT versus NOMT

In the 1990s, the National Renewable Energy Laboratory (NREL) developed a set of PV test parameters designed to more closely mimic real-life atmospheric conditions than STC. The **Photovoltaics for Utility Scale Application Test Conditions**, or **PTC**, varies significantly from STC conditions.

(PVUSA Test Conditions (PTC))

The PTC tests panels at 1,000 W/m^2 (same as STC), at an air mass of 1.5 (same as STC), but at a different temperature (20° C ambient air temperature and 45° C cell temperature, rather than 25° C cell temperature). It also assumes the panel is 10 meters off the ground (simulating the panels placed up on a roof) with a breeze of 1 meter per second (which will tend to cool the panels).

The PTC designation (pushed largely by US-based manufacturers) was never widely adopted by other nations. But this designation can still be found occasionally referenced in the specification sheets of some panels. In the state of California, PV system components for buildings (modules, inverters, etc.) must be certified through the **California Energy Commission's (CEC)** PV system certification program. The CEC requires panels installed within that state to have a PTC rating as part of their certification process.

(California Energy Commission (CEC))

(NOCT)

Another condition often referred to in panel specifications is **NOCT (nominal operating cell temperatures)**. Measurements at NOCT are made at 800 W/m^2, ambient air temperature of 20° C (68° F), and wind speeds at 1.0 m/s.

This measurement is often included with panels to give a better "real world" indication of the power the panel will produce once installed.

In 2012 the IEEE began modifying the testing criteria solar panel manufacturers must use when testing NOCT values. During this process they also introduced the term **Nominal Module Operating Temperature**

(NMOT) to replace the NOCT term, distinguishing the new modified testing procedures from the old.

NMOT

While this may be important to panel manufacturers, from the installer's point of view these terms are interchangeable. More and more panel specification sheets reference NMOT values rather than NOCT values.

Panel IV (current-volt) Curves

Every solar panel has unique power output characteristics that can be diagrammed in a chart by plotting how many watts of power are generated at each and every level of current and at every voltage. This is called an **IV curve**. Figure 2-10 shows a typical IV curve when measured at STC.

IV CURVE

The watts generated by a PV panel are equal to amps produced x volts produced. So, if amps = 0 or volts = 0, then watts will also equal zero (anything times zero equals zero).

At some point along the curve, the panel will produce the most energy (watts). But this point will not be at the maximum voltage the panel can produce (known as the **Voc**, or **open circuit voltage**) nor will it be at the maximum current the panel is capable of producing (referred to as the **short circuit current** or **Isc**). This point along the curve when the most power is produced by the panel is referred to as its **maximum power point (Pmax or Mpp)**.

OPEN CIRCUIT VOLTAGE (Voc)

SHORT CIRCUIT CURRENT (Isc)

MAXIMUM POWER POINT (Pmax)

The maximum power point when measured under standard test conditions is known as the panels **rated power**. When advertised as a 380 W panel, for example, this is the power rating referenced.

RATED POWER

When operating at the maximum power point (which only happens under STC), the voltage at which the panel will operate is referred to as the **maximum power voltage (Vmp or Vmax)**. The amperage at this point is

MAXIMUM POWER VOLTAGE (Vmax)

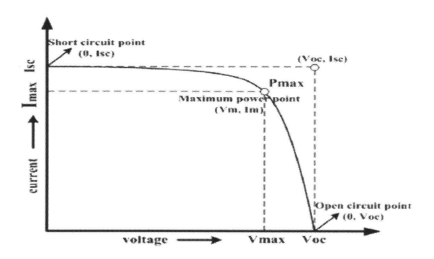

FIGURE 2-10: TYPICAL IV CURVE FOR PV PANEL AT STC

MAXIMUM POWER CURRENT (IMAX)

referred to as the **maximum power current (Imp or Imax)**. In all cases, Imp x Vmp = Pmax.

Standard test conditions represent a very specific set of environmental conditions and are used to rate solar panels. But in the real world, there are an infinite number of condition measurements that will affect how much a panel will produce.

Changing levels of sunlight result in varying levels of irradiance. Figure 2-11 shows how changes in irradiance affect the number of watts a panel will produce.

Changes in temperature also affect a panel's power production, as illustrated in Figure 2-12. The colder the cell temperature, the more power a panel will generate.

Factors that Impact Solar Modules

A number of physical factors will affect how well a solar module generates power.

Physical Size

In the world of PV panels, size does matter. There is a one-to-one relationship between the surface size of a panel and the amount of energy that panel will produce. A panel that is two square meters in surface area will produce twice as much power as a similarly manufactured panel that is only one square meter in size.

FIGURE 2-11: TYPICAL IV CURVE FOR PV PANEL AT VARIOUS LEVELS OF IRRADIATION

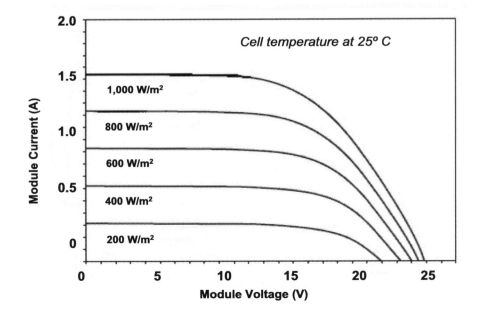

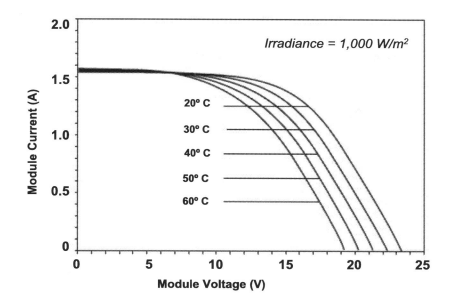

FIGURE 2-12: TEMPERATURE ADJUSTED IV CURVE.

Reflectivity

Silicon by its very nature is shiny. If left untreated, about 35% of the Sun's energy would simply bounce off or be reflected back into the atmosphere. For that reason, most panels have an anti-reflective coating applied to the surface of the solar module.

Size and Placement of the Electrical Contacts

The small electrical conductors that provide a circuit for the energy from the top layer of the panel (busbars) can also block sunlight from reaching the panel. So manufacturers must be cautious and clever about how these contacts are placed to maximize the efficiency of the panel. The smaller the wire, the more energy will reach the photovoltaic cells. Often as much as 5% of the panel's surface is blocked by wires.

Temperature

For a typical solar panel, there will be around a 0.5% decrease in performance expected for every 1° C increase (from a starting place of 25° C as tested under STC) in the air temperature. This will result in lower voltages generated by the panel as the temperature rises. The exact number is known as the panel's **temperature coefficient.**

TEMPERATURE COEFFICIENT

Conversely, as temperatures fall, the voltage output of the system will increase. On an extremely cold but clear winter day, voltage output might increase to such a degree as to exceed safe levels for the other parts of the system. So temperature corrections must be designed into the system.

FIGURE 2-13: SIX PV CELLS CONNECTED IN SERIES.

FIGURE 2-14: SIX PV CELLS CONNECTED IN SERIES WITH FINAL CELL SHADED.

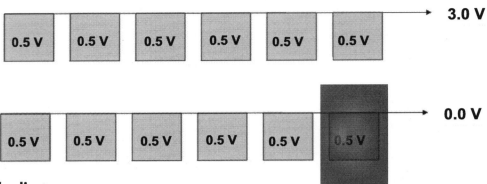

Shading

A simple leaf-covered branch, shading only a portion of a solar panel, can have a profound impact on how much energy that panel produces. To understand why, it is necessary to understand how panels are constructed.

A typical solar panel is comprised of a number of solar cells, each with the potential to generate about 0.5 V of energy. How these cells are connected together will determine the overall voltage rating of the solar panel. They are most often wired in a combination of series and parallel connections.

Assume there are six PV cells connected together in series (see Figure 2-13). Under normal conditions, the voltage will rise from one cell to the next (in the direction of the arrow), generating at the end of the series a three-volt string.

But if the cell furthest to the right is shaded (as demonstrated in Figure 2-14), it will resist the flow of electricity through it, effectively shutting off all six of the cells.

BYPASS DIODE

In order to avoid this significant loss of power due to shading, panel manufacturers will place **bypass diodes** within the panel wiring system. A diode is typically made of semi-conductor material and allows energy to flow in only one direction.

With the bypass diode in place, the flow of energy can bypass the shaded cell, carrying the remaining power through the system (as shown in Figure 2-15).

FIGURE 2-15: SIX PV CELLS CONNECTED IN SERIES WITH BYPASS DIODE IN PLACE.

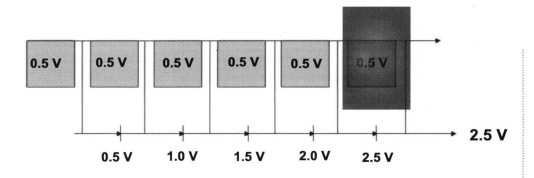

Ideally there would be one bypass diode for each solar cell, but this can be rather expensive and inserts a great deal of resistance into the panel, so normally one diode is used per group of series cells.

How diodes are incorporated into the design of the panel will affect how well a specific PV module deals with shading. Most panels currently on the market incorporate three bypass diodes into the junction box, as depicted in Figure 2-16.

The same concepts can be incorporated when connecting multiple panels together in series. Bypass diodes can be incorporated to bypass entire panels or strings that may fall into shade, or perform poorly for other reasons.

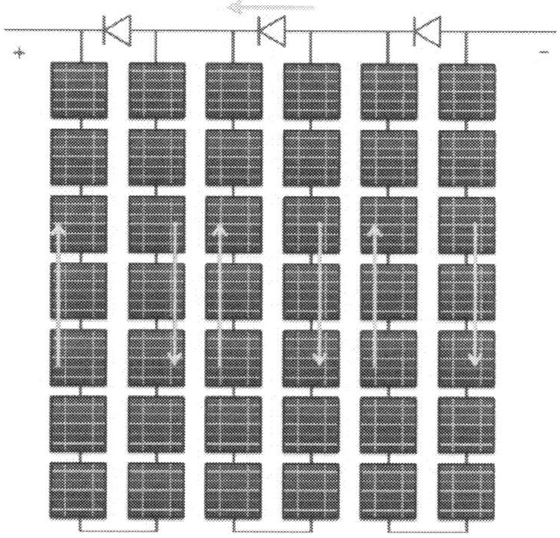

FIGURE 2-16: STRINGS OF CELLS BYPASSED WITHIN A PANEL IF SHADING BECOMES AN ISSUE

Age

As solar panels age, they become slightly less effective in generating power. The industry has long assumed that crystalline panels will degrade about 0.5% per year, meaning that at the end of the 25-year expected life span, the panel will generate 12.5% less power than it did on the day it was installed. Thin film panels are assumed to degrade at a rate about double that (1.0% per year) of crystalline panels.

Recent studies, however, have found that the assumed rate of degradation may be significantly overstated - finding a loss of only about 0.1% annual loss in installed crystalline solar arrays (tracked over a 10-year period).

Regardless of the rate of degradation, it can safely be assumed that a solar panel will produce less power as it ages.

Blocking Diodes

Another system incorporated in most panels (certainly those rated 60 watts or higher) are **blocking diodes**. Under normal conditions, electricity is flowing from the solar panels to the batteries. But at night, or if the panels are shaded, the energy stored in the batteries can flow backwards, from the battery to the panel. This would effectively drain the stored energy within the battery.

BLOCKING DIODE

To prevent this from happening, a blocking diode is placed within the system that only allows the energy to flow in one direction. These diodes are generally placed within the charge controller or inverter, or can be purchased and installed (Figure 2-17) into the wiring system on the load side of the array.

Chapter 2: Solar Cells & Solar Modules

FIGURE 2-17: BLOCKING DIODES INCORPORATED INTO A PV SYSTEM.

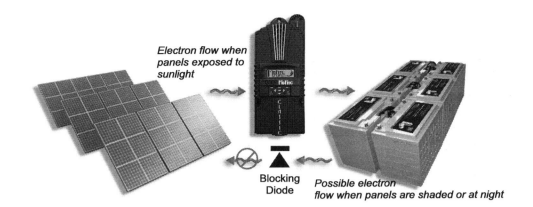

Blocking diodes may also prevent the "backfeeding" of electricity from one string of panels to another. For example, if a panel or a string of panels is shaded or damaged, other strings or panels connected to it in parallel may try to service this false load. A blocking diode would prevent this **backfeed**.

BACKFEED

In some countries, blocking diodes are allowed to replace overcurrent protective devices (fuses and breakers). This is an effective method of overcurrent/fault prevention, provided the blocking diode is reliable.

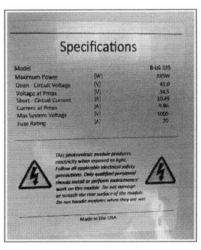

FIGURE 2-18: RATING INFORMATION STICKER ON BACK OF A SPECIFIC PV PANEL.

Typical Panel Ratings

Look on the back of a typical commercial PV panel and you will find a label that outlines that particular panel's ratings (an example of which can be found in Figure 2-18).

While the information displayed will vary from manufacturer to manufacturer - critical information needed in selecting the right panel for any given application includes:

- Maximum Power (Pmax): The power (watts) the panel will produce when operating at its maximum power point (MPPT).
- Open Circuit Voltage (Voc): Theoretically the most voltage the panel can produce (without a load) under STC.
- Maximum Power Voltage (Vmp): The voltage the panel will produce when operating at its maximum power point. This should not be confused with the **nominal voltage** of the panel. For example, a panel with a Vmp of 36.36 V (such as the 385 W panel detailed in Figure 2-19) might have a nominal voltage of 24 V, meaning it is compatible with a nominal 24-volt battery bank and charge controller.
- Short Circuit Current (Isc): Theoretically the most current the panel can produce (without a load) under STC.
- Maximum Power Current (Imp): The current the panel will produce when operating at its maximum power point.

NOMINAL VOLTAGE

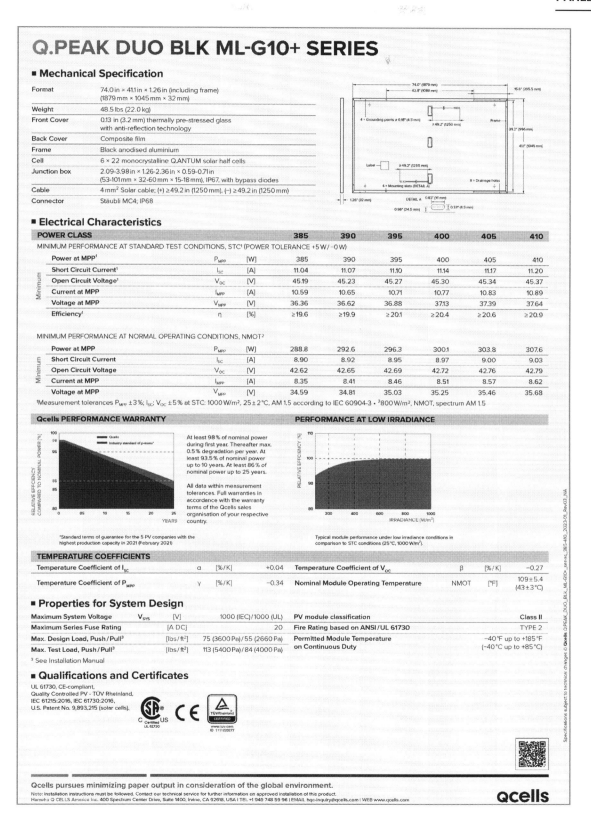

FIGURE 2-19: SOLAR PANEL DATA SHEET

- Temperature Coefficient: How will voltage be affected for every degree Celsius above or below standard test conditions.

Other information which may or may not be included directly on the panel (but will certainly be contained in the supporting documentation) includes:
- type of cell (monocrystalline, polycrystalline, etc.),
- cell configuration (how many in series, parallel, etc.),
- module efficiency,
- maximum system voltage,
- series fuse rating,
- connector type,
- operating temperatures,
- warranty information,
- standards the panel was tested against,
- and mechanical characteristics such as length, weight, etc.

Resistance to Hail Damage

In 1996, Underwriters Labs developed the first test standard to assess the impact resistance of roof coverings – **UL 2218** *Impact Resistance of Prepared Roof Covering Materials*. This standard has subsequently been used to measure the effectiveness of solar panels and solar shingles in resisting damage from hail.

This test uses steel balls ranging in size from 1.25 - 2.0 inches (32 - 51 mm) in diameter to simulate damage that might occur in a hail storm. The steel balls are dropped from heights of 12 ft (3.66 m) for the 1.25 inch (32 mm) ball to 20 ft (6 m) for the 2 inch (51 mm) ball. The test assembly is struck with the steel ball twice in the same location.

To meet the acceptance criteria of UL 2218, the solar panel's exposed surface and back surface must show no evidence of tearing, fracturing, cracking, splitting, rupture, or other evidence of opening.

Qualifying products are given a class rating depending upon successful performance of the assembly under impacts as noted in Table 2-1.

As weather events grow increasingly severe, resistance to hail damage becomes more and more important. GCube Insurance, an underwriter for renewable energy, said about 54% of all incurred solar insurance losses from 2018-2023 can be attributed to hail, despite being only 1.4% of the total number of claims filed.

Hailstorms are often unpredictable in nature and very destructive to solar panels. In 2021, hail related losses exceeded $1 billion across the United States.

Rating	Description
1	Sample did not crack when hit twice in the same spot with a steel ball measuring 1.25 inches in diameter.
2	Sample did not crack when hit twice in the same spot with a steel ball measuring 1.50 inches in diameter.
3	Sample did not crack when hit twice in the same spot with a steel ball measuring 1.75 inches in diameter.
4	Sample did not crack when hit twice in the same spot with a steel ball measuring 2.00 inches in diameter.

TABLE 2-1: HAIL RESISTANCE RATINGS AS MEASURED BY UL 2218

Chapter 2 Review Questions

1. Solar Cells rely on a materials such as silicon that allow some, but not all of the energy to flow through them. These materials are known as:
   A) p-n junction compounds
   B) semi-conductors
   C) doped conductors
   D) a diode

2. The energy from a photon striking a solar panel must be at least as much as is required to "knock" an electron across the space where the top wafer of a solar cell and the bottom wafer meet. Silicon has a relatively low _____ energy level (1.1 electron volts - or 1.1 eV).
   A) transference zone
   B) convergence zone
   C) PV junction
   D) band gap

3. A _____ is typically made of a semi-conductor material and allows energy to flow in only one direction.
   A) diode
   B) capacitor
   C) conductor
   D) insulator

4. A _____ is necessary for electrons to flow from the negatively charged portion of the solar panel to the positively charged portion of the panel.
   A) permit
   B) circuit
   C) diode
   D) charge controller

5. Silicon works well as a material from which to construct solar cells because:
   A) it is doped
   B) it is an n-type material
   C) it is an excellent conductor
   D) it is a semiconductor

6. _____ are made by slicing a single silicon crystal into wafers. They are generally more expensive, but are also more efficient in converting the energy of the sun into electricity.
   A) Monocrystalline cells
   B) Multi-junction photovoltaic cells
   C) polycrystalline cells
   D) Thin-film

7. _____ are constructed by pouring molten silicon into bars, and then slicing wafers from the hardened material. This type of PV cell is less expensive to produce, but is also a bit less efficient.
   A) Monocrystalline cells
   B) Multi-junction photovoltaic cells
   C) Polycrystalline cells
   D) Thin-Film

8. _____ employ multiple layers of thin films. Each layer can more efficiently absorb light within a very narrow band. In combination, these cells can more efficiently absorb light over a wider portion of the electromagnetic spectrum.
   A) Monocrystalline cells
   B) Multi-junction photovoltaic cells
   C) Polycrystalline cells
   D) Thin-film

9. One advantage of _____ technology (also referred to as amorphous, meaning "without shape") is that they are flexible and can be more easily integrated into the building design - such as solar shingles.
   A) monocrystalline cells
   B) multi-junction photovoltaic cells
   C) polycrystalline cells
   D) thin-film

10. Which of the following is **NOT** a common type of thin film solar cell?
    A) Amorphous Silicon (a-Si)
    B) Lithium-ion Polymer (LiPo)
    C) Cadmium Telluride (CdTe)
    D) Copper Indium Gallium Selenide (CIGS)

11. What emerging PV technology results in transparent or many colored cells, making them ideal for solar windows or other building integrated applications?
    A) multi-junction thin film
    B) flat-plate modules
    C) two-dimensional solar cells
    D) perovskite solar cells

12. Which of the following is **NOT** a relatively recent innovation in solar cell design?
    A) bifacial cells
    B) PERC cells
    C) half-cut cells
    D) encapsulated cells

13. A solar cell comprised of one layer of silicon and one layer of perovskite is an example of a:
    A) bifacial cell
    B) polycrystaline cell
    C) split cell
    D) tandem cell

14. Which of the following is the smallest, most basic component?
    A) a solar photovoltaic array
    B) a solar photovoltaic module
    C) a solar photovoltaic cell
    D) a solar photovoltaic panel

15. "A mechanically integrated assembly of modules or panels with a support structure and foundation, tracker, and other components, as required, to form a dc or ac power-producing unit" is known as a:
    A) solar array
    B) solar cell
    C) solar panel
    D) solar module

16. Historically, solar panels sold in the United States were manufactured and marketed in accordance with _____.
    A) ANSI/UL 1703
    B) IEC 62446-1
    C) IEEE 1547.1
    D) NFPA 70

17. In 2017 the US and the International manufacturing standards for PV module safety were harmonized. The new standard now referenced for the construction and testing of flat-plate solar modules is:
    A) ANSI/UL 1703
    B) UL 61730
    C) IEEE 1703
    D) NFPA 1703

18. Under Standard Test Conditions (STC) for solar panels, the amount of sunlight hitting the panel (known as irradiance) is assumed to be
    A) 1,000 watts per square meter
    B) 364 watts per square foot
    C) higher in cold weather than in warm weather
    D) dependent upon the relative humidity

19. Standard Test Conditions also assume two other conditions (other than irradiance) be present. These include:
    A) wind speeds of less than 5 mph, and ambient air temperatures of 77 degrees Celsius.
    B) wind speeds greater than 5 mph, and ambient air temperatures of 25 degrees Celsius.
    C) atmospheric pressure equal to 1.5, and cell temperatures of 77 degrees Celsius.
    D) atmospheric pressure equal to 1.5, and cell temperatures of 25 degrees Celsius.

20. One main difference between STC and PTC (PV USA Test Conditions) is that:
    A) PTC is only used in the USA.
    B) PTC does not concern itself with atmospheric pressure.
    C) PTC addresses wind speed while STC does not.
    D) PTC measures in Imperial units (U.S. based) rather than metric.

21. Every solar panel has unique power output characteristics that can be diagrammed in a chart by plotting how many watts of power are generated at each and every level of current and at every voltage. This is called:
    A) the maximum efficiency diagram.
    B) the I-V curve.
    C) multi-junction display pattern.
    D) manufacturer's reliability function.

22. At some point along the diagram, the panel will produce the most power (watts). But this point will not be at maximum voltage (the Voc) nor will it be at maximum current (Isc). This point along the curve when the most power is produced by the panel is referred to as its _____.
    A) nexus
    B) matrix
    C) p-n junction
    D) maximum power point

23. The _____ is the rating on a solar panel that indicates the theoretical maximum voltage a panel can produce when measured under standard test conditions.
    A) maximum power (Pmax)
    B) maximum power voltage (Vmp)
    C) open circuit voltage (Voc)
    D) short circuit voltage (Isc)

24. The _____ is the rating on a solar panel that indicates the voltage a panel will produce when generating its rated power when measured under standard test conditions.
    A) maximum power (Pmax)
    B) maximum power voltage (Vmp)
    C) open circuit voltage (Voc)
    D) short circuit voltage (Isc)

25. In space (at the outside edge of the Earth's atmosphere), the energy of the sun (irradiance) will measure about:
    A) 1.5 atmospheres
    B) 1.0 atmospheres
    C) 0.0 atmospheres
    D) 1,300 watts/m2

26. NOCT stands for:
    A) nominal operating cell temperatures.
    B) normal operating test conditions.
    C) national official certified test.
    D) normal observed conditional testing.

27. The cell temperature of a panel is measured and found to be 30°c. This panel will produce _____ energy compared to a similar panel measured at 25°C.
    A) more
    B) less
    C) the same
    D) twice the

28. How much a panel's energy production increases or decreases for every degree Celsius the cell temperature varies from STC is known as the panel's:
    A) IV curve.
    B) open circuit voltage.
    C) nominal operating cell temperature.
    D) temperature coefficient.

29. Which of the following is **NOT** a factor that will affect the performance of a solar cell?
    A) the band gap energy characteristics
    B) the internal load resistance
    C) the size of the cell
    D) its Ah capacity

30. In order to avoid a significant loss of power due to shading, panel manufacturers will often place _____ in the panel wiring system.
    A) bypass diodes
    B) blocking diodes
    C) rectifying diodes
    D) amplifying diodes

29. At night, or if the panels are shaded, the energy stored in the batteries can flow backwards, from the battery to the panel. To prevent this from happening, manufacturers place _____ within the charge controller.
    A) bypass diodes
    B) blocking diodes
    C) rectifying diodes
    D) amplifying diodes

30. Which of the following is **NOT** a factor that affects how much power a solar panel will produce?
    A) the size of the panel
    B) the reflective coating on the panel
    C) the air temperature
    D) the relative humidity

31. Which of the following will likely create the greatest **REDUCTION** in a PV module's output?
    A) measured cell temperature of below zero degrees Celsius
    B) the age of the panel is 10 years old
    C) approximately one-quarter of the panel is covered in snow
    D) the 200 W panel is half the size of another 200 W panel

Lab Exercises : Chapter 2

Lab Exercise 2-1:

Supplies Required:
- Monocrystalline module
- Polycrystalline module
- Thin Film module

Tools Required:
- Multimeter
- Solar Panel Tester Photovoltaic Multimeter (such as Frogbro)
- Screwdriver

2-1. **Examination of Solar Modules**

Select one of the PV modules. Turn it over and examine the label on the back. Note the various specifications listed.
a. Note the Voc, Isc, Vmp, Imp, and Pmax.
b. Note the electrical contacts on the face of the panel.
c. Remove the cover of the junction box on the back of the panel (if one is in place).
d. Locate the bypass diode(s) if any.
e. Set the multimeter to DC voltage. Attach the leads from the meter to the negative and positive leads of the panel. Note the voltage output of the panel.
f. Discuss whether the readings make sense, given the panel specifications and the available sunlight.
g. Reverse the polarity of the leads (negative to positive, etc) and note the new reading. What has changed?
h. Repeat for all panels.
i. Connect solar panel tester to a panel and obtain a reading (in Watts) in the classroom. Take the panel outside and test it again in direct sunlight. Compare the results to Voc and discuss whether the panel is producing as expected.

Chapter 3
Types of Photovoltaic Systems

Chapter Objectives:

- Identify the various types of photovoltaic systems and how they function.
- Determine what constitutes an AC-coupled, DC-coupled and hybrid PV system.
- Be aware of the voltage limits imposed upon PV systems.
- Evaluate system loads and how that effects the size and composition of a PV system.
- Become familiar with the various components of a PV system and where they fall within the total system design.
- Understand concepts such as kilowatt hour, system load, maximum power draw, power rating affect system design.
- Determine how energy efficiencies can reduce the system load demands.
- Conduct a load analysis.
- Understand maximum power draw versus daily load demand.

The purpose, and perhaps the location, of the PV system will in large measure determine how that system is configured. In broad terms, there are four major types (or configurations) of PV systems.

Stand-Alone or Off-Grid

In the early days of photovoltaics, systems were developed to provide power where no other power source was easily or economically available.

These systems were not connected to any form of utility power, and relied on **battery banks** to service the electrical load (and the solar panels were used to keep the batteries charged).

Examples of a **stand-alone PV system** (also referred to as **off-grid PV system**) as illustrated in Figure 3-1 may include a very remote mountain cabin, an isolated railroad signaling device, or even a small pocket calculator. Recreational vehicles and boats are also popular locations for stand-alone PV systems.

A stand-alone PV system is the configuration most people think of when they first begin to explore the world of photovoltaics. The dream of living independent of society's infrastructure appeals to the pioneer spirit.

However, batteries are expensive and require a

(BATTERY BANK)

(STAND-ALONE PV SYSTEM)

(OFF-GRID PV SYSTEM)

FIGURE 3-1: DIAGRAM OF A STAND-ALONE PV SYSTEM

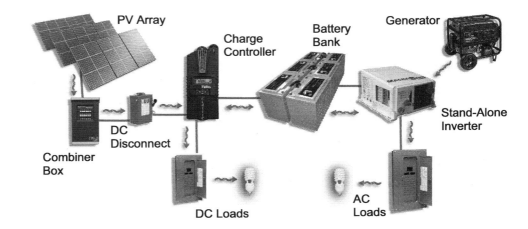

great deal of maintenance. As a result, stand-alone systems today represent only a very small minority of the systems installed in the U.S.

In areas that rely on diesel generators for power, stand-alone PV systems are increasingly an attractive alternative as prices for systems and batteries continue to fall. The industry expects to see a 7.5% year-on-year growth in off-grid systems worldwide over the coming decade.

In rural locations, the cost of connecting to the electric grid can be quite expensive. Installation of wire to connect the home to the grid generally runs between $10-$25 per foot. So if a home is located 1,000 feet from the point where it connects to the grid, it may cost up to $25,000 to make the connection. At these prices, installing a stand-alone PV power system may be the most economical option.

Grid-Tied Systems

In recent years (specifically in developed countries), most residential PV systems (around 90%) are connected to the **electrical grid**. These basic **grid-tied PV systems** contain no battery bank for energy storage, but rely on the

ELECTRICAL GRID

GRID-TIED PV SYSTEM

FIGURE 3-2: DIAGRAM OF A GRID-TIED PV SYSTEM

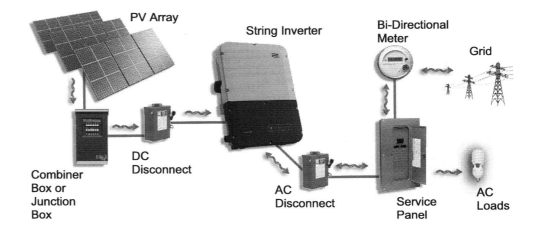

grid as a "virtual" storage system. The NEC refers to this type of solar PV system as an **interactive system**.

INTERACTIVE SYSTEM

In a simple grid-tied PV system, when there is enough sunlight available, the home runs off the power generated by the PV system. Any excess energy generated is fed into the electrical grid (in effect, the home becomes a very small power plant). When the home needs more power than it is generating from the PV system (such as at night, or during a storm), the required power that the PV system is not producing at that moment is supplied from the grid.

In effect, the electrical grid serves as a battery bank for the system. This avoids the cost of a battery system (often more than half of the total system cost), and also avoids the maintenance issues involved in dealing with batteries. As a result, grid-tied systems are far and away the most common and lowest cost type system installed in developed countries today.

A simple grid-tied system, such as the one depicted in Figure 3-2, has the advantage of requiring fewer components than a stand-alone PV system configuration. It is therefore normally less expensive to design and install. Installed prices for the average residential grid-tied system have, in recent years within the U.S. Market, stabilized at around $3 per watt, as shown in Figure 3-3.

One disadvantage, however, is that grid-tied photovoltaic systems are designed to shut down when the grid loses power. This avoids a situation where workers repairing the lines may assume there is no power present on the grid, only to find that a PV system is pumping electricity onto supposedly "dead" lines (known as **islanding**). Workers have been killed or severely injured when a distributed energy system (such as a generator) continued to operate while connected to a grid under repair.

ISLANDING

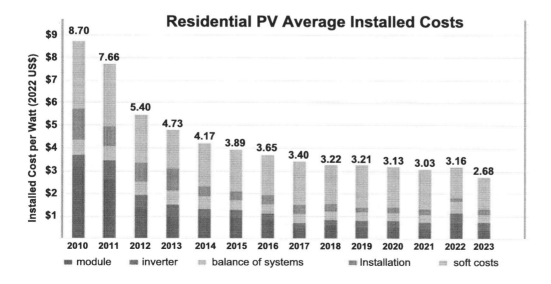

FIGURE 3-3: AVERAGE INSTALLED COST OF RESIDENTIAL SOLAR GRID-TIED SYSTEM IN US

(FROM NREL)

Chapter 3: Types of Photovoltaic Systems

As a result, a typical grid-tied system will only be as reliable as the electrical grid to which it is connected.

Grid Interactive with Battery Backup

To avoid the problem of loss of power when the electrical grid goes down, some grid-tied systems incorporate a battery backup system into their design.

This is a small but rapidly growing design option for residential solar. In 2020, a study by Berkeley Labs found that only about 6% of installed residential solar systems incorporated batteries. One reason for this is that incorporating batteries adds about $1.20/W to the cost of the system (about a 35% increase over a simple grid-tied system).

Lower battery costs, as well as changes in utility pricing schemes, are motivating many consumers to consider adding batteries to their systems. According to BloombergNEF, in 2023, over 70% of residential solar systems in Germany and Italy, as well as 20% in Australia and 13% within the US, had batteries attached.

AC-coupled or DC-coupled systems may be installed at a location for a number of reasons.

In residential settings, these normally include:

- the grid is unreliable. The battery bank serves as a backup power supply when the grid goes down.
- net-metering isn't permitted at the location. A battery bank can be set to store excess energy (since it cannot be exported to the grid), and then offset power that might otherwise be purchased from the grid when the array is not producing sufficiently to service the loads (such as at night, for instance).
- net-metering policies are inadequate. Many utilities are shifting the way they compensate for power that is exported to the grid. Rather than paying retail (rolling the meter backwards), they have shifted to simply crediting the "**avoided cost**" or the wholesale price they would have to pay to purchase power from another energy generator. The result is that the cost of energy the homeowner pays the utility may be $0.20/kWh (for

(AVOIDED COST)

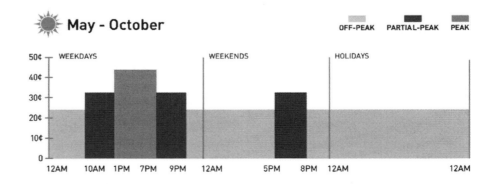

FIGURE 3-4: TIME-OF-USE PRICING STRUCTURE PUT IN PLACE IN 2016 BY PACIFIC GAS & ELECTRIC

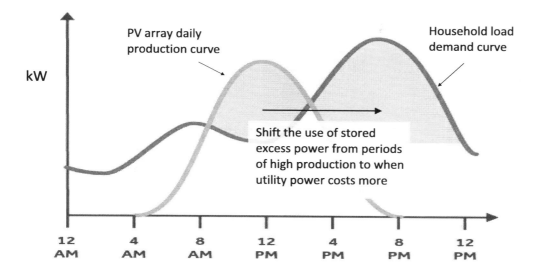

Figure 3-5: Load Shifting for household located where evening power from the utility costs more than afternoon power

example), but energy sent back to the grid would only receive a credit of $0.05/kWh.

As a result of this compensation scheme, it makes financial sense for the homeowner to avoid sending energy back to the grid, but rather store on site when there is excess energy being generated by the array, and then using it when the array is not producing enough to service the loads. This practice is known as **self consumption**.

- **time of use pricing.** If the utility pricing scheme charges more for power during periods of high demand (Figure 3-4), the customer can shift their consumption by storing excess power generated by the PV array (rather than selling it back to the utility) and then using the stored power during

(Self Consumption)

(Time of Use Pricing)

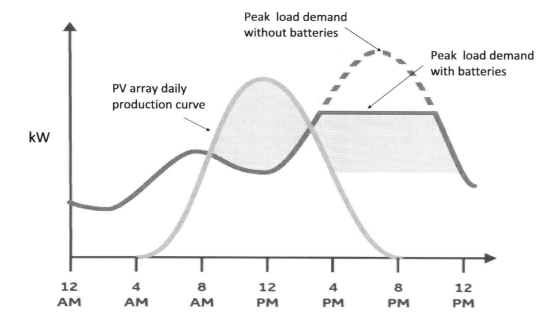

Figure 3-6: Peak shaving affect on peak load demand by incorporating battery bank

(LOAD SHIFTING)

(DEMAND CHARGES)

(PEAK SHAVING)

(SELF-CONSUMPTION GRID-TIED)

periods of high-cost power (rather than purchasing it from the utility). This process is also known as **load shifting,** as illustrated in Figure 3-5.

Utilities often charge commercial customers a rate based on the 15-minute interval within the month where the most power was used. The higher the use, the higher the bill. These are referred to as **demand charges.**

Customers can reduce their peak usage, offsetting it with power from the battery bank, as illustrated in Figure 3-6. This can dramatically reduce their monthly bill from the utility company. Peak demand reduction, commonly called **peak shaving**.

PV systems that incorporate batteries designed to deal with load shifting and/or peak shaving are often referred to as **self-consumption grid-tied** systems.

Grid-Coupled Battery Sizing

The sizing of self-consumption battery banks often involves obtaining a detailed load-versus-time profile of home energy consumption profile using energy-monitoring equipment.

The data from the solar array must also be collected, determining how much excess power is sent back to the grid during peak generation times. Most grid-tied inverters collect and report this data.

This information can be used to size a PV array to offset all of or the desired portion of the home's energy consumption. The battery bank is sized to store the portion of the PV array's daily energy that would normally be exported to the grid in a net-metered system. Designers often oversize battery banks, assuming power outages will last longer than they actually do.

Eaton's *2015 Blackout Tracker* report found that of the more than 3,500 outages in the U.S., the average duration was only 49 minutes. The Electric Power Research Institute also estimated that 57% of U.S. power outages last less than five minutes. Depending on the client's system availability requirements, a relatively small battery bank can handle most grid outages. For example, a typical home in the U.S. consumes about 30 kWh per day. This power is not consumed at a uniform rate (30 kWh / 24 hrs). An hourly load assessment could be conducted, but for this example assume during waking hours the client consumes 2 kWh per hour.

A lithium ion battery bank designed to provide one (1) hour of emergency backup would be sized:

load demand / system voltage / depth of discharge / inverter efficiency, or

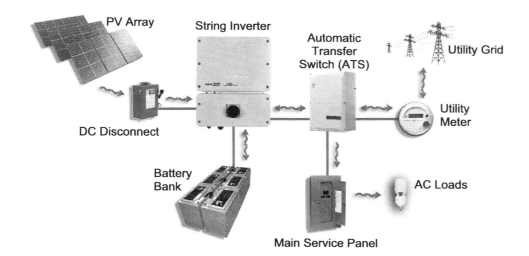

Figure 3-7: Diagram of a DC-Coupled Multimode system with whole home backup

2 kWh / 48 volts = 41.66 Ah / 95% DOD = 43.85 Ah / 97% = 45.2 Ah

As most grid-coupled systems are marketed in terms of kWh rather than the more traditional Ah rating of battery systems, and the DOD and inverter efficiency is already calculated into this rating, then this system would need a 2 kWh battery bank to deal with any power outages less than one hour in duration. Effectively, this system acts as an **uninterruptible power source (UPS)** for the entire house.

DC-Coupled

The most common design, a **DC-coupled multimode system** (as seen in Figure 3-7), might be desirable if the installation is located in an area with a dodgy grid, or if there are critical loads within the building that absolutely must have power all the time (such as within a hospital).

In a DC-coupled system, the multimode inverter is connected to both the solar array (which produces DC power) and a battery bank (which stores and produces DC power). When the grid is operating normally, the inverter keeps the batteries fully charged, and feeds power to the loads and to the grid, just like a simple grid-tied inverter.

But when the grid goes down, the inverter or more commonly a **"smart switch"** or **automatic transfer switch (ATS)**, senses this and disconnects the home from the grid (avoiding islanding). It then draws on the array and the battery bank to service all the loads in the home.

However, servicing all the loads during a power outage may require a very large battery bank and may not be necessary (not all the loads are critical during a power emergency).

So it is typical that the system is designed so only specific circuits (like a refrigerator, selected outlets, furnace fan, sump pump, etc). receive power

Figure 3-8: Diagram of an AC-Coupled Multimode PV system

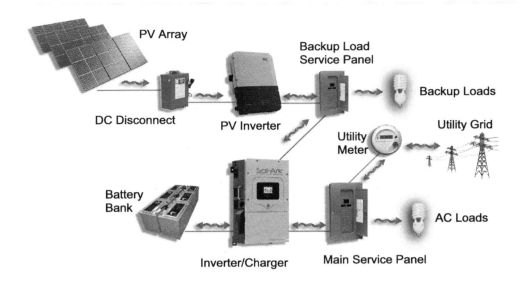

from the battery bank (rather than the entire building) when the grid is down. These emergency loads are normally separated from the main service panel and wired through a **back-up load panel.**

(BACK-UP LOAD PANEL)

Incorporating a battery bank into a typical grid-tied system will add substantially to the overall cost and maintenance requirements of the system.

AC-Coupled

Systems that convert the DC power of an array into AC power, and then convert it back to DC to charge a battery bank, are gaining in popularity. At first glance, this may seem an unnecessary step and added complexity, but there are a number of advantages.

(AC-COUPLED MULTIMODE SYSTEM)

AC-coupled multimode systems, such as the one illustrated in Figure 3-8, allow owners of existing PV systems (both string inverter systems as well as micro inverter systems) to add a battery backup without changing their installed system configuration.

In an AC-coupled system, there is usually already an inverter present that converts the DC power from the array to AC power that can be used in the home. This may be micro inverters that are connected directly to the solar panels, or a string inverter that is connected to the entire array (often called the **PV inverter**).

(PV INVERTER)

This AC power must be fed to the battery bank, which only receives DC power. So it must undergo a conversion, typically through a charger that is either built into the battery bank storage system (more properly referred to as an **energy storage system (ESS)**) or incorporated into an AC-coupled **inverter/charger**.

(ENERGY STORAGE SYSTEMS (ESS))

(INVERTER/CHARGER)

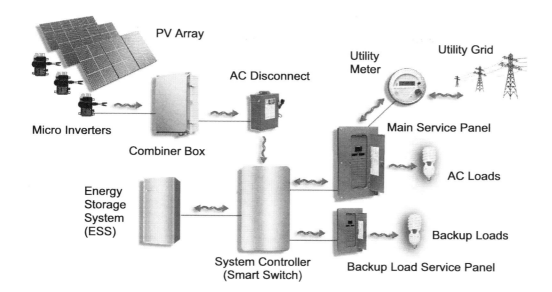

Figure 3-9: Diagram of an Enphase AC-Coupled System

In some systems, the inverter/charger is not a separate piece of equipment, but built into the ESS. In these cases the power flow is controlled by a smart switch, illustrated in Figure 3-9.

When the grid is operating normally, power passes through the inverter/charger (or smart switch) from the array and services the energy storage system (keeping the batteries full), the main service panel, and the back-up load panel.

But when the grid is down, a relay automatically disconnects the system from the main service panel (which is where the home connects to the grid) and only powers the items wired into the back-up loads panel.

The inverter/charger also sends a signal to the existing inverter (or micro inverters) that "fools" them into thinking the grid is still operational, avoiding shutdown due to the anti-islanding feature within the inverter.

Integrating Multiple Generating Sources (Hybrid Systems)

Distributed energy systems (such as PV systems) can integrate multiple generating sources into the system.

For example, integrating a wind turbine with a PV array can (in many instances) provide complimentary power sources, as the wind typically blows during periods when the sun is not shining. These are referred to as **hybrid systems.**

HYBRID SYSTEM

Far and away the most common form of a hybrid system incorporates a generator as backup to the PV array so the system can be sized to handle normal operations and the generator provides backup during extreme energy emergencies.

PV System Voltage Limits

Photovoltaic systems must be designed to operate within the voltages limits established by the local authorities, as well as the capacity limitations of the equipment used.

In North America, the NEC and the CEC limit the voltage of solar PV systems installed in residential buildings (one or two-family dwellings) to under 600 Vdc. For multi-family and commercial units, the maximum voltage in the US is 1,000 Vdc. Commercial ground mounted systems can be installed up to 1,500 Vdc.

Europe's electrical standards body, the IEC, considers 1,500 Vdc the low-voltage limit and enables certification to that voltage. Although most PV systems installed in Europe (and pretty much everywhere outside of North America) use equipment limited to 1,000 Vdc.

Battery Bank Voltage Limits

It has long been understood that the NEC imposed a 50 Vac /60 Vdc limit on battery banks installed within buildings. But in the 2017 NEC, a new article was added, Article 706: Energy Storage Systems. This article applies to all permanently installed energy storage systems (this was changed in the 2020 NEC to include even temporary systems) that operate above the 50 Vac/ 60 Vdc limit.

An energy storage system (ESS) is defined as *"one or more components assembled together capable of storing energy for use at a future time. ESS(s) can include but is not limited to batteries, capacitors, and kinetic energy devices (e.g. flywheels and compressed air). These systems can have AC and DC output for utilization and can include inverters and converters to change stored energy into electrical energy."*

In other words, the term applies to battery banks, as well as assembled units such as a Tesla Powerwall that incorporate an inverter within the unit. A qualifying ESS must be listed as complying with UL1973, a safety standard for stationary battery systems.

An accessible ESS installed in a residential dwelling is limited to 100 volts or less. However, if the live parts of the ESS are not accessible during routine maintenance (in other words, they can only be accessed by a trained and authorized technician), then the unit can exceed this voltage limitation (no upper limit is stated, but since Article 690 limits voltages to 600 Vdc for residential systems, that is the effective limit).

Many such systems are available. For example, the LG Chem 9.8 kWh lithium ion battery (RESU10H) operates at a nominal 400 Vdc (with a voltage range of 430 Vdc - 550 Vdc).

Advantages of Higher Voltage

Higher-voltage systems are less expensive than similarly sized low-voltage options, because fewer and smaller materials are needed.

At its most basic, higher voltage means lower amps. Higher voltage and lower amps means smaller wire, as well as smaller and fewer components - which translates into significant cost savings. Higher-voltage systems enable longer strings, which translates into fewer combiner boxes, fewer circuit breakers, smaller and less wiring, and therefore less labor.

Industry studies show that installing a 1,000 Vdc system versus a 600 Vdc system will result in an approximately 40 percent BOS wiring savings and up to 2 percent performance improvement, resulting in a cost savings of about $0.10 per watt.

Increasing the voltage to 1,500 Vdc would result in additional savings of approximately $0.05 per watt.

Determining the System Load

The first step in designing a solar PV system is to determine the load to be serviced (which will in turn determine the size of the the system).

When adding a PV system to an existing home that is already connected to the electrical grid, determining the **system load** requirements is quite easy and straightforward. Simply review the past year's utility bills. Averaging them over the course of a year should provide a pretty good idea of the electric needs of the household.

Of course the size of the system could and should be minimized by instituting a number of **energy efficiencies** within the home. As they say, "eat your energy efficiency vegetables before deciding how much renewable energy dessert you should take".

For homes not yet built, budget or norms may be the best determiners of how much electricity will be required within the structure. Electricity is sold in units known as a **kilowatt hour (kWh).** One kWh is equal to one thousand watts of electricity consumed over a period of one hour.

According to U.S. government statistics, the average home in the U.S. consumes about 890 kilowatt-hours (kWh) of electricity each month (as of 2021). The average detached home in the United Kingdom uses about 346 kWh per month, while the average home in China uses about 125 kWh per month. Iceland holds the record as the nation that consumes the most electrical energy per person.

For stand-alone systems, it may be necessary to create an inventory of each electrical device to be used in the home. Then calculate the energy consumption (in watts) of each device, the anticipated amount of time each device will be used – and total them all to determine the anticipated load requirements.

Most appliances will specify the amount of energy (in watts) that they draw when in use. If this information is not available, a handy and inexpensive tool is widely available that can assist in measuring a specific electrical device's power consumption. The Kill A Watt® meter plugs into any standard outlet. The appliance is then plugged into the front of the meter. When in use, the meter will display the watts consumed. It can also be set to display the volts, amps and hertz feeding the device.

FIGURE 3-10: KILL-A-WATT METER HOOKED UP TO A TOASTER

Figure 3-10 demonstrates a toaster plugged into the meter, indicating that it is drawing 807 watts. To determine the daily use of this appliance, simply multiply the watts used by the amount of time the appliance will be used each day. For example, if it is anticipated that this toaster will be used on average about 15 minutes per day, then the daily use is calculated by multiplying 807 watts x .25 hours (15 minutes) = 202 Wh per day. This process is referred to as a **load analysis**.

LOAD ANALYSIS

Calculating the daily load for an appliance that consumes energy at a constant rate is fairly straightforward (energy demand x hours used each day). But some appliances (such as a refrigerator) consume more power when first turned on, then less as they operate. They also operate in cycles, making a daily prediction of consumption difficult. With such a load, a good method of assessing the daily load demand is to plug into the watt meter for a typical 24-hour period and then read the kW consumed indicator.

Load Analysis Spreadsheet

Example: The client has decided to build a small cabin on a five-acre farm located in a remote area - and as a result, far from any utility hookup. The first step is to determine the load requirements of this stand-alone system.

Assume the following electrical requirements:
- refrigerator/freezer (runs 10 hours each day)
- radio/CD-player (used 3 hours per day)

Load Description	Quantity	Operating Watts	# Hours/Day	Daily Watt hrs
Refrigerator/freezer	1	Variable		7500
Radio/CD-player	1	70	3	210
Cistern pump	1	1100	1	1100
fluorescent lights	8	20 x 8 = 160	3	480
Ceiling fan	2	100 x 2 = 200	5	1000
Radio/alarm clock	1	10	24	240
Microwave	1	900	0.5	450
Laptop Computer	1	50	8	400
Solar array electronics	1	40	5	200
Odds and ends	N/A	100	5	500
TOTAL:				12,080 Wh

TABLE 3-1: LOAD ANALYSIS FOR SMALL ONE-BEDROOM CABIN

- cistern pump (operates 1 hour per day)
- 8 compact fluorescent lights (average 3 hours per day)
- ceiling fan (5 hours per day)
- radio/alarm clock (on all the time)
- microwave (30 minutes per day)
- laptop Computer (8 hours per day)
- solar array electronics (inverter and such - 5 hours per day)
- unanticipated odds and ends (things we may plug in over time)

As indicated in Table 3-1, the cabin will use 12,080 Wh or 12.08 kWh each day. Now make a few energy efficiency changes. How will this affect the daily usage (and as a result, reduce the required size of the PV system)?

Changes to the household:
- smaller, more energy efficient refrigerator/freezer (125 operating watts/hr)
- more efficient water pump (500 operating watts/hr)
- only install one ceiling fan
- reduce light usage from 3 hours per day to 2 hours per day

How will these changes affect the electrical usage?

Energy Use after Energy Efficiency Measures

As seen in Table 3-2, by making a few minor energy efficiency changes, the required size of the PV system can be reduced to almost one-third of the original estimate (reducing the cost of the system by at least half).

Maximum Power Draw versus Daily Load

The **daily load** of a system is the total power (measured in watt-hours or kilowatt-hours) that a household will use during a typical 24-hour period.

DAILY LOAD

It accounts for the electrical consumption rate of all the appliances, lights, motors, etc. within the household, times the number of hours those items will draw electricity during a typical day. A PV system (both stand-alone and

TABLE 3-2: LOAD ANALYSIS FOR SMALL ONE-BEDROOM CABIN AFTER MAKING ENERGY EFFICIENCY ADJUSTMENTS

Load Description	Quantity	Operating Watts	# Hours/Day	Daily Watt hrs
Refrigerator/freezer	1	Variable		1250
Radio/CD-player	1	70	3	210
Cistern pump	1	500	1	500
fluorescent lights	8	20 x 8 = 160	2	320
Ceiling fan	1	100	5	500
Radio/alarm clock	1	10	24	240
Microwave	1	900	0.5	450
Laptop Computer	1	50	8	400
Solar array electronics	1	40	5	200
Odds and ends	N/A	100	5	500
TOTAL:				**4,570 Wh**

grid-tied) must be sized to meet this daily load if the system is to provide all the electrical needs of the household.

As a result, the daily load will be critical in determining the size of a PV array. However, some components of the system (such as an inverter in a stand-alone system) are not impacted by how much power is used over the course of a day. They are sized based on how much power is used at any one time (**continuous power rating**).

CONTINUOUS POWER RATING

So it is not only necessary to determine how much power is required to operate the household on a daily basis, but for a stand-alone system it is also necessary to calculate how much power may be required at any one time.

For example, a refrigerator may draw 400 watts (its power rating). If it runs for one hour, it will consume 400 watt-hours of power. If it runs for six hours during a typical day, it will contribute 2,400 watt-hours of consumption (400 watts x 6 hours) to the daily load of the system. But it will never draw more than 400 watts at any one time.

Or a home may have 10 light fixtures (each rated at 20 watts). However no more than six are ever operated at one time. So the **maximum power draw** for the lights would be 120 watts (6 x 20 watts).

MAXIMUM POWER DRAW

It is possible to determine the maximum power draw for the small one-bedroom cabin in this example, as demonstrated in Table 3-3. Assume no more than six of the fluorescent lights are used at any one time, but that all the other electrical devices might be used at the same moment (simultaneously).

With these assumptions, it is clear that at any one moment, the cabin may draw as much as 2,015 watts of power. Devices such as a stand-alone inverter used within the PV system in this example must be rated to handle at least 2,015 watts of continuous power.

Load Description	Quantity	Operating Watts	Power Draw
Refrigerator/freezer	1	125	125
Radio/CD-player	1	70	70
Cistern pump	1	500	500
fluorescent lights	6	20	120
Ceiling fan	1	100	100
Radio/alarm clock	1	10	10
Microwave	1	900	900
Laptop Computer	1	50	50
Solar array electronics	1	40	40
Odds and ends	N/A	100	100
TOTAL:			2015 W

TABLE 3-3: MAXIMUM POWER DRAW ANALYSIS FOR SMALL ONE-BEDROOM CABIN

Appliances that use motors will often draw more power than their rating when they first start up. This is why in many homes the lights briefly dim when the water pump or the refrigerator "kicks on."

Stand-alone inverters can accommodate brief periods when the energy draw exceeds their continuous rating. This **surge rating** can dramatically exceed the continuous rating of the inverter.

SURGE RATING

For example, an inverter rated to provide 2,000 watts of power on a continuous basis might be able to supply, 2,400 watts for up to 30 minutes, 3,100 watts for up to five minutes, and even up to 3,700 watts for up to five seconds.

Calculating Unusual Loads

Sometimes calculating daily load demand is not as straightforward as simply calculating the number of watts an appliance uses and then multiplying it by the number of hours that appliance will operate during a day (watts x hours = watt-hours).

Variable or Cycling Loads

Many common appliances, such as a refrigerator, experience a surge of power demand when first entering their power cycle. The energy demand quickly settles to a much lower number (after a couple of seconds, or so).

However, the refrigerator may then demand less and less electricity as it cools. At the end of the cycle, it may have a small but constant draw (such as required for an ice maker or an automatic defrosting system). When the door opens, a light turns on, which draws some power. And depending on the temperature of the room and the number of times the door is opened, the cycle frequency will vary.

Determining the daily load draw for this appliance can prove quite complex. A practical way to accomplish this is to plug the appliance into a Kill-a-Watt meter. Allow it to remain plugged in for a typical day of use. Then read the

FIGURE 3-11:
EXAMPLES OF
WEIGHTED LOADS

Kiln:
3,000 W
6 hours,
twice a
week

Mixer:
4,000 W
1.5 hours,
once a
week

Daily Load Demand = 857 Wh
Maximum Draw = 4,000 W

Daily Load Demand = 5,143 Wh
Maximum Draw = 3,000 W

meter's watt-hour setting. This will indicate the number of watt-hours the appliance consumes over a typical day.

Weighted Loads

Another common situation occurs when a load is used infrequently, yet is a significant portion of the system's electrical load. These are referred to as **weighted loads**.

(WEIGHTED LOAD)

For example, an artist might use a kiln twice a week and a mixing unit once each week. Each draws a significant amount of power when in use, so must be factored in to the daily load demand (when sizing the array and battery bank for a stand alone system), and the maximum power draw (when selecting an inverter).

For example (Figure 3-11), assume that the kiln draws 3,000 watts when operating (no surge or variation) and runs for 6 hours each time it is used. Also assume that the mixer uses 4,000 watts and runs for 90 minutes (again, no surge or variation).

So the impact on daily load will be:

watts used x hours x days per week operated / days per week

So the kiln's impact on the daily load demand = 3,000 watts x 6 hours x 2 days per week / 7 days per week = 5,143 watt-hours

The mixer's impact on the daily load demand = 4,000 watts x 1.5 hours (90 minutes) x 1 day per week / 7 days per week = 857 watt-hours

Total impact on the daily load demand of the system from both appliances will be: 5,143 watt-hours (for the kiln) + 857 watt-hours (for the mixer) = 6,000 watt-hours

The array will have to be sized to produce 6,000 watt-hours of energy each day to handle the weekly demands of these loads. In a stand alone system, this energy would be stored in the battery bank to be used when needed, then completely replaced over the seven day week. In a grid-tied system, the daily production would be "stored" on the grid, then used on the days when the kiln and/or mixer are in operations. The result over the seven-day-period would be net zero.

When calculating the maximum power draw (required when sizing a stand-alone system's inverter), the system does not care how often the load is operating, just the maximum draw during any one second.

In this example, assume the designer will have to determine if the two appliances ever operate at the same time.

It could be that the system is set up so that the artist cannot physically operate both the mixer and the kiln at the same time. If this is the case, then the impact of the two loads on the maximum power draw will be the greater of the two loads (or 4,000 watts as the mixer draws more power than the kiln when operating).

If the two units can operate at the same time, then their combined impact on the maximum power draw of the system will be the sum of their draw when operating.

In this example, 4,000 watts (draw of mixer) + 3,000 watts (draw from kiln) = 7,000 watts (impact on the maximum power draw of the two loads on the system).

Chapter 3 Review Questions

1. A PV system that uses batteries and is **NOT** connected to the local utility is referred to as a:
   A) grid-tied system.
   B) grid-interactive system.
   C) grid-fallback system.
   D) off-grid system.

2. Which of the following is **NOT** an example of a load demand that is often powered by a stand-alone PV system?
   A) a pocket calculator
   B) a remote hunting cabin
   C) an urban traffic light
   D) a recreational vehicle

3. A PV system that **DOES NOT** use batteries and IS connected to the local utility is referred to as a:
   A) grid-tied system.
   B) grid-interactive system.
   C) grid-fallback system.
   D) off-grid system.

4. The most common type of residential home PV system installed in developed countries today is the:
   A) grid-tied system.
   B) grid-interactive system.
   C) grid-fallback system.
   D) off-grid system.

5. The situation where a PV system is generating power to the electrical grid while the grid is "down" (no electricity flowing from the utility) is known as:
   A) islanding.
   B) grid-tied.
   C) a utility interface.
   D) band gap energy.

6. A PV system that uses batteries and IS connected to the local utility is referred to as a:
   A) grid-tied system.
   B) grid-interactive system.
   C) grid-backup system.
   D) off-grid system.

7. A distributed energy system that incorporates both solar (PV) and wind power would be an example of a:
   A) grid-tied system.
   B) grid-fallback system.
   C) hybrid system.
   D) redundant system.

8. Which of the following will likely be the least expensive (per watt) solar PV system configuration to install ?
   A) grid-tied
   B) grid-interactive
   C) hybrid
   D) off-grid

9. The device that senses the grid has lost power and disconnects a grid-interactive solar PV system so that it can function in a stand-alone mode is known as a/an:
   A) automatic transfer switch.
   B) anti-islanding switch.
   C) interconnection switch.
   D) utility inter-tie switch.

10. Which of the following PV systems will need to incorporate a utility interface?
    A) stand-alone
    B) grid-tied
    C) solar thermal
    D) direct current

11. According to the NEC, in the US the maximum voltage the DC portion of a residential PV system can be configured to is:
    A) 50 V
    B) 240 V
    C) 600 V
    D) 1,000 V

12. According to the IEC, in Europe the maximum DC voltage a residential PV system can be configured to is:
    A) 50 V
    B) 600 V
    C) 1,000 V
    D) 1,500 V

13. According to the NEC, in the US the maximum DC voltage a roof mounted commercial or multi-family PV system can be configured to is:
    A) 50 V
    B) 600 V
    C) 1,000 V
    D) 1,500 V

14. Generally the first step in designing a PV system is to determine the:
    A) load requirements.
    B) customer's budget.
    C) cost of solar panels.
    D) the size of the array.

15. According to government statistics, in 2021 the average U.S. home consumed About _____ of electricity per month.
    A) 460 kWh
    B) 890 kWh
    C) 1,250 kWh
    D) 2,675 kWh

16. Electricity is generally purchased in units referred to as _____.
    A) watts (W)
    B) kilowatts per hour (kW/h)
    C) watt-hours (Wh)
    D) kilo-amps (kA)

17. The most cost-effective way of reducing the price of a photovoltaic system in a residential setting is:
    A) to increase the number of hours of insolation at the site.
    B) to increase the size of the array by adding additional panels.
    C) to connect the system to the electrical grid
    D) reduce the load demand through energy efficiencies.

18. The electricity consumed by all the lights and appliances in a structure is referred to as:
    A) the system load.
    B) the grid fallback.
    C) the irradiance of the site.
    D) the net-meter.

19. Sally is conducting a load analysis for a small business, a bakery. Their dough mixer runs for 3 hours, two times a week. Each time it is operated, it consumes 1,400 Wh during that three-hour period. How much would you calculate is the mixer's contribution to the DAILY load requirements of the business?
    A) 2.8 kWh
    B) 467 Wh
    C) 400 Wh
    D) 1.4 kWh

20. The most power (measured in watts) a household might use at any particular moment is referred to as the household's:
    A) maximum power draw.
    B) continuous power rating.
    C) daily load demand.
    D) power surge rating.

21. The amount of energy (measured in watt-hours) a household might use in an average day is known as the household's:
    A) maximum power draw.
    B) continuous power rating.
    C) daily load demand.
    D) power surge rating.

For Questions 22 - 24:
Oscar is conducting a load analysis on a barn. There are three critical loads that must be serviced by the PV system. A water heater uses 1,700 Wh/day and runs for four hours each day. A milking machine uses 900 Wh/day and runs for three hours. A grain mixer uses 1,200 W when operating, but only operates for two hours, twice a week. Assume the rate of energy used by all the machines is constant when running (no surges or cycles).

22. What is the daily load demand if these three critical loads are the only loads in the barn?
    A) 3,286 Wh
    B) 2,600 Wh
    C) 3,800 Wh
    D) 543 Wh

23. Assuming all the machines can be operated at the same time, what is the maximum power draw that must be accommodated?
    A) 3,800 W
    B) 1,925 W
    C) 1,200 W
    D) 1,025 W

24. Assuming only one of the machines can be operated at a time, what is the maximum power draw that must be accommodated?
    A) 3,800 W
    B) 1,925 W
    C) 1,200 W
    D) 1,025 W

Lab Exercises : Chapter 3

Lab Exercise 3-1:

Supplies Required:
- Copy of Load Analysis Worksheet

Tools Required:
- Kill A Watt meter
- Internet access

3-1. Conduct a Load Analysis
1. Design a two bedroom isolated cabin. Populate this cabin with all the appliances and electrical devices that might make this a comfortable home environment.
2. Research the power consumption of each of the electrical devices. If available, use the Kill A Watt meter to test the power consumption of a specific device. For devices that are not available, research their power consumption online.
3. Estimate the average daily use of each appliance or device.
4. Enter all the findings on the Load Analysis Worksheet.
5. Calculate the average daily load requirements of the fictional two-bedroom cabin.
6. Make changes (either by replacing, eliminating or reducing the time of use of selected devices) that will result in a 25% reduction in average daily load requirements.

Lab Exercise 3-2:

Supplies Required:
- Copy of Maximum Power Draw Worksheet

Tools Required:
- Internet Access

3-2. Determine Maximum Power Draw
1. Using the results from step 6 of Exercise 3-1 (above), determine the maximum power draw for the cabin.
2. Determine what size stand-alone inverter will be required to meet this draw.
3. Locate an example of a suitable unit online.

Chapter 4
Basic Electrical Concepts

Chapter Objectives:
- Understand basic electricity and how it is measured.
- Become aware of safety issues while working with electricity.
- Understand how the national electrical grid operates.
- Comprehend circuits, and how electrical characteristics change when connected in series and parallel.
- Become familiar with common voltages, single phase and 3 phase power.
- Note the differences and similarities of AC and DC current.
- Understand Ohm's Law and resistance, and ways to minimize its impact on voltage drop.
- Appreciate how testing for resistance can determine continuity within a system or component.
- Become familiar with bonding and grounding issues as they relate to PV systems.

Before getting too deeply into the world of photovoltaics, it is necessary to have a very basic understanding of **electricity** and how it works.

ELECTRICITY

It should be assumed that to design or install PV systems, it is necessary to study electrical theory in much more detail than is covered in this text, or to work with people who have a strong electrical background. This text only provides a very basic introduction into electricity.

Electricity improperly handled can kill (both those who work on the systems or those who may later come into contact with them) – so it is very important to understand how it works.

It should be noted that most fatal accidents involving electricity take place while the victim is working on common household electrical circuits. So always exercise caution when working with electricity, regardless of the system.

What is Electricity?
Everyone is familiar with electricity, but few people really understand how it works. Plug a toaster into an electrical outlet and it heats bread. Turn on a light switch, and suddenly there is light. Turn it off, and the light is gone.

But where did the power go? Where did it come from? How does it work? What does it look like?

Electricity is actually made up of several distinct physical effects.

These include:

- An **electrical charge.** This is built up electrical energy. An electrical charge can be seen when a thunderstorm rolls in and lightning strikes. Or when a metal door handle discharges a static charge after feet slip across a dry carpet. For a discharge to occur, there must be an imbalance of positive and negative forces. The spark seen and felt when a metal door handle is touched is that imbalance of charge trying to find a balance once more through a conductive material.
- An **electrical current**. This is the flow of an electrical charge. If it has a proper pathway (a conductor), an electrical charge will flow along this pathway. The amount of flow or current is measured in **amps (amperes)**.
- **Electrical potential.** When there is an imbalance between positive and negative charges (such as at the terminals of a battery) there is a difference in electrical potential. This difference is measured in **volts**. The more the difference in potential, the greater the voltage.
- **Electromagnetism.** Early experimenters found that there is a relationship between electricity and magnetism. They found that electricity could be generated from mechanical devices (such as a generator) and conversely, electricity could drive mechanical devices (such as a motor).

How Electricity is Measured?

It may be useful to think of electricity that flows through a wire as being similar to water flowing through a pipe. The pressure, or how fast the water flows can be thought of as the voltage of the system. Volts, in the world of electrical equations and diagrams are represented with the symbol "V" or "E" (for **electromotive force**).

How much water (or electricity) flows past a certain point in the system over a given period of time is the system's **current**. In electricity, this current is measured in amps, represented with the symbol "I" (from the French *"intensité de courant"* or current intensity).

The total amount of water (or electricity) flowing through the pipe at any one moment is the **power** generated by the system. In electricity, this is measured in **watts**, with the symbol "W".

These three forces are related to each other. Just as with the amount of water flowing through a pipe can be measured by multiplying its intensity (how fast the water is flowing) by the size of the pipe (its current), electrical power can be similarly measured:

Watts (power) = Volts (intensity) x Amps (current)
or more simply, **W=V x I**

Energy is essentially power over time. So energy is measured as:

watts (power) x hours (time) = **watt-hour (Wh)**

Understanding this **power equation** is critical for those hoping to design and/or install PV systems. Understanding the relationship between watts, volts and amps is necessary to properly size PV arrays, determine the size of wires and protection devices, select the proper system components - fundamental to all aspects of designing and installing the system.

> ENERGY
> WATT-HOUR (WH)
> POWER EQUATION

Electrical Safety

Electricity can be quite dangerous. A few safety tips should always be followed when working with electricity. Some of these include:

- ✓ Always make sure the electrical conductors (wires) are not energized. Turn off circuits (breakers and switches) and test them with a voltmeter to make sure no current is present. Photovoltaic panels produce energy when exposed to sunlight, so it is a good idea to always cover them with a blanket or tarp when working on the system.
- ✓ Transient (stray) power can flow through the system, even when normal power sources are turned off. These stray power sources may include lightning, or down power lines, or short circuits. Inspect the system prior to working on it, test it with a voltmeter, and do not work on systems during severe weather.
- ✓ Always use protective gloves and wear insulated shoes when working on electrical systems. Use protective mats when working with higher voltage systems.
- ✓ Never assume that voltage is not present simply because you have turned off the switch. Check it with a voltmeter. And test the voltmeter to make sure it is working properly.
- ✓ Don't work alone. Make sure someone else is present who knows how to disconnect the power or remove you from danger in an appropriate and safe manner.
- ✓ Always replace fuses and breakers with properly rated replacements.
- ✓ Make sure your tools are well insulated, with plastic or rubber handles.
- ✓ Never stand on a damp surface while working with electricity.
- ✓ Do not touch pipes or other metal surfaces while working with electricity.
- ✓ Avoid using aluminum ladders, or ladders made of any conductive material. Always look up, there may be exposed wires overhead.
- ✓ Remember that low voltage does not mean low danger. No amount of electricity should be considered "safe."
- ✓ Protective eyewear should be worn when working with power tools, batteries and while working with electricity. OSHA (Occupational Safety and Health Administration) requires that all eye protection used in the workplace conform with ANSI Z87.1 requirements.
- ✓ Always inspect your tools and meters for damage and/or wear. Replace or repair as appropriate.
- ✓ In many workplace environments, hard hats are required (and nearly always a good idea, even where they are not required). Class A hard hats offer good impact protection, but limited voltage protection. Class B hard hats are suitable for electrical work, offering protections against shocks and burns. Class C hard hats are generally more comfortable, but offer only limited protection against falling objects.
- ✓ When working in attics or other enclosed spaces, a breathing mask may be appropriate.

The power equations is sometimes referred to as Watt's Law and may be expressed as P (power) = I (amps) x E (electromotive force, or volts). Just remember that "electricians love PIE".

If two of the variables are known, it is easy to calculate the third. For example, a 60 watt light bulb that is being powered by a 12-volt battery will need a 5 amp circuit to make it work. (60 watts = 12 volts x 5 amps, or 5 amps = 60 watts / 12 volts, or 12 volts = 60 watts / 5 amps).

If the voltage of the system is increased to 120 volts, then only ½ amp is required to power the same light bulb (60 watts = 120 volts x .5 amp).

The Difference between Watts, Watt-Hours and Watts/Hour

The terms watts (W), watt-hours (Wh) and watts per hour (W/hr) are often used interchangeably, but they are not the same thing. This can lead to confusion. A useful analogy to demonstrate the difference can be found in the following statements.

- "1,000 miles" (watts, just a simple statement of power - doesn't indicate how much energy, if any, was used)
- "I traveled 1,000 miles." (watt-hours, or how much power was consumed)
- or "I traveled 1,000 miles per hour." (watts per hour - or the rate at which the power was consumed).

Clearly these statements each mean completely different things.

A watt is a unit of measure (defined as one **joule** per second). A watt-hour is a convenient way of measuring electrical consumption. It refers to the amount of energy that is consumed or produced at a continuous rate for a period of one hour.

For example, when referring to a 60-watt light bulb, this reference is really incorrect and misleading. It should be referred to as a 60 watt-hour light bulb, indicating that when the bulb is operating, it will consume 60 watts of power over the course of an hour.

But 60 watt-hours can also be consumed at a rate of 120 watts/hr for 30 minutes, or 30 watts/hr over a two-hour time period.

What is the Grid?

Most photovoltaic systems designed and installed in today's market will be connected in some way to the electrical grid. It is therefore important to understand how the grid works, and how it will impact the decisions made in designing the right PV system for the client.

Getting power to a home or business requires a number of systems.

First, the power must be generated in some way. The power within the **generation** portion of the grid can come from turbines fueled by coal, natural gas, oil, hydro, wind, or even nuclear energy. Or increasingly it might come from utility-scale solar farms. A power plant may have more than one generator, and some generators may use more than one type of fuel.

> GENERATION

According to US government statistics, as of 2022 there are over 11,000 utility-scale power plants in operation feeding power onto the grid.

Once the power is generated, it must somehow make its way from the power plant to where it will be used. This is done through a network of wires. This network is normally organized into two parts, **transmission** and **distribution**.

> TRANSMISSION
> DISTRIBUTION

There are over 450,000 miles of high-voltage power lines and about 160,000 miles of overhead transmission lines in the United States. These lines connect electrical power plants to sub-stations located closer to the home or business where it will be used to power loads.

The maximum distance for a high voltage transmission line is about 300 miles. The electricity on these lines is normally in the range of between 115,000 to 765,000 volts. The higher the voltage, the less energy is lost as it travels long distances.

FIGURE 4-1: A TYPICAL ELECTRIC UTILITY DISTRIBUTION SUB-STATION

The high voltage transmission lines terminate at a **distribution sub-station**, as shown in Figure 4-1. At this point the electricity enters the distribution system. At the sub-station several things happen. The voltage is stepped down with the use of **transformers**. The sub-station also may feed the electricity into multiple distribution systems, as well as provide a method to disconnect that section of the grid in case of problems.

> DISTRIBUTION SUB-STATION

> TRANSFORMER

FIGURE 4-2: TRANSFORMER DRUM AT CUSTOMER'S SITE

Power leaving the distribution sub-station is considered part of the primary distribution system. Voltages within this part of the system range from 4 kV to 35 kV. Only large consumers (such as industrial facilities) are fed directly from distribution voltages.

Chapter 4: Basic Electrical Concepts

The power from the sub-stations is distributed to more than 145 million homes and businesses through millions of miles of local electrical lines. At each home, the power is reduced in voltage again within the transformer drum that normally is attached to a power pole, like the one pictured in Figure 4-2.

Residential Voltages

Most of the world (Europe, Africa, Asia, Australia, New Zealand and much of South America) use electricity from the grid that is within 6% of 230 V (a range of 216 V to 244 V).

Mainland Western Europe has traditionally used electricity rated nominally at 220 V, 50 Hz. The UK used 240 V, 50 Hz power from the grid. During European "harmonization," in an effort to require a single voltage standard across Western Europe (including UK and the Irish Republic) a few minor modifications took place.

Harmonization was achieved by changing the tolerances of previously existing standards. UK voltage was changed to 240 V + 6% and - 10% (or a range of 216 V to 254 V) and Europe to 220 V +10% and -6% (or a range of (207 V to 242 V), thereby creating a manageable overlap. These voltages are often referred to as 230 V (its **nominal voltage**), despite the fact that no system was intentionally generating at 230 V.

(NOMINAL VOLTAGE)

The United States and Canada use a supply voltage of 120 volts ± 6% (or a voltage range of 113 V to 127 V). Japan, Taiwan, Saudi Arabia, Mexico, Central America and some parts of northern South America use a voltage between 100 V and 127 V. Brazil is unusual in having both 127 V and 220 V systems at 60 Hz and also permitting interchangeable plugs and sockets.

Single Phase Power

For residential service in North America (Canada, the US and Mexico) - power is typically stepped down to 240 volts at the transformer before it enters the home. The service from the transformer to the home is normally provided at either 100 or 200 amps of current. The larger the amps, the larger the wire required to safely transmit the power from the final transformer to the home's **service panel**. Within the service panel, the voltage is then split into two 120 Vac circuits.

(SERVICE PANEL)

FIGURE 4-3: STANDARD US SINGLE-PHASE WALL OUTLET

(NEUTRAL)

(GROUND)

(SINGLE-PHASE)

It is the nominal 120 V power that supplies a standard US electrical outlet as shown in Figure 4-3. There is a "hot" conductor (black) a **neutral** (white) and a bare copper wire known as the **ground**.

The power coming from a typical transformer drum to a home is referred to as **single-phase** electric. This refers to the waveform of the AC power. As illustrated in Figure 4-4, a two-wire single-phase signal (in the US) will have a voltage of 120 volts, and a frequency of 60 hertz (cycles per second).

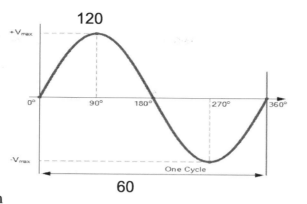

FIGURE 4-4: SINGLE-PHASE WAVEFORM

But the service coming into a typical home has three, rather than two wires. These carry two 120 volt power lines that share a common neutral. The first 120 volt AC power line runs exactly ½ cycle, or 180° out of phase with the second 120 Volt AC power line (as shown in Figure 4-5). This is called **split phase** power, because the 240 volt signal from the transformer is split into two 120-volt circuits, that operate 180 degrees apart.

SPLIT-PHASE

In the main electrical panel of the building (where the circuit breakers are located), phase 1 and phase 2 are connected within the panel to separate busbars that essentially "zigzag" and interlock. As a result, each single-pole circuit breaker (on the left side, and the right side) is powered by a different phase. In other words, the top right breaker is powered by phase 1, the one below it by phase 2, the one below that by phase 1, below that by phase 2, and so on.

By doing this, a double-pole breaker (that fits into two slots, rather than one, will receive power from both phases - or at 240 volts rather than 120 volts as shown in Figure 4-6.

Appliances that require 120 volts are connected to only one leg (either phase 1 or phase 2) through a single-pole breaker in the service panel. Appliances that receive 240 volts must be wired to both of the hot wires, as well as the common neutral, as illustrated in Figure 4-7. In a residential setting, the color coding will typically be black and red, for the "hot" wires, and white for the common neutral.

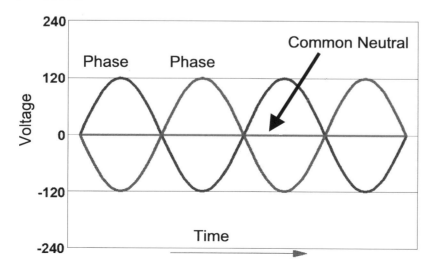

FIGURE 4-5: SPLIT PHASE WAVEFORM

Chapter 4: Basic Electrical Concepts   Page 83

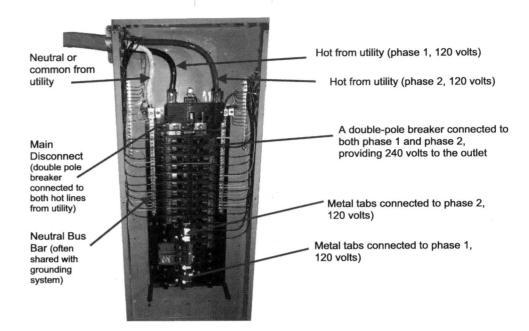

Figure 4-6: Typical Single-Phase Split Phase Service Panel with double-pole breaker feeding a load

Three Phase Power

THREE-PHASE

Many commercial and industrial customers receive their electricity in the form of **three-phase** power. In a three-phase system, the power received or generated is in the form of three sine waves, each 120° out of phase with the other, as illustrated in Figure 4-8.

Three-phase power is most effective in driving large motors. Normally, three-phase motors generate about 1.5 times the power output of similarly sized single-phase motors.

Three-phase power is also more constant. In single-phase circuits, the power pulsates. Due to the nature of the waveform, the power falls or rises to zero

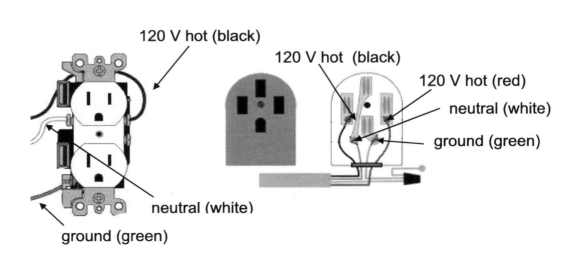

Figure 4-7: US Outlet wiring for 120 volt outlet (left) and 240 volt outlet (right)

Page 84   Understanding Photovoltaics: Design & Installation of Residential Solar PV Systems

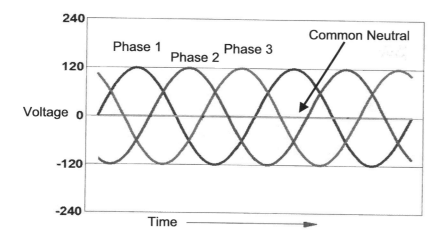

FIGURE 4-8: THREE-PHASE POWER SINE WAVE FORM

three times in each cycle. Whereas, in **polyphase systems** (three-phase), the power delivered is almost constant when the loads are balanced, and never falls to zero.

POLYPHASE SYSTEMS

When designing a PV system, it is important that the selected inverter match the utility power system feeding the property.

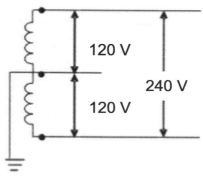

FIGURE 4-9: UTILITY PROVIDED SINGLE-PHASE POWER DIAGRAM

For standard split-phase power, the two phases are exactly out of phase (shifted by 180 degrees), so that the voltage of the waveform becomes twice that of the phases: 2 x 120 Volts = 240 Volts (Figure 4-9) when both hot conductors are connected to the load.

But in a three-phase system, it is not quite so simple. Two selected phases are only 1/3 of the way out of phase (shifted by 120 degrees), so their peaks do not line up. The voltage of the resulting waveform is only 1.73 (the square root of 3) larger than the voltage of each of the phases:

1.73 x 120 volts = 208 Volts or 1.73 x 277 volts = 480 volts

Who Owns the Grid?

The electric grid, which includes generation, transmission, and distribution, is owned and operated quite differently in different nations.

Often the grid is owned and operated exclusively by the national government - as it is in Mexico and France, for example. Or it may be in part publicly owned and in part owned by private companies - as it is in Australia. Or, as in the U.S. and Germany, it may be owned by private corporations with oversight by a number of governmental bodies.

INVESTOR-OWNED UTILITY (IOU)

As of 2022, in the U.S. there were 166 **investor-owned utilities (IOU)** that owned 33% of all net generation, 80% of the transmission infrastructure, and 50% of the distribution infrastructure. An IOU is typically what one thinks of as a utility. It is a for-profit corporation owned by one or more investors. This type of utility nationwide serves 72% of all customers connected to the grid.

MUNICIPAL SYSTEM

RURAL ELECTRIC COOPERATIVE

There are about 2,900 publicly-owned utilities and cooperatives (**municipal systems** and **rural electric cooperatives**) that account for 20% of net generation, 12% of transmission, and nearly 50% of the nation's electric distribution lines.

INDEPENDENT POWER PRODUCER

Approximately 2,800 **independent power producers** account for about 45% of net generation (these would include large wind farms as well as utility-scale solar arrays). An independent power producer is defined as a corporation, person, agency, authority, or other legal entity (other than a utility) that owns or operates facilities for the generation of electricity for use primarily by the public.

The federal government owns nine power agencies (such as the Tennessee Valley Authority) that own 7% of net generation and 8% of the transmission lines.

The 6,000 or so "owners" each control a small portion of a vast electrical system that we call the "grid." Clearly there must be a substantial amount of oversight from some central authority to make sure it all operates smoothly.

State Regulation

Those who own the generating, transmission and distribution systems are regulated by individual state's **Public Utility Commission (PUC)** or the **Public Service Commission (PSC)**. These are regulatory agencies appointed by the state legislatures. These agencies approve the construction of generation facilities and also set electricity rates within their state.

PUBLIC UTILITY COMMISSION (PUC)

PUBLIC SERVICE COMMISSION (PSC)

Because it is more local in nature, the state PUC or PSC is largely responsible for regulating the generation and distribution portion of the grid within the borders of that state.

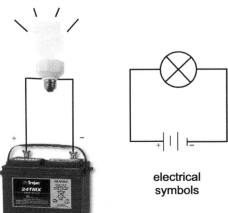

FIGURE 4-10: DIAGRAM OF A SIMPLE ELECTRICAL CIRCUIT

electrical symbols

Electrical Circuits

Whenever there is an electrical charge (an imbalance of electrical potential - the difference between the positive and negative charges) built up between two

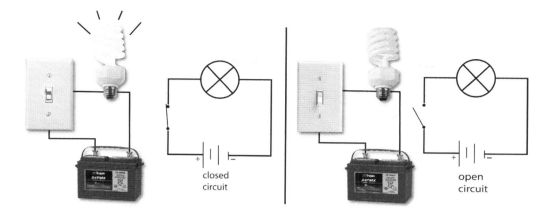

FIGURE 4-11: OPEN AND CLOSED ELECTRICAL CIRCUITS

points, the different charges are attracted to each other (like the negative and positive poles of a magnet) - seeking a neutral balance.

If these two points are connected with a conductor (typically a wire, but it could be anything that will allow electricity to pass through it) as illustrated in Figure 4-10, a **circuit** will be created and the **electrons** (with a negative charge) will flow towards the positively charged location.

As the electrons flow through this circuit, they can be used to power a **load** (any device that consumes electricity, such as an electric light).

While the circuit is intact, it is said to be "closed" (referred to as a **closed circuit**). If the circuit is interrupted (an **open circuit**), then the flow of electricity along it stops.

When a light switch is turned off, as indicated in Figure 4-11, the circuit is opened and the light bulb turns off, no electricity is now flowing along that circuit.

Connecting in Series and Parallel

If the battery in a circuit is a nominal 12-volt battery, the energy flowing around the circuit will measure approximately 12-volts, more or less depending on its state of charge. A fully charged battery might measure 13.4 volts, and the same battery after long use might measure 11.2 volts. But the battery is still said to be a 12-volt (nominal) battery.

However, if there are multiple batteries (power sources), then the voltage of the circuit can be altered depending on how the batteries are hooked together.

When hooking a battery bank (multiple batteries) together in **series** (connecting the negative

CIRCUIT

ELECTRON

LOAD

CLOSED CIRCUIT

OPEN CIRCUIT

FIGURE 4-12: FOUR 12-VOLT, 90 AH BATTERIES CONNECTED TOGETHER IN SERIES

SERIES

Chapter 4: Basic Electrical Concepts

terminal of one battery to the positive terminal of the next battery – as shown in Figure 4-12), the voltage of the system increases - simply adding the individual voltages of each connected battery together (while the amps stay the same).

PARALLEL

FIGURE 4-13: FOUR 12-VOLT, 90 AH BATTERIES CONNECTED TOGETHER IN PARALLEL

When connecting this system together in **parallel** (connecting all the positive terminals together and all the negative terminals together as shown in Figure 4-13), the voltage of the system remains the same but the amps (the current) increase.

The same principles apply when PV modules are hooked together into a solar array. They are simply a different energy source (as opposed to the more familiar batteries).

Six nominal 12-volt solar panels hooked together in series (as shown in Figure 4-14) will create an array that operates at a nominal voltage of 72 volts (6 x 12).

If each panel had a maximum power current (Imp) of seven amps, then the system connected together in series would still operate with an Imp of seven amps (voltage increases, but amps remain the same when they are connected in series).

For panels connected in series, the amps do not change. But if the amps generated by the panels are not the same (either because of shading, mismatched panels, damage or aging), then the current from the string will be equal to the lowest amps coming from any one panel.

FIGURE 4-14: SIX 12-VOLT SOLAR PANELS CONNECTED TOGETHER IN SERIES

FIGURE 4-15: SIX 12-VOLT SOLAR PANELS CONNECTED TOGETHER IN PARALLEL

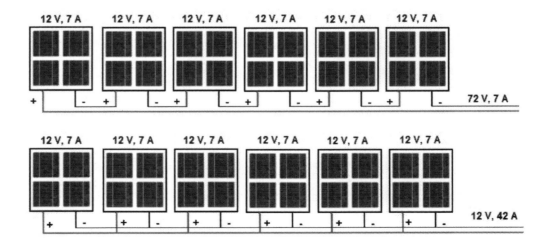

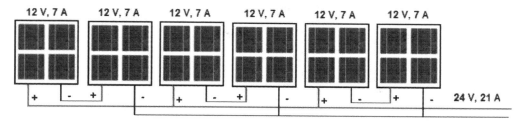

Figure 4-16: Six 12-volt solar panels connected together in series and parallel

If these same panels are connected together in parallel as illustrated in Figure 4-15, then the voltage will remain the same (at a nominal rating of 12 volts), but the amps (current) will increase. In this case, the circuit must be designed to carry a current of 42 amps (6 panels x 7 amps each).

Systems can also be configured to incorporate both series and parallel connections (as shown in Figure 4-16). In a configuration connected as shown, both the voltage and amps flowing through the circuit will be modified.

Direct and Alternating Current

Electricity can flow along a conductor (wire) in one of two ways. The simplest to understand is referred to as **direct current (DC)**.

DIRECT CURRENT (DC)

The unmodified power which flows from a battery or from a PV panel is in the form of a direct current. Technicians generally view this power as negatively charged electrons flowing along the wire at a constant voltage, from the negative terminal of a battery back to the positive terminal. This can be visualized as a steady state flow, as shown in Figure 4-17.

Typical household electricity flows in the form of **alternating current (AC)**.

ALTERNATING CURRENT (AC)

As the AC current flows, it is constantly reversing direction. It does this in a regular pattern as it flows down the circuit. This pattern is called a **sine wave**, as shown in Figure 4-18. The height of the wave is an indication of its voltage.

SINE WAVE

CYCLES PER SECOND

The number of peaks and valleys (**cycles per second**) is referred to as the signal's **frequency**. Frequency is measured in units called **hertz (Hz)**. In the

FREQUENCY

HERTZ (HZ)

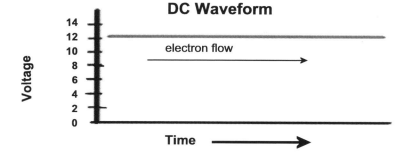

Figure 4-17: DC Power Waveform

Chapter 4: Basic Electrical Concepts

FIGURE 4-18: AC POWER SIGNAL

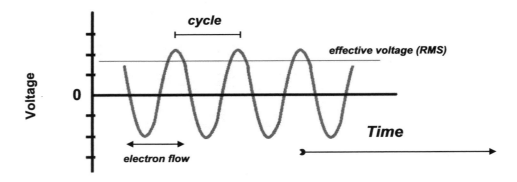

United States, normal AC electrical systems operate at 60 hertz (or 60 cycles per second). This frequency is not standard around the globe. In the United Kingdom, for example, the standard for typical household current (**mains current**) is 50 Hz.

System voltages are also not uniform around the globe. In the United States, typical residential voltage is 120/240 Vac (split phase / single phase). The voltage coming into the home from the utility is 240 Vac, but is split into two 120 Vac branches within the electrical service panel.

Commercial systems often use three-phase voltages, that, in the USA, can be at 120/208 Vac, 277/480 Vac, 240 Vac or 480 Vac. Three-phase power systems are generally beyond the scope of this text, but it should be noted that such systems exist (this can be confusing for beginning installers when selecting components for PV systems).

The effective voltage of an AC circuit is not the peak voltage of the wave, but is the **RMS (root mean square)** voltage of the wave. This is the point on the sine wave equivalent to the power dissipation in the load that would occur if the AC circuit were a DC circuit.

Resistance

As current flows through a conductor, the material that makes up the conductor will offer some **resistance** (blocking just a bit of the flow). Using the water analogy, as the water flows through a pipe, it will rub up against the sides of the pipe, the friction causing the flow to slow just a bit. The water will, over time, flow just a bit slower because of this friction (resistance).

In the world of electricity, this friction, or resistance, will result in a drop in voltage (logically referred to as **voltage drop**). Resistance is measured in **ohms**, designated with the symbol Ω. The relationship between volts, amps and ohms can be calculated through the equation known as **Ohm's Law**. This equation states:

Volts = Current x Resistance (or, V = I x R)

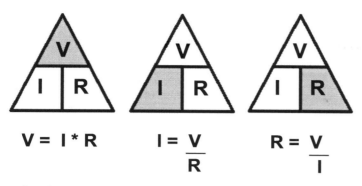

FIGURE 4-19: OHM'S LAW TRIANGLE

This equation is often depicted graphically using the **Ohm's Law Triangle**, a shorthand method (Figure 4-19) of visually seeing the resulting equation while solving for one of the variables.

(OHM'S LAW TRIANGLE)

Minimizing the Impact of Resistance on a PV System

Generally accepted practices in the PV industry seek to limit voltage drop within the wiring system to between 2-5%.

In order to reduce the impact of resistance (and the resulting voltage drop) on a PV system, the designer/installer can do a number of things:

- *Select the appropriate wire.* The material making up the wire will affect its resistance to the flow of electricity. Copper is an excellent conductor, and for this reason it is typically the standard material used in wires appropriate for PV systems.

- *Shorten the distance.* Resistance increases as the length of the wire increases. For example, a wire that has a resistance of 1.588 ohms at 1,000 ft (305 m) of length, will have twice the resistance if the length of the run is increased to 2,000 ft (610 m) or 3.176 ohms. Reducing the length of the circuit will help in avoiding unwanted voltage drop.

- *Adjust the size of the wire.* The larger the wire, the less resistance this wire has to the energy flowing through it. Larger wires will create a system less prone to voltage drop. But larger wires cost more money. So a balance between price and performance must be reached when designing a PV system.

In the U.S., wires are measured using the **American Wire Gauge (AWG)** system. Contrary to logic, the smaller the AWG number, the larger the wire. Throughout most of the rest of the world, conductors are referenced by the cross sectional diameter of the conductor, measured in square millimeters (mm^2).

(AMERICAN WIRE GAUGE (AWG))

Table 4-1 reflects the resistance characteristics of the various available copper wire sizes (in AWG) per 1,000 feet (kFT). This table would be used when calculating voltage drop within the US.

Table 4-2 lists the resistance characteristics using mm² per 1,000 meters (km). This table would be used in most nations around the globe.
In order to compare "apples to apples," the NEC in Chapter 9, Table 8, references the resistance of various wire sizes and types in terms of its resistance per 1000 feet or resistance per kilometer.

- *Increase the system voltage.* Even though the actual voltage drop does not change when changing the system voltage, the effective percentage of voltage lost decreases as the system voltage increases.

For example: a voltage drop of 8 Vdc would be significant if the system was operating at 12 Vdc (67% loss), but not so much if the system was operating at 600 Vdc (only about a 1% loss).

Voltage Drop Equation

The voltage drop equation is essentially Ohm's Law adjusted for distance (the length of the cable). When calculating resistance for a cable run within a PV system (for DC circuits and single-phase AC circuits), the modified **voltage drop equation** is:

$$V_d = I \times R \times 2\, d/1{,}000\ \text{ft}$$

where... V_d = the voltage drop of the cable run

TABLE 4-1: US TABLE UNCOATED COPPER CONDUCTOR RESISTANCE CHARACTERISTICS MEASURED IN OHMS (Ω) PER 1,000 FEET MEASURED AT 75°C (167°F)

AWG	mm²	R/ 1,000 feet Solid Copper	R/ 1,000 feet Stranded Copper
18	0.823	7.77	7.95
16	1.31	4.89	4.99
14	2.08	3.07	3.14
12	3.31	1.93	1.98
10	5.261	1.21	1.24
8	8.367	0.764	0.778
6	13.3	—	0.491
4	21.15	—	0.308
3	26.67	—	0.245
2	33.62	—	0.194
1	42.41	—	0.154
0 (1/0)	53.49	—	0.122
00 (2/00)	67.43	—	0.0967
000 (3/00)	85.01	—	0.0766
0000 (4/00)	107.2	—	0.0608

I =	the amps on the circuit, normally in PV this would be the I_{mp} of the string
R =	resistance from the NEC table for the selected wire
2d =	twice the distance of the cable run (multiplied by 2 since there are two conductors in the cable to complete the circuit). This adjusts resistance for actual distance rather than the standard 1,000 feet/ 1,000 meter rating provided by the table.

For example, if…

I_{mp} = 7.2 amps
R = 1.24 Ω / 1000 ft (10 AWG stranded copper)
d = 250 ft cable run

So… V_d = 7.2 amps x (1.24 Ω / 1000 ft) x 2 (250 ft)
V_d = 7.2 amps x (1.24 Ω / 1000 ft) x 2 (250 ft)
V_d = 7.2 amps x 0.00124 Ω x 500 = 4.464 volts

Therefore, over a distance of 250 feet, using an AWG #10 stranded copper cable, transmitting 7.2 amps of current, there will be a loss of 4.464 volts.

mm²	R/ 1,000 meters Solid Copper	R/ 1,000 meters Stranded Copper
0.823	25.5	26.1
1.31	16	16.4
2.08	10.1	10.3
3.31	6.34	6.5
5.261	3.984	4.07
8.367	2.506	2.551
13.3	—	1.608
21.15	—	1.01
26.67	—	0.802
33.62	—	0.634
42.41	—	0.505
53.49	—	0.399
67.43	—	0.317
85.01	—	0.2512
107.2	—	0.1996

TABLE 4-2: METRIC UNCOATED COPPER CONDUCTOR RESISTANCE CHARACTERISTICS MEASURED IN OHMS (Ω) PER 1,000 METERS MEASURED AT 75°C (167°F)

If the system is operating at 12 volts and 4.464 volts are lost due to voltage drop, then 4.464/ 12 or 37.2% of the system voltage has been lost.

Clearly this is not an acceptable design. In fact, it is generally accepted that voltage drop for the entire wiring system from array to load be limited to between 2% - 5% (ideally less than 2%). To reduce the affect of voltage drop on the PV system, the designer could increase the voltage of the system.

If the nominal voltage of the system were raised to 200 volts, with the same voltage drop, then the loss due to voltage drop would be 4.464/200 or 2.23% of the available power. So while the amount of voltage drop due to resistance did not change, by increasing the voltage of the system, the percentage of the power lost was reduced from 37.2% to 2.23%.

For three-phase systems the calculation is slightly different, due to the different voltage factors that affect these systems as compared to a single-phase system.

For standard single-phase power, the two phases are exactly out of phase (shifted by 180 degrees), so that the voltage of the waveform becomes twice that of the phases: 2 x 120 Volts = 240 Volts.

But in a three-phase system, it is not quite so simple. Two selected phases are only 1/3 of the way out of phase (shifted by 120 degrees), so their peaks do not line up. The voltage of the resulting waveform is only 1.73 (the square root of 3) larger than the voltage of each of the phases: 1.73 x 120 volts = 208 Volts.

So in a three-phase circuit, the voltage drop is calculated:

$$V_d = I \times R \times 1.73 \, d/1{,}000 \text{ ft}$$

AMBIENT AIR TEMPERATURE

Another factor that can affect resistance is the **ambient air temperature**. The hotter the cable gets, the more the resistance. Therefore protecting the cables from extreme heat will also help in system efficiency.

Measuring Resistance as a Continuity Test

A short bit of wire or a small well-terminated circuit will present a resistance reading of near zero when tested with a multimeter. Conversely, an open circuit will show a resistance reading of infinite, typically depicted as "1" on a meter set to test for resistance.

CONTINUITY TEST

As a result, the measuring of resistance can be used as a **continuity test** (as demonstrated in Figure 4-20), checking a circuit to see if a wire has become disconnected, if a fuse is good or bad, or if an unmarked switch is in the off (open) or on (closed) position.

FIGURE 4-20: A CONTINUITY TEST OF A SWITCH USING A STANDARD MULTIMETER

Open Switch (infinite resistance)   Closed Switch (near zero resistance)

Technicians often use this continuity test to check switches, fuses and circuits during the troubleshooting of PV systems.

Voltage Variations

The voltage of power supplied by the grid is often referred to by its nominal voltage - for example, 240 Vac for single-phase power supplied to a residential home in the US. But the actual voltage varies within a range (+/- 6%). So the actual measured voltage at the main disconnect might measure anywhere from 226 Vac to 254 Vac when the grid is operating normally.

But the voltage may occasionally fall outside this normal operating range. If the voltage falls below acceptable levels, it is referred to as a **voltage sag**. Voltage sags may be caused by system design flaws such as undersized wire or excessively long cable runs - which result in unacceptable voltage drop. But more commonly it is the result of sudden high load demands or overloaded transformers.

(VOLTAGE SAG)

It may help to again think of electricity flowing through wires like water flowing through a pipe. If you suddenly turned on all the taps in a building, the water pressure would drop until the pressure from the supply could be increased to compensate.

A **voltage swell** occurs when the voltage on the system exceeds the normal operating range. These typically occur close to substations (where the voltage is highest in anticipation of voltage drop as the power moves through the power lines). A voltage swell may occur if there is a sudden decrease in system demand, or immediately after a voltage rise if the generators overcompensate by putting too much power onto the grid.

(VOLTAGE SWELL)

Voltage sags and voltage swells normally cause the PV system to shut down until the grid has returned to acceptable voltage limits. This usually happens within the inverter. The inverter senses that the grid is no longer operating within normal limits and shuts down. It monitors voltages within the grid and turns back on after a preset period of time.

A **transient** is a very short (up to a few milliseconds) voltage spike on the system, normally caused by a lightning strike, arcing, or an extremely high current load being turned on and off.

Transients differ from voltage sags and voltage swells in that they are shorter in duration, greater in magnitude, and more erratic. Equipment that is unprotected (through the use of a surge protector, for example) can be permanently damaged by transients.

Bonding and Grounding

Most items constructed of metal can conduct electricity, whether they were designed to or not. Clearly the copper wire of a PV system is designed to carry an electric current. But the frame holding the solar panel, or the metal box containing a disconnect switch can also carry any random charge that might come in contact with its metallic surface.

Because the unexpected nearly always happens, the National Electrical Code (NEC) requires that all metal components within PV systems be **bonded** together and **grounded**. This is to protect the life and safety of people who may come in contact with the system.

There are two distinct components to a properly configured bonding and grounding system.

The first is **equipment grounding**, in which bonding and grounding is accomplished by connecting (bonding) all exposed non-current-carrying metal parts of the system together as seen in Figure 4-21, using a bare copper wire or a wire with green insulation, clips or other conductive devices. The size of the wire will vary depending on the amperage of the system.

Once all the metal bits are connected together, the circuit is then connected to a metal rod buried in the earth. This metal rod or plate is called the **grounding electrode (GE)**, often an 8-foot long copper rod (typically called a **ground rod**). The wire used to connect the grounding rod to the equipment

FIGURE 4-21: EQUIPMENT GROUNDING CONDUCTOR BONDING THE METAL FRAMES OF TWO PV PANELS TOGETHER

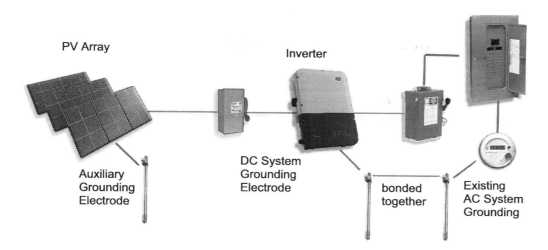

FIGURE 4-22: BONDING AND GROUNDING SYSTEM

grounding system is called the **grounding electrode conductor (GEC)**. The grounding electrode conductor is typically a #4 AWG or #6 AWG bare or green copper wire.

Taken together, the grounding electrode and the grounding electrode conductor are referred to as the **system ground**.

The purpose of a properly configured grounding system (Figure 4-22) is to ensure that all the conductive materials (typically metal) within the PV system have the same **voltage potential** as the earth.

Should any "live" electrical source accidentally touch the metal surface (such as a fallen wire, a short circuit in the system, lightning, etc), the electricity has a low-resistant path to ground (the grounding system conductors). This makes it much less likely that a person touching an energized system becomes the conductor that provides a pathway to ground (and is injured or killed).

(GROUNDING ELECTRODE CONDUCTOR (GEC))

(SYSTEM GROUND)

(VOLTAGE POTENTIAL)

Chapter 4 Review Questions

1. The amount of flow or current is measured in _____.
   A) watts
   B) volts
   C) amps
   D) ohms

2. When there is an imbalance between positive and negative charges (such as at the terminals of a battery) there is a difference in electrical potential. This difference is measured in _____.
   A) watts
   B) volts
   C) amps
   D) ohms

3. Which of the following is **NOT** a safety recommendation when working with electric wires?
   A) Always check and test exposed wires, ensuring there is no current present before handling.
   B) Wear safety goggles when working with electrical systems.
   C) Always use protective gloves and wear insulated shoes when working on electrical systems.
   D) Always wear a respirator when working with exposed wiring.

4. More people are killed each year in the U.S. working on _____ systems than any other type of electrical system.
   A) photovoltaic electrical
   B) high voltage electrical
   C) regular household electrical
   D) hybrid electrical

5. Electricity is generated when _____ move along conductive wires.
   A) electrons
   B) photons
   C) neutrons
   D) protons

6. The power equation is defined as:
   A) $V = I \times R$
   B) $V = I \times R \times 2d$
   C) $W = I \times R$
   D) $W = I \times V$

7. Watts equals:
   A) volts minus resistance.
   B) amps times volts.
   C) volts divided by amps.
   D) resistance plus amps divided by volts.

8. A kilowatt is:
   A) 10 watts.
   B) 100 watts.
   C) 1,000 watts.
   D) 10,000 watts.

9. 60 watt-hours is equal to:
   A) 6 kilowatts consumed over a 10-hour period of time.
   B) 240 watts consumed for 15 minutes.
   C) 60 watts consumed at a constant rate for a 24-hour period.
   D) 120 watts consumed for 2 hours.

10. Which of the following is NOT considered a major component of the electrical grid?
    A) generation
    B) harmonization
    C) transmission
    D) distribution

11. The device that steps up or steps down the voltage in an AC power distribution system is know as a:
    A) transformer.
    B) resistor.
    C) capacitor.
    D) harmonizer.

12. In the US and Canada, the voltage that enters a typical residential home at the main disconnect is:
    A) single-phase power .
    B) 240 nominal volts (+ or - 6%).
    C) 60 hertz.
    D) all of the above.

13. In the US and Canada, the waveform of the power that enters a typical residential home on line one and line two are _____ out of phase with each other.
    A) 90°
    B) 120°
    C) 180°
    D) 360°

Chapter 4: Basic Electrical Concepts   Page 99

14. Which of the following is **NOT** generally true with regards to three-phase power?
    A) it will produce less vibration in motors than single-phase power
    B) it is often used in commercial and industrial systems
    C) the waveforms in each leg of the system are 180° out of phase with each other
    D) three-phase motors generate about 1.5 times the power output of similarly sized single-phase motors

15. In the United States, which type of utility serves the majority of customers?
    A) investor-owned utility
    B) municipal system
    C) rural electric cooperative
    D) independent power producers

16. Within each state, utilities are generally regulated by the:
    A) Energy Information Agency (EIA).
    B) Public Utility Commission (PUC).
    C) Federal Energy Commission (FEC).
    D) Power Siting Board (PSB).

17. When a string of batteries are connected in series:
    A) the voltage and amps remain the same.
    B) the voltage increases but the amps remain the same.
    C) the voltage remains the same, but the amps increase.
    D) both voltage and amps increase.

18. When a number of solar panels are connected in parallel:
    A) the voltage and amps remain the same.
    B) the voltage increases but the amps remain the same.
    C) the voltage remains the same, but the amps increase.
    D) both voltage and amps increase.

19. When a circuit is closed:
    A) it is disconnected from the utility grid.
    B) power cannot flow through it.
    C) a fuse has blown or a circuit breaker has tripped.
    D) electrons can flow through it.

20. Six solar panels are connected in series. Each panel has a Vmp rating of 18 Vdc and an Imp rating of 8.33 amps. The output **CURRENT** of this string will be:
    A) 108 Vdc
    B) 108 Vac
    C) 18 Vdc
    D) 8.33 amps

21. Six solar panels are connected in series. Each panel has a Vmp rating of 18 Vdc and an Imp rating of 8.33 amps. The output **VOLTAGE** of this string will be:
    A) 108 Vdc
    B) 108 Vac
    C) 18 Vdc
    D) 8.33 amps

22. Which of the following is NOT relevant to DC circuits?
    A) resistance
    B) hertz
    C) Ohms Law
    D) voltage drop

23. Cycles per second are referred to as:
    A) hertz
    B) ohms
    C) joules
    D) intensity

24. When seeking to decrease the effect of voltage drop on a circuit, which of the following is **NOT** an effective solution?
    A) Shorten the length of the circuit.
    B) Use a smaller (in diameter) wire size.
    C) Increase the voltage of the system.
    D) Select a wire with better conducting properties.

25. The loss of an electrical signal over distance as it travels along a conductor due to resistance is referred to as:
    A) its root mean square.
    B) voltage drop.
    C) the ampacity rating.
    D) a ground and bonding system.

26. When referencing the resistance characteristics of various conductors or wires, the NEC measures in the following units.
    A) volts per watt
    B) amps per volt
    C) watts per square meter
    D) ohms per 1,000 feet

27. Ohm's Law is expressed as:
    A) volts = current x resistance.
    B) watts = amps x volts.
    C) voltage drop = amps x resistance x twice the distance.
    D) resistance = amps x volts.

28. When measuring the terminals of a blown fuse with a multimeter set to perform a continuity test, the reading will show:
    A) resistance = 0
    B) voltage = 0
    C) resistance = 1 (or infinite)
    D) amps = 1 (or infinite)

29. What is the voltage drop of a 400 foot cable run, using #10 AWG stranded copper wire on a circuit designed to carry 8.3 amps and 48 volts of electricity?
    A) 10.29 V
    B) 8.23 V
    C) 4.216 V
    D) 1.24 V

30. What percentage of the power from the circuit in Question 23 will be lost due to voltage drop?
    A) 21.44%
    B) 17.15%
    C) 8.78%
    D) 2.58%

31. If you have a wire length of 100' in a 90 volt, 8.5 amp derated DC circuit, what is the **SMALLEST** wire size you would need to use to get less than a 2% voltage drop?
    A) #6 AWG with a resistance of .510
    B) #8 AWG with a resistance of .809
    C) #10 AWG with a resistance of 1.29
    D) None of the above

32. The system designed to eliminate all unwanted current from non-current carrying metal components is referred to as _____ .
    A) net metering
    B) a blocking diode
    C) transient flow resistance
    D) bonding and grounding

33. Bare wire or green insulation will always indicate the wire is part of _____ .
    A) a photovoltaic system
    B) a hybrid system
    C) a DC system
    D) a ground and bonding system

34. The eight-foot long metal rod, commonly called the "ground rod" is more accurately referred to as:
    A) the grounding electrode.
    B) the system ground.
    C) the equipment ground.
    D) the grounding conductor.

Lab Exercises : Chapter 4

4-1. Crimping Connectors on Wires

The purpose of this exercise is to practice crimping wires and making solid electrical connections.

a. Select a 6-inch length of wire. Strip about ¼ inch of insulation off both ends.
b. Select the appropriately sized connector, based on the wire size selected. Crimp a connector on one end of the wire.
c. Slide shrink tape over the connector and wire.
d. Use heat gun or lighter to shrink tape over connection.
e. Repeat for connector on other end of the wire (so both ends now have a connector and shrink tape.
f. Check that all connections are secure and that no copper is exposed outside of the connectors.
g. Set the multimeter to the 20k Ω test setting.
h. Test the wire (place probes on the metal part of each connector) for continuity.

> **Lab Exercise 4-1:**
>
> *Supplies Required:*
> - Assortment of wire, #10 AWG - #22 AWG
> - Assortment of wire connectors
>
> *Tools Required:*
> - Wire stripper
> - Wire crimper
> - Shrink tape
> - Multimeter
> - Heater or lighter
> - Safety glasses

4-2. Continuity and Voltage Testing

The purpose of this exercise is to practice continuity tests and ensure the components are in working order.

a. Set the multimeter to the 20 DC voltage setting.
b. Test the voltage of the battery to ensure it is working properly.
c. Set the multimeter to the 20k Ω test setting.
d. Test the switch, light and fuse for continuity.

> **Lab Exercise 4-2:**
>
> *Supplies Required:*
> - 12-volt battery
> - 12-volt DC switch
> - 12-volt DC light
> - fuse
>
> *Tools Required:*
> - Multimeter
> - Safety glasses

4-3. Create a Simple Circuit

The purpose of this exercise is to create a simple 12-volt circuit that includes a power source (battery), switch, overcurrent protection (fuse) and load (light bulb).

a. Select the appropriate connectors and terminate short lengths of cable.
b. Connect in the proper sequence the battery, fuse, switch and light.
c. Test to see that the circuit operates properly by turning on and off the light with the switch.

> **Lab Exercise 4-3:**
>
> *Supplies Required:*
> - 12-volt battery
> - 12-volt DC switch
> - 12-volt DC light
> - Fuse and holder
> - #12 AWG wire
> - Assorted connectors
>
> *Tools Required:*
> - Wire stripper
> - Wire crimper
> - Safety glasses

Lab Exercise 4-4:

Supplies Required:
- (4) 12-volt batteries
- #12 AWG wire
- Assorted connectors

Tools Required:
- Wire stripper
- Wire crimpers
- Multimeter
- Safety glasses

4-4. Connecting Batteries in Series/Parallel

The purpose of this exercise is to connect four batteries in various configurations and test the effect on the voltage of the system.

a. Connect all four batteries together in series.
b. Set the multimeter to 200 DC voltage setting.
c. Test the battery bank with the multimeter. Note the result.
d. Connect two batteries together in parallel.
e. Test the battery bank with the multimeter. Note the result.
f. Connect the four batteries together in a combination of parallel and series.
g. Test the battery bank with the multimeter. Note the result.

Lab Exercise 4-5:

Supplies Required:
- NEC resistance table

Tools Required:
- Calculator

4-5. Calculate Voltage Drop for a Cable Run

The purpose of this exercise is to practice calculating the voltage drop of a wire length given a number of parameters.

The cable run is connected to string of ten (10) 390 W solar panels. Each panel has a Vmp measured at 30.4 Vdc and an Imp rating of 12.84 amps. The cable run is 290 feet, using #12 AWG stranded copper wire.

a. Look up the resistance of a 1000-foot length of #12 AWG stranded copper wire.
b. Using the voltage drop equation, determine the voltage drop for this example.
c. What would be the voltage drop if #10 AWG stranded copper wire were used instead?

Chapter Objectives:

- Explore how solar panels are rated and how they connect to the system.
- Understand the function of a combiner box / junction box.
- Identify the various types of conductors used in a PV system and how environmental conditions affect their performance.
- Describe the function and selection of AC and DC disconnects.
- Understand how overcurrent protection functions within the system.
- Note the various charge controllers and how they interact with the battery bank.
- Describe inverter options and how they converter DC waveform to AC.
- Comprehend the concept of Rapid Shutdown and how that affects PV system design.
- Become familiar with module level power electronics.
- Explore how the PV system is connected to the existing AC wiring system of a home.
- Identify various tools used when working on a PV system.

Chapter 5

Parts of the PV System

Before a complete photovoltaic system can be designed and installed, it is important to understand how all of the system components function and interact with each other.

This chapter will define each system component, beginning at the solar panel and working down through the system to the load (the device that actually uses the power, such as a light or a radio). Bear in mind that not every component is used in every system.

Solar Panels

Solar panels can be constructed in various ways (monocrystalline, polycrystalline, thin film), but most are simply arrays of solar cells connected together.

How these cells are connected (in a combination of series and parallel connections) will affect the output voltage of the module (the terms solar panel and solar module, such as is illustrated in Figure 5-1, are typically interchangeable in today's market).

FIGURE 5-1: SOLAR PANEL (ALSO CALLED A SOLAR MODULE)

Commonly, a solar panel for a stand-alone PV system will be described as a 12-volt or 24-volt panel. This designation refers to the panel's nominal voltage. This generally means it is compatible with a nominal 12Vdc battery bank.

But such a reference does not mean that the panel always generates 12 or 24 volts of energy. As a cloud covers the sun, for example, voltage generated from any panel will drop. As full sunlight returns, the panel will generate more power. In fact a panel with a nominal rating of 12 volts will typically operate between 14 to 18 volts in full sunlight when the system is **under load** (powering something).

(UNDER LOAD)

The nominal rating is a shorthand way of letting designers and installers of stand-alone PV systems know how to select the proper panel to go with the rest of the components. A nominal 12-volt panel will be compatible with a nominal 12-volt battery, and a nominal 12-volt charge controller, and so on.

In fact, testing a nominal 12-volt solar module by touching the leads of a multimeter to the connections at the back of the panel with no load attached (measuring the panel's open circuit voltage (Voc) as shown in Figure 5-2) will generally indicate a voltage much higher than the panel's nominal rating. A nominal 12-volt panel may generate 21.5 volts when tested at the back of the panel under standard test conditions.

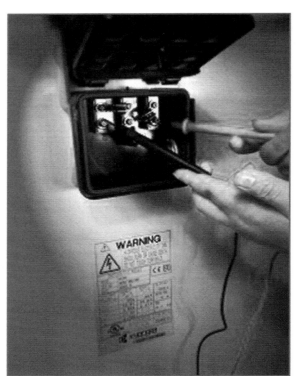

FIGURE 5-2: TESTING A SOLAR PANEL

(RATED VOLTAGE)

When the same panel is then hooked up to a load (again under standard test conditions), the voltage reading will immediately fall to 16.9 volts, which is the maximum power voltage (Vmp) the panel will produce under load. This measurement is sometimes referred to as the panel's **rated voltage**.

There are a number of other specifications commonly associated with solar panels with which the designer/installer will need to be familiar. The short circuit current (Isc) is theoretically the highest current (amps) that the panel can generate under standard test conditions (although a panel can generate higher current and voltage as temperatures fall). Once again, if the

▪ Electrical Characteristics

POWER CLASS			405	410	415	420	425
MINIMUM PERFORMANCE AT STANDARD TEST CONDITIONS, STC[1] (POWER TOLERANCE +5W/-0W)							
Power at MPP[1]	P_{MPP}	[W]	405	410	415	420	425
Short Circuit Current[1]	I_{SC}	[A]	13.35	13.39	13.42	13.46	13.49
Open Circuit Voltage[1]	V_{OC}	[V]	38.56	38.58	38.61	38.64	38.67
Current at MPP	I_{MPP}	[A]	12.62	12.68	12.75	12.82	12.88
Voltage at MPP	V_{MPP}	[V]	32.10	32.32	32.55	32.77	32.98
Efficiency[1]	η	[%]	≥21.1	≥21.4	≥21.6	≥21.9	≥22.2
MINIMUM PERFORMANCE AT NORMAL OPERATING CONDITIONS, NMOT[2]							
Power at MPP	P_{MPP}	[W]	306.3	310.0	313.8	317.6	321.4
Short Circuit Current	I_{SC}	[A]	10.76	10.79	10.82	10.84	10.87
Open Circuit Voltage	V_{OC}	[V]	36.58	36.61	36.63	36.66	36.69
Current at MPP	I_{MPP}	[A]	9.91	9.97	10.03	10.09	10.15
Voltage at MPP	V_{MPP}	[V]	30.90	31.09	31.29	31.48	31.66

[1]Measurement tolerances P_{MPP} ±3%; I_{SC}; V_{OC} ±5% at STC: 1000 W/m², 25±2°C, AM 1.5 according to IEC 60904-3 • [2]800 W/m², NMOT, spectrum AM 1.5

FIGURE 5-3: EXAMPLE OF SOLAR PANEL SPECIFICATIONS

panel is hooked up to a load, this current will drop. The maximum current a panel can generate while under load and tested at standard test conditions is known as its maximum power current (Imp), or also sometimes referred to as its **rated current**.

(RATED CURRENT)

Specifications often reported in addition to those listed in Figure 5-3 include:
- physical dimensions,
- weight of the unit,
- type of cells in the module (monocrystalline, polycrystalline, etc),
- cell efficiency (how much of the sun's energy is actually converted into electricity by an individual cell),
- and more.

Panel Junction Box

The **junction box** (Figure 5-4) is typically located on the back of a PV panel, allowing for easy connecting and disconnecting of the wiring system. These boxes also often house bypass diodes to enhance the efficiency of the panel when a cell is shaded or damaged.

(JUNCTION BOX)

Most modern junction boxes accept standard connectors, creating a weather tight "plug and play" system of wiring the panel to the rest of the system.

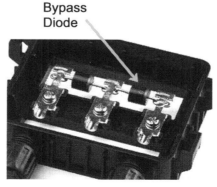

Bypass Diode

FIGURE 5-4: JUNCTION BOX WITH BYPASS DIODES

Chapter 5: Parts of the PV System

Solar Panel Connectors

MC CONNECTORS

MC3 CONNECTOR

Most solar panels on the market today are pre-wired, fitted with **MC connectors** designed to fit easily together and provide a weatherproof connection. Early versions of pre-wired solar panels used a connector known as the **MC3** or Solarline 1 (pictured in Figure 5.5).

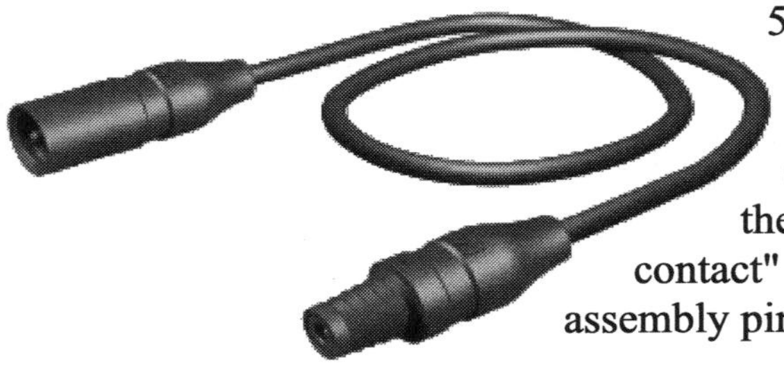

FIGURE 5-5: MC3 (ALSO REFERRED TO AS A SOLARLINE 1) CONNECTOR

First introduced in 1996, MC3 connectors took their name from the fact that they were a "multi contact" unit with a 3mm² contact assembly pin.

These types of connectors have been phased out in favor of the newer locking connectors and are no longer supported by the National Electrical Code (NEC).

MC4 CONNECTOR

The **MC4 connector** is actually a registered trade name of the German firm Stäubli International AG, but is generally used to refer to an entire class of connectors. Similar (and compatible) products are available from other manufacturers, such as the H4 connector that is manufactured by Amphenol and the T4 connector from Canadian Solar.

The connector is self-locking to ensure a tight and lasting connection. The MC4 connector, as shown in Figure 5-6, incorporates a flexible water tight seal and is supplied with "male" and "female" terminations to minimize the chance of incorrect connections.

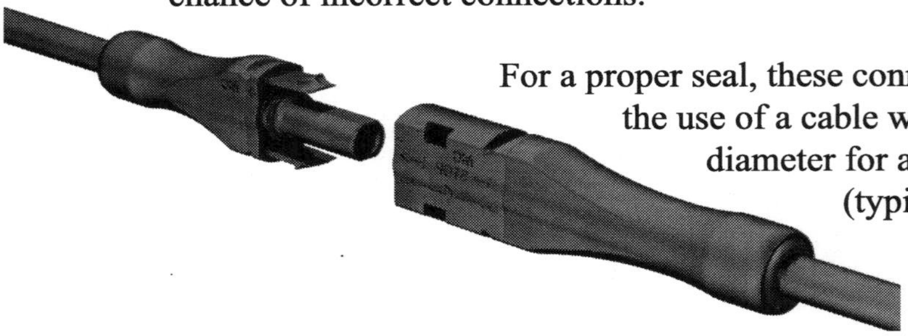

FIGURE 5-6: MC4 (ALSO REFERRED TO AS A SOLARLINE 2) CONNECTOR

UV RESISTANT

For a proper seal, these connectors require the use of a cable with the correct diameter for a tight fit (typically PV wire). Normally MC4 connectors are double-insulated (wire insulation plus a black sheath) and **UV (ultra-violet light) resistant** (many cables deteriorate if used outdoors without protection from sunlight).

MC4 connectors from different manufacturers may not be compatible with each other. Loose connections may cause arching and mismatched connectors are a leading cause of fires in installed arrays. So care must be taken to ensure that connectors on site are matched to those that come installed on the selected solar panel.

Combiner Box

For larger PV installations, quite a number of solar panels will be connected together in an array. Those panels connected in series, are referred to as **strings** (when hooking panels together in series, the voltage increases, but the amps remain the same). Once the maximum voltage of the system is reached, additional strings must be added to provide enough power to service the home or business.

> STRINGS

Rather than run a wire from each string to the rest of the system (which might be some distance away), the wiring for these strings are typically connected together in a **combiner box** (Figure 5-7).

> COMBINER BOX

Multiple wires enter the box from the array strings, and are connected together in parallel within the box. Only one circuit (the exiting conductors will need to be larger to handle the increased amps of the system) leaves the box to connect to the rest of the system.

FIGURE 5-7: COMBINER BOX WITH OVERCURRENT PROTECTION

(FROM MIDNIGHT SOLAR)

While these combiner boxes offer a weatherproof location to connect the strings together, they also normally provide **overcurrent protection** (fuses and/or breakers) for each string – protecting the modules, as well as the wiring system and equipment connected to it.

> OVERCURRENT PROTECTION

Many combiner boxes are offered pre-wired as shown in Figure 5-8. Connections are made simply by snapping the MC4 connector into the proper connection point on the box.

FIGURE 5-8: PRE-WIRED COMBINER BOX

(FROM ECO-WORTHY)

Some units may also offer LED monitor indicators that demonstrate at a glance if there is a problem with one or more of the strings.

Increasingly, **data monitoring** is also incorporated into the combiner box, connected to the Internet so that the system performance can be monitored remotely.

> DATA MONITORING

Chapter 5: Parts of the PV System

The combiner box may also house a DC disconnect switch (to isolate the array from the rest of the PV system). If a disconnect is not incorporated into the combiner box, it is still good practice to locate one as close as practical to the array.

Junction Box

More and more systems are now designed and installed with multiple strings running from the panels to the inverter (not combined in parallel within the combiner box). This allows each string to be monitored and controlled independently, reducing losses due to shading.

FIGURE 5-9: SOLADECK JUNCTION BOX INSTALLED ON ROOFTOP SYSTEM

However, most systems still need a transition point where the loose wires coming from the array strings can be fed into conduit. The junction box also serves as a transition point where the PV wire that is attached to most solar panels is transitioned to another wire type (such as USE-2 or THHN).

Junction boxes, such as the unit illustrated in Figure 5-9, serve as a transition point within the system. The conduit containing the wires leaving the junction box then continue on to a combiner box, or more commonly, to a DC disconnect located at or near the inverter.

Junction boxes normally do not contain overcurrent protection (fuses or circuit breakers), but they can if the situation warrants it.

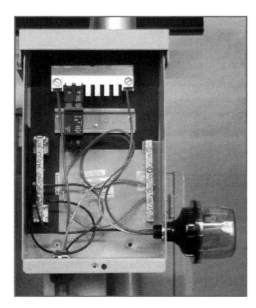

FIGURE 5-10: SURGE PROTECTION ATTACHED TO COMBINER BOX

Lightning & Surge Protection

Lightning is the number one cause of catastrophic failures in solar electric systems and components. A properly designed and installed bonding and grounding system will help minimize damage (by providing a clear pathway to ground), but is not a guarantee against damage.

An average lightning strike has a current of about 30,000 amps, however, NASA has recorded strikes during the Apollo launches as high as 100,000 amps. Fuses and breakers will not help in the event of a lightning strike. They are simply not designed for this purpose and do not react quickly enough.

A **surge protection** device such as the unit shown in Figure 5-10 is designed to "clamp down" on extreme surges (as high as 115,000 amps) to a current and voltage that the system can withstand.

SURGE PROTECTION

Multiple units can be installed and should be located as close as possible to the source of the surge (the array or the grid connection, for example).

Surge protection is commonly installed in combiner boxes (to protect the system from lightning that may strike the array) or in the main AC load panel (to protect from surges that may come in from the utility grid connection).

Wiring the System

Electrical wire (conductors) in the U.S. is measured using the American Wire Gauge (AWG). Typical wire sizes for PV installations in the US range from AWG 0000 (the largest) to AWG 40 (smallest). Most nations simply measure the cross sectional area of the wire in square millimeters (mm^2).

The larger the wire, the lower the resistance (resulting in less voltage drop). Thicker wire is also capable of carrying more current without fusing (becoming damaged). The **NEC Table 310-16** (a portion reproduced in

TABLE 310-16

Size AWG	60°C (140°F)	75°C (167°F)	90°C (194°F)	Size mm²
	Types TW, UF	Types RHW, THHW, THW, THWN, USE	Types PV, THHN, THHW, USE-2	Equivalent (nearest available size)
COPPER				
18	–	–	14	0.823 (1.0)
16	–	–	18	1.31 (1.5)
14	15	20	25	2.08 (2.5)
12	20	25	30	3.31 (4.0)
10	30	35	40	5.261 (6.0)
8	40	50	55	8.367 (10)
6	55	65	75	13.3 (16)
4	70	85	95	21.15 (25)
3	85	100	110	26.67 (30)
2	95	115	130	33.62 (35)
1	110	130	145	42.41 (50)
1/0	125	150	170	53.49 (70)
2/0	145	175	195	67.43 (70)
3/0	165	200	225	85.01 (95)
4/0	195	230	260	107.2 (120)

TABLE 5-1:
ALLOWABLE AMPACITIES FOR INSULATED CONDUCTORS

FROM NEC TABLE 310-16

Table 5-1) has established maximum current levels allowed for the various size wires when placed in conduit or directly buried. Conductors in free air (like PV wire in the PV string circuit) are capable of slightly higher ampacities which can be found in NEC Table 310.17.

Temperature Ratings of Conductors and Terminations

Note that different types of wire are rated for different temperatures. Conductors coated with insulations in column 1 (such as TW or UF) are rated for 60°C (140°F). Conductors in column 2 (RHW, THW, etc) are rated to operate in temperatures up to 75°C (167°F). Conductors in column 3 (PV, USE-2, THHN, etc) are rated to operate in temperatures up to 90°C (194°F). Because they have better heat characteristics, they can carry more amps.

Conductors must be sized by considering where they will terminate as well and how that termination is rated. The rating must apply to the entire cable run, which includes the wire as well as the terminations on both ends.

If the termination within the equipment (or breaker) is rated for only 75°C, the maximum temperature rating of the cable run is 75°C, even if 90°C wire has been used (lowest rating prevails).

If the temperature ratings of the terminations on both ends of the cable run are known, then an appropriately rated conductor can be selected and the ampacity ratings in the appropriate column can be used.

However it is often the case that when selecting the wire, the terminating equipment is unknown. Or it may be that at some point in the future the equipment will be changed to a unit with a lower temperature rating.

TABLE 5-2:
TEMPERATURE CORRECTION FACTORS FOR INSULATED CONDUCTORS AT 30°C

FROM NEC TABLE 310-15 (B) (1)

Temp (°C)	TEMPERATURE CORRECTION FACTORS		
	60°C (140°F)	75°C (167°F)	90°C (194°F)
21-25	1.08	1.05	1.04
26-30	1	1	1
31-35	0.91	0.94	0.96
36-40	0.82	0.88	0.91
41-45	0.71	0.82	0.87
46-50	0.58	0.75	0.82
51-55	0.41	0.67	0.76
56-60	–	0.58	0.71
61-65	–	0.47	0.65
66-70	–	0.33	0.58
71-75	–	–	0.5
76-80	–	–	0.41
81-85	–	–	0.29

So in practice it is best to always refer to column 1 (60°C) of the table when selecting wire. Doing so will always ensure that the conductor selected will have the ampacity rating required for the cable run regardless of the rating of the terminations.

Effect of Temperature on Wire

Temperature affects the way wire conducts electricity. So in order to give a point of reference, the NEC **ampacity** ratings found in Table 310-16 are determined at an ambient air temperature of 30° C (86° F).

> AMPACITY

If the anticipated operating temperature of the wire is expected to be greater than 30° C (86° F), larger wire may be required. Correction factors to adjust for this increased temperature can also be found in the NEC Table 310.15(B) (1) (1). These correction factors are reproduced in part in Table 5-2.

For example, if the high temperature of a rooftop installation has been measured and found to be 46° C (115° F), then the ampacity of the wire must be adjusted (as the table assumes a temperature of 30° C (86° F)).

Assuming the installer is using USE-2 cable (high heat rated at 90° C), then the ampacity must be adjusted by a factor of 0.82.

Assume the circuit was to be designed to handle a maximum of 38 amps. The installer was planning to use #10 AWG USE-2 wire, which has a maximum rating of 40 amps. But correcting for temperature, #10 AWG USE-2 wire only has a maximum ampacity rating of 32.8 amps (40 amps x 0.82 = 32.8 amps). So #8 AWG must be selected instead, which has a temperature adjusted ampacity rating of 55 amps x 0.82 = 45.1 amps, which is more than enough to handle the 38 amps maximum current for the design.

Wires in Conduit

The NEC suggests that wires placed in conduits that are exposed to sunlight be placed with the bottom of the **conduit** or **raceway** no closer than 23 mm (7/8 inch) to the deck of the roof to avoid excess heat. If placed closer than this, 33° C must be added to the highest measured ambient air temperature.

> CONDUIT

> RACEWAY

FIGURE 5-11:
CABLE MOUNT OFFSET FROM IRONRIDGE

Using the previous example, if the same wire (#10 AWG USE-2) was placed in conduit that lay directly on the roof surface, then the corrected temperature would be 46° C (the highest recorded temperature) plus 33° C (added for exposed conduit less than 23 mm

Chapter 5: Parts of the PV System

TABLE 5-3: AMPACITY ADJUSTMENT REQUIRED WHEN PLACING MORE THAN THREE CONDUCTORS IN A RACEWAY OR CONDUIT

FROM NEC TABLE 310.15(C)(1)

Number of wires in a single conduit or raceway	Percent of ampacity values (inclusive of temp adjustment, if necessary)
4 to 6 wires	80%
7 to 9 wires	70%
10 to 20 wires	50%
21 to 30 wires	45%
31 to 40 wires	40%
41 or more wires	35%

from the roof surface). The combined temperature of 79° C requires a temperature correction factor of 0.41 (from Table 5-2).

The #10 AWG USE-2 that would normally have an ampacity rating of 40 amps can now only safely carry 16.4 amps (40 amps x 0.41 = 16.4 amps).

To avoid these ampacity problems due to increased temperature, installers seek to place cable in the shade (under the solar panels in most cases) and also offset them from the roof using mounting brackets similar to the one shown in Figure 5-11 or the rails that support the solar panels.

When many wires are placed into a conduit, the heat generated by the wires can increase the air temperature within the conduit. The air space within the conduit or raceway is trapped and therefore does not easily allow for cooling.

When placing more than three conductive wires in a single conduit or raceway (the equipment ground conductor does not count), adjustments must be made to the ampacity rating of the wire. In other words, the wire will likely be capable of conducting less current than its normal rating due to the build up of heat within the conduit.

NEC Table 310.15(C)(1) outlines the adjustments required when placing multiple wires in a single raceway.

As shown in Table 5-3, if 18 wires were placed in a single conduit or raceway, each wire would only be capable of transmitting half (50%) of the current they would be able to transmit if three or fewer wires had been placed in that same conduit.

Types of Wire

Aluminum or Copper

The conductors used in PV installations are generally made from either copper or aluminum. Copper is a better conducting material than aluminum. It can therefore carry more current through the same sized wire.

Aluminum wire is generally cheaper than copper, but it can become weakened and break (during bending). Aluminum wire is not permitted inside of residential homes.

Solid or Stranded

Solid conductors are made from one single piece of metal. **Stranded conductors** consist of many smaller wires twisted together. Solid conductors tend to be more robust and rigid, while stranded conductors are more flexible.

With AC current, the electrons tend to flow on the outside of the wire due to a phenomenon known as **skin effect**. The higher the frequency, the more pronounced this effect. At 60 hertz, the skin effect is negligible.

DC current tends to use the entire conductor, flowing evenly over the surface as well as through the core. In the photovoltaic world, both types of cable are perfectly acceptable.

Type	Name	Max. Temp.	Environment	Insulation
THHN	Heat Resistant Thermoplastic	90 C (194 F)	Indoor Dry or Damp Locations	Flame Retardant, Heat Resistant Thermoplastic
THW	Moisture & Heat Resistant Thermoplastic	75-90 C (167-194 F)	Indoor Dry or Wet Locations	Flame Retardant, Moisture and Heat Resistant Thermoplastic
THWN	Moisture & Heat Resistant Thermoplastic	75 C (167 F)	Indoor Dry or Wet Locations	Flame retardant, moisture and heat resistant thermoplastic
TW	Moisture resistant thermoplastic	60 C (140 F)	Indoor Dry or Wet Locations	Flame Retardant, Moisture Resistant Thermoplastic
UF and USE	Underground Feeder & Underground Service Entrance	60-75 C (140-167 F)	Outdoor Service Entrance	Moisture and Heat Resistant
USE-2 and RHW-2*	Underground Service Entrance	90 C (194 F)	Outdoor Dry or Wet and Service Entrance	Moisture and Heat Resistant
PV Wire	Photovoltaic cable	90 C (194 F) wet, 150 C (302 F) dry	Dry or Wet and Service Entrance	Moisture and Heat Resistant

TABLE 5-4: COMMON TYPES OF CONDUCTORS USED IN PV SYSTEMS

TABLE 5-5:
INSULATION COLOR-CODING OF AC AND DC WIRES

AC (Alternating Current)		DC (Direct Current)	
Color	Application	Color	Application
Black, Red or other	Ungrounded Hot	Black, Red or other	Positive
White or Gray	Grounded Conductor (Neutral)	White or Gray	Solidly Grounded Conductor (Negative)
		Black, Red or other	Functionally Grounded Conductor (Negative)
Green, Green with Yellow Stripes, or Bare	Equipment Grounding Conductor	Green, Green with Yellow Stripes, or Bare	Equipment Grounding Conductor

While there is some debate as to which type of cable is superior (some people insist that stranded must be used for DC), both will work. Some terminations are easier with solid wire. Stranded wire is often easier to pull through conduit. Select the type of cable based on which is easier to work with, easier to terminate, and attractively priced.

Insulation

INSULATION

There are many different types of **insulation** cladding (or covering) the metal conductors contained in cable. The different types of insulation serve different functions.

The insulation covering the conductor can protect it from heat, moisture, ultraviolet light or chemicals. Various insulation styles commonly used in PV systems are outlined in Table 5-4. Some insulation materials are manufactured so as not to emit toxic gases when burned (important when installed indoors).

A wire with the designation "-2" such as THW-2, generally indicate that this wire is permitted in environments experiencing continuous 90° C (194° F) operating temperatures.

Wires located in conduit should be assumed to be in a wet environment. Wires located on rooftops are in a high heat environment. Wires run inside a building must be rated for indoor use.

Color Coding of Wires

The color of the insulation on a conductor indicates its function or use.

Note that the **color coding** for AC and grounded DC applications, as indicated in Table 5-5, are similar, but not the same. The terminology used to describe the wires are also different between AC and DC systems.

> COLOR CODING

The color code used may be slightly different for DC systems depending on whether the system is grounded through the negative wire, through the positive wire (although most systems in the U.S. are grounded through the negative wire), or ungrounded.

Systems that incorporate a **solid ground** (which are rare in today's market) require white or gray insulation on the grounded conductor. **Functionally grounded** systems (which are now the norm) are not permitted to use white or gray insulation for the grounded conductor (usually the negative).

> SOLID GROUND

> FUNCTIONAL GROUND

A solidly grounded system connects the grounded conductor directly to the grounding system. This is normally done within the inverter. No breakers, fuses, or any other means of disconnection is permitted anywhere along the solidly grounded conductor.

But in today's market, all inverters have some means of disconnect incorporated into the inverter on the grounded conductor (arc fault protection or ground fault protection). This protective device is designed to disconnect the PV system in case of a problem.

It was determined in the 2017 NEC that the negative wire in a functionally grounded system should not be considered a grounded conductor. So the ungrounded conductor color scheme applies.

The NEC does not specify what color (other than they cannot be white, gray or green) should be used for ungrounded conductors. But in practice, most installers select red for the positive leg of the DC circuit and black for the negative leg. Typical color coding conventions are detailed in Table 5-6.

Typical Color Coding (US)	
Application	Color
DC Wiring	Positive - red, Negative - black
120V AC Wiring (single phase)	Hot - black, Neutral - white
240V AC Wiring (single phase)	Line 1 - black, Line 2 - red, Neutral - white
120V, 208V, 240V Three Phase AC Wiring	Line 1 - black, Line 2 - red, Line 3 - blue, Neutral - white
277V, 480V Three Phase AC Wiring	Line 1 - brown, Line 2 - orange, Line 3 - yellow, Neutral - gray

TABLE 5-6: COLOR CODING CONVENTIONS FOR DC, AC AND THREE-PHASE AC CURRENT CARRYING CONDUCTORS

Installers must be consistent with the color coding throughout the installation. These color codes are strictly enforced in the field.

Sizes #4 AWG or Larger

For wire sizes #4 AWG or larger, the NEC allows colored tape to be placed around the wire near each termination point rather than requiring that the entire wire have the appropriate color. This is because larger wire sizes are often only commercially available in black.

If, for example, an installer wishes to run #2 AWG wire, but can only find it in black, it is okay to use the black wire for both the positive and the negative, so long as the positive circuit is identified with red tape wrapped around the conductor near each termination (assuming black for negative and red for positive).

For wire sizes smaller than #4 AWG, the properly colored insulation must be used.

Identification and Grouping

All conductors must be identified and labeled at all accessible points of termination, connection and/or splices.

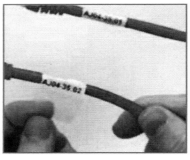

FIGURE 5-12: LABELS ON CONDUCTORS

(FROM HELLERMANN/TYTON)

This identification includes the proper color coding as well approved labels and/or tags, as shown in Figure 5-12. Typical identification codes include the polarity (in DC circuits), line number (in AC circuits), string number, etc. Where wires of more than one PV system occupy the same junction box, raceway or equipment, the conductors of each system must be grouped separately and labeled.

DC Disconnect

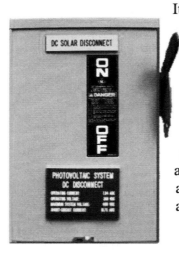

FIGURE 5-13: TYPICAL LABELED PV ARRAY DC DISCONNECT

DC DISCONNECT

It is not only a good idea, but also required that DC system power sources (such as a solar array or a battery bank) have the ability to be easily isolated from the rest of the system.

This is accomplished through the use of a **DC disconnect.** At a minimum, these disconnects should be located between the array and the inverter (or charge controller, when one is present) and between the battery bank and the inverter. It is also a good idea to locate an additional disconnect as close to the PV array as practical.

Modern inverters generally incorporate a DC disconnect within or attached to the inverter unit. If this is not the case, a DC disconnect must be installed in a "readily accessible location" - what this means exactly is to be determined by the local code official.

DC disconnects designed to isolate equipment (like charge controllers and inverters) must be located either within the equipment or within sight of it and no more than 10 feet (3 m) from the unit.

Make sure that the disconnect is indeed rated for DC power (such as the unit shown in Figure 5-13), since DC-rated units work differently than AC disconnects (switches, fuse boxes and circuit breakers must also be rated for DC current).

The disconnect must be externally operable and be lockable (although a lock does not have to be placed on the unit).

All current-carrying conductors must be switched (both the positive and negative). In the case of a solidly grounded system (which is rare), only the ungrounded conductor should be switched (typically the positive).

The NEC also requires that disconnects for the PV array, battery circuits, and AC circuits that are used in the system be grouped together and located near the inverter. This is to allow people who are perhaps unfamiliar with the installation (such as utility workers, emergency responders, etc.) to easily locate the method to shut down or disconnect the system.

Local utilities may have other requirements that also must be accommodated.

Service Disconnect

The design of large PV systems can become quite complex, especially when incorporating multiple power sources into a facility. These power sources (generators, PV array, the utility grid, batteries) may operate independently or in tandem with each other. Each of these power sources is considered a separate **service**.

(SERVICE)

As of the 2020 NEC, each service can have no more than one means of disconnecting it from its power source, with a few exceptions (outlined in NEC Article 230.71) that do not generally apply to residential installations.

These disconnects must all be grouped together. Not just the disconnects from one service, but the disconnects from all services feeding power to the home or business. This is usually interpreted to be within 10 feet (3 m) of each other, easily accessible and visible at one location. The location is typically at or near the site's utility meter.

FIGURE 5-14: COMMON BACK-FED SINGLE-POLE, TWO-POLE AND THREE-POLE CIRCUIT BREAKERS

If for some reason the disconnects cannot be grouped together, directories or a plaque must be in place indicating where all the disconnects are located.

Overcurrent Protection

The wiring and equipment that comprise a PV system are designed to only accept a certain amount of current. If current in excess of this maximum amount flows through the system, it could very likely damage or destroy the various components (and perhaps cause a fire).

FUSE

CIRCUIT BREAKER

OVERCURRENT PROTECTION DEVICES (OCPD)

For this reason, **fuses** and/or **circuit breakers** are incorporated into the system, designed to "blow" or trip off should excess current find its way onto the circuit. Collectively these are known as **overcurrent protection devices (OCPD)**.

Often this overcurrent protection is incorporated into combiner boxes, disconnects, charge controllers and/or inverters. But regardless of where it is housed, overcurrent protection must be part of a well-designed system.

While short circuits in the PV output system will likely not be much more than the rated operating output current of the system, short circuits from batteries can be thousands of amps.

AMPERE INTERRUPT RATING (AIR)

The rating on breakers and fuses is known as its **ampere interrupt rating (AIR)**. If current flowing through the OCPD exceeds the rating of the unit, it will open the circuit, preventing current flow that exceeds design parameters.

Available OCPD sizes were modified in the 2014 NEC and now include: 1, 3, 6, 10, 15, 20, 25, 30, 35, 40, 45, 50, 60, 70, 80, 90, 100, 110, 125, 150, 175, 200, 225, 250, 300, 350, 400, 450, 500, and 600 A ratings.

Fuses and breakers used in the DC circuits must be specifically rated for DC photovoltaic systems. Underwriters Labs (UL) has specific certifications for both fuses (UL 2579) and circuit breakers (UL 489B) specifically designed for use in DC PV systems.

Circuit breakers normally are available as single-pole, two-pole, or three-pole, as illustrated in Figure 5-14.

Back-Fed and Ground Fault Breakers

Any circuit breaker that does not specifically state on it that it can only be fed from the line or the load side is considered a **back-fed breaker**. This means that it will open (turn off) when current passing through it exceeds its rating - regardless of the direction from which the power originates.

Some breakers are designed to protect the circuit they are connected to should a ground fault occur. These are typically referred to as **Ground-Fault Protection of Equipment (GFPE) breakers**, or more commonly simply as GFPE breakers.

FIGURE 5-15: GROUND-FAULT CIRCUIT INTERRUPTER (GFCI) CIRCUIT BREAKER

Smaller GFPE breakers designed to protect a single load circuit can be easily identified by their white pigtail connection, as seen in Figure 5-15. These incorporate electronic ground-fault detection circuitry, or a trip solenoid, within the unit. These smaller breakers are called **ground-fault circuit interrupter (GFCI)** circuit breakers.

Most people are much more familiar with the receptacle type GFCI, as shown in Figure 5-16, which protects a single outlet. The NEC (210.8(A)) now require that GFCI outlets be placed anywhere moisture may be present (in the kitchen, bathrooms, basements, garage, outdoors, etc.).

FIGURE 5-16: STANDARD US GFCI WALL OUTLET

GFCI breakers are designed to provide that same protection to the entire circuit. GFCI works by sensing and comparing the amount of current flowing through the conductors. If there is a sudden imbalance (a ground fault), it then opens the circuit.

When a ground fault is detected, the breaker opens. But if it is back-fed, then the detection circuitry can be damaged. For this reason, the terminals on these circuit breakers are identified with "line" and "load" markings.

Some larger GFPE breakers (Figure 5-17) have been designed to operate while back-fed. Unless marked as line and load, the breaker is assumed to be back-fed.

FIGURE 5-17: LARGER GFPE FAULT POWERED CIRCUIT BREAKER THAT MAY ALLOW FOR BACK-FED CIRCUITS

Charge Controller

In any PV system that includes batteries (stand-alone or grid-interactive), it is necessary to include a **charge**

CHARGE CONTROLLER

controller, such as the one pictured in Figure 5-18, within the system. The charge controller acts a bit like a traffic cop. It is connected to the solar array, to the battery bank, and sometimes to the load. The controller electronically determines how much power (modifying both amps and volts) will be directed to the various locations. The most basic function of a charge controller is to prevent the over charge and the over discharge of the batteries.

SERIES CHARGE CONTROLLER

Early **series charge controllers** functioned a bit like a thermostat, charging the battery when its level fell below a certain point, and stopping when the bank exceeded a certain preset voltage level.

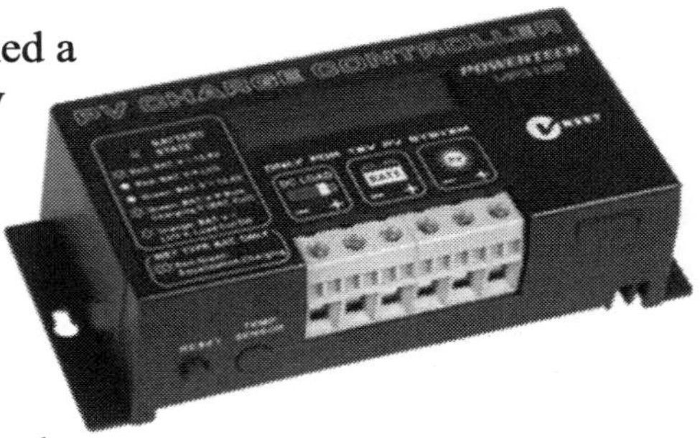

FIGURE 5-18: TYPICAL SMALL CHARGE CONTROLLER

For example, a nominal 12-volt battery might begin charging when it dropped to 11 volts, and stop charging when it reached 14 volts. While this got the job done, it was not very efficient as there were long periods of time when power from the array was not being used to charge the system.

Most multi-stage charge controllers for lead acid batteries on the market today employ **Pulse Width Modulation (PWM).** PWM charge controllers function as an electrical switch between batteries. The switch can be quickly switched on and switch off. As the desired voltage is obtained within the battery as it charges, the charge current slowly decreases.

PULSE WIDTH MODULATION (PWM)

MPPT Charge Controllers

To increase system efficiency, most high-quality charge controllers now employ **maximum power point tracking (MPPT)**, which electronically monitors both the output from the array and the state-of-charge of the battery bank, then automatically adjusts the charging voltage coming from the panels to best match what is needed by the battery bank.

MAXIMUM POWER POINT TRACKING (MPPT)

If an array is generating 800 watts, for example, at 28 volts the battery bank is being charged with 28.57 amps (800 W = 28 V x 28.57 A). If the MPPT reduced the voltage from the array to 19 volts, then the battery bank would receive a charge of 42.1 amps from the same energy output of 800 watts (800 W = 19 V x 42.1 A). By increasing or decreasing the voltage of the system, an MPPT can dramatically affect how much current is received by the battery bank.

MPPT systems will typically increase the effective useable power from the array by 20-45% during winter months, and 10-15% during summer months. They are, however, usually more expensive than the less sophisticated PWM controllers.

Remember that charge controllers are **DC-to-DC voltage converters**. They specify a maximum DC input voltage rating, as well as a maximum DC output current rating. In a MPPT charge controller, the DC voltage is subtly altered (from input to output) to more efficiently charge the battery bank.

> DC-TO-DC VOLTAGE CONVERTER

Traditional charge controllers require that the nominal voltage of the array matches the nominal voltage of the charge controller, which then matches the nominal voltage of the battery bank. In other words, if the battery bank was rated at 48 nominal volts, then the charge controller must be rated for 48 volts, and the array must be configured to generate 48 nominal volts.

FIGURE 5-19: MORNINGSTAR TRISTAR MPPT 60-AMP CHARGE CONTROLLER, WITH 600-VOLT INPUT AND 48-VOLT OUTPUT

But with the advent of MPPT charge controllers (Figure 5-19), manufacturers have been able to extend the input voltage range of the charge controllers to accept array output as high as 600 volts DC, the maximum allowed by the NEC for residential systems.

Through a DC-to-DC voltage converter, the charge controller can "step down" to voltage that matches the battery bank's nominal voltage. This allows for maximum flexibility in the design of the array and can result in substantial cost savings. This type of charge controller is often referred to as a **buck/boost charge controller**, as they can boost (raise) and buck (lower) voltage levels.

> BUCK/BOOST CHARGE CONTROLLER

Many charge controllers can also serve as a **load diverter**. This prevents energy produced by the solar array from being wasted when the batteries are fully charged and all load demands are being met - yet additional power is still being produced by the array.

> LOAD DIVERTER

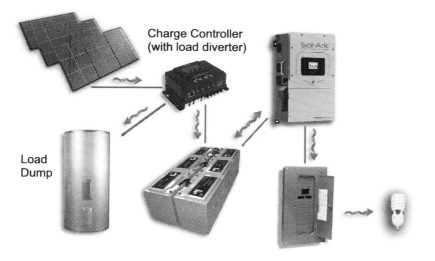

FIGURE 5-20: STAND-ALONE PV SYSTEM WITH LOAD DIVERTER AND LOAD DUMP INCORPORATED

Chapter 5: Parts of the PV System

LOAD DUMP

In such an instance, the charge controller can direct the excess power to a **load dump**, such as a hot water heater or perhaps even a heated swimming pool. The excess energy from the array is then captured and stored in the form of hot water as depicted in Figure 5-20.

This is generally only an issue in a stand-alone configuration, since the utility system will serve as a load dump within systems that are connected to the grid.

Battery Management Systems

Lithium ion batteries charge differently than lead acid batteries. For this reason, the system designer must make sure the charge controller is rated for lithium ion batteries if they are to be incorporated into the PV system. Many charge controllers that work with lead acid batteries are also designed to properly charge lithium ion batteries as well.

BATTERY MANAGEMENT SYSTEM

Most lithium-based batteries are equipped with a **battery management system (BMS)** due to their high energy density.

A battery management system is an electronic control unit that monitors and manages the performance of the battery bank. The BMS ensures the battery operates within safe limits, maximizes its lifespan, and maintains optimal performance. The BMS also often manages communication between the various systems (the array, backup loads, vehicle charging, etc).

FIGURE 5-21:
EXPLODED VIEW OF ENPHASE IQ BATTERY 3T ESS UNIT

(FROM ENPHASE)

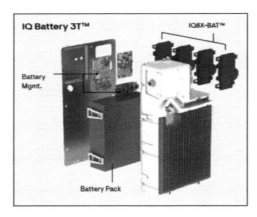

Increasingly, batteries integrated into PV systems are not just simple isolated batteries. They are complete **energy storage systems (ESS).**

ENERGY STORAGE SYSTEM (ESS)

POWER CONVERSION SYSTEM (PCS)

In addition to the battery management system, the ESS may also include a **power conversion system (PCS)** that handles AC to DC conversions (to charge the battery) and DC to AC conversions (to power the loads), as shown in Figure 5-21.

Charging Lithium Ion Batteries

When installing a smaller lithium ion battery that does not include a battery management system, it will be necessary to incorporate a charge controller.

SET POINT VOLTAGE

Lithium ion batteries charge at a constant voltage (usually referred to as the **set point voltage**). Since solar power varies with the intensity of the sun, the charge controller must adjust the amps coming from the array while holding the voltage constant. Once the battery is fully charged, the controller then draws only as many amps as required to hold the voltage of the battery steady.

The set point voltage is critical and must match the requirements of the battery bank. If it is set too high, the number of charge cycles the battery can complete will be reduced (shortening the battery life).

Lithium ion batteries do not have to function at their maximum voltage, and doing so actually adds stress to the unit and will limit its life, as indicated in Table 5-7. But if the voltage is too low, the battery will not be fully charged.

The lithium ion battery charging cycle is divided into two separate segments. The first is the **current limit** (sometimes called constant current) phase of charging. During this phase the maximum charging current is flowing into the battery since the battery voltage is below its set point.

> CURRENT LIMIT

However, the charge controller must limit the current coming from the solar array to the maximum allowed by the manufacturer to prevent damaging the batteries.

When the battery reaches its set point voltage (usually at about 65% of its full charge), the charging process enters its second phase. This phase is referred to as the **constant voltage** phase.

> CONSTANT VOLTAGE

Even though the battery has reached the desired voltage, the charge has not fully saturated the battery. The charge controller holds the voltage constant while reducing the current over time as the battery charge reaches full saturation.

Inverters

The role of an **inverter** (Figure 5-22) is to convert the DC current from a solar array or battery bank into AC current (that can then be used by conventional AC loads).

> INVERTER

Charge V/cell	Capacity at cut-off voltage	Charge time	Capacity with full saturation
3.8	60 %	120 min	~65%
3.9	70 %	135 min	~75%
4	75 %	150 min	~80%
4.1	80 %	165 min	~90%
4.2	85 %	180 min	100%

TABLE 5-7: TYPICAL CHARGE CHARACTERISTICS OF A LITHIUM ION BATTERY CELL

FIGURE 5-22: FRONIUS SYMO ADVANCED GRID-TIE INVERTER

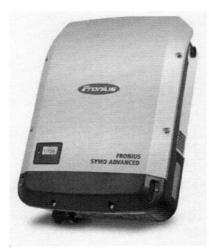

In the early days of solar electricity, inverters often were not incorporated into the system. Without the inverter's power conversion, the entire system needed to be wired to handle the DC current, and DC fixtures and appliances had to be purchased and installed. This added greatly to the cost of a PV system and often made the installation in an existing structure (already wired for AC) impractical.

Today, nearly all home and commercial PV systems incorporate an inverter, so that conventional AC wiring and loads (appliances, lighting, etc) can be serviced by the system. This makes converting an existing home to run off a PV system much easier, since the existing wiring and appliances do not need to be replaced. It also provides a maximum degree in flexibility, allowing the homeowner to use either grid-supplied power (AC power), or power from the solar array (DC converted to AC).

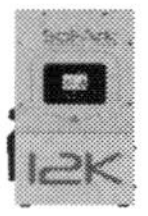

Inverters are available in many sizes and with many features. However, they can be divided into two broad categories: those that service systems that are not attached to the electrical grid, and those that must be able to communicate with the existing electric utility provider.

Sol-Ark

Sol-Ark 12K-P Spec Sheet

Solar	Input Power 12000W
Max Allowed PV Power	6500W + 6500W = 13000W
Max PV Power Delivered to Battery & AC Outputs	12000W
Max DC Voltage (Voc)	500V @ 18A, 450V @ 20A
MPPT Voltage Range	150-425V
Starting Voltage	125V
Number of MPPT	2
Max Solar Strings Per MPPT	2
Max DC Current per MPPT (Self Limiting)	20A
Max AC Coupled Input (Micro/String Inverters)	9600W

AC	Output Power 9kW On-Grid & Off-Grid
Connections	120/240/208V Split Phase
Continuous AC Power to Grid (On-Grid)	9000W 37.5A-L (240V)
Continuous AC Power to Load (Off-Grid)	9000W 37.5A-L (240V)
Surge AC Power 10sec Surge AC Power 100ms	16,000VA L-L (240V) 25,000VA L-L(240V)
Parallel Stacking	Yes
Frequency	60/50Hz
Continuous AC Power with Grid or Generator	15120W 63A L-L (240V)
CEC Efficiency	96.5% (Peak 97.5%)
Idle Consumption Typical ---No Load	60W
Sell Back Power Modes	Limited to Household/Fully Grid-Tied
Design (DC to AC)	Transformerless DC
Response Time (Grid-Tied to Off-Grid)	4ms
Power Factor	+/- 0.9 - 1.0

Battery (optional)	Output Power 9000W
Type	Lead-Acid or Li-Ion
Nominal DC Input	48V
Capacity	50 — 9900Ah
Voltage Range	43.0 — 63.0V
Continuous Battery Charging Output	185A
Charging Curve	3-Stage w/ Equalization
Grid to Batt Charging Efficiency	96.0%
External Temperature Sensor	Included
Current Shunt for Accurate % SOC	Integrated
External Gen Start Based on Voltage or %SOC	Integrated
Communication to Lithium Battery	CanBus & RS485

General	
Dimensions (H x W x D)	30.0" x 18.3" x 10.0"
Weight	78 lbs
Enclosure	NEMA 3R
Ambient Temperature	-25-55°C, >45°C Derating
Installation Style	Wall-Mounted
Wi-Fi & LAN Communication	Included
Standard Warranty (verified by HALT Testing)	10 Years

Protections & Certifications	
Electronics Certified Safety by SGS Labs to NEC & UL Specs - NEC 690.4B & NEC 705.4/6	Yes
Grid Sell Back — UL1741-2010/2018, IEEE1547a-2003/2014, FCC 15 Class B, UL1741SA,	Yes
PV DC Disconnect Switch — NEC 240.15	Integrated
Ground Fault Detection — NEC 690.5	Integrated
PV Rapid Shutdown Control — NEC 690.12	Integrated
PV Arc Fault Detection — NEC 690.11	Integrated
PV Input Lightning Protection	Integrated
PV String Input Reverse Polarity Protection	Integrated
AC Output Breakers - 63A	Integrated
250A Battery Breaker / Disconnect	Integrated
Surge Protection	DC Type II / AC Type II

FIGURE 5-23: STAND-ALONE INVERTER SPECIFICATIONS SHEET

(FROM SOL-ARK)

STAND-ALONE INVERTER

Stand-Alone Inverters

Stand-alone PV systems are independent of the electrical grid. Often they are installed in remote locations where power from utilities is not available.

These systems rely on battery banks to store power for use when the sun is not shining. **Stand-alone inverters** connect to the battery bank, converting the DC power of the batteries into AC power to use within the home. When the batteries run low on power, the inverter shuts off – and no power is available to service loads within the building. The solar array is required to recharge and/or keep the battery bank adequately charged.

When selecting a stand-alone inverter (sometimes referred to as off-grid or **battery-based inverter)**, specifications to be considered (Figure 5-23) include:
- the battery bank voltage (the inverter should be compatible with the system voltage). The input voltage tolerance should be able to accept the maximum voltage generated by your battery bank or solar array. Most residential stand-alone inverters will be rated at a nominal 48 Vdc. Smaller inverters may be rated at 12 Vdc or occasionally 24 Vdc. Output voltages in the U.S. should be 120 Vac or 240 Vac for typical household systems.
- the **power rating**, or the watts the system is designed to handle. Power ratings will often be listed as **continuous** power (the amount of power the unit can safely produce for an ongoing or indefinite period of time), or a **surge rating** (the maximum or peak amount of power the unit can handle for a very brief period of time). The amount of power produced by a stand-alone inverter is determined by the load. Turn on more appliances, and the inverter will produce more power. If the load (maximum power draw) exceeds the surge rating of the inverter, the system will shut down (and perhaps be damaged). The inverter should have enough capacity to be able to handle the anticipated load of the system (as determined by a load analysis) on a continuous basis.
- the output frequency, measured in hertz. In most of the western hemisphere (including the U.S) and a few other nations, it is around 60 Hz. Much of the rest of the world operates at 50 Hz.
- the **waveform** of the system.
- the efficiency of the unit. Some power is lost as the signal is converted from DC to AC. The efficiency of various inverters can range from 80 - 98% percent. The more efficient the unit, the more power from the battery bank will be available for use to service the loads.

Inverter Waveforms

Inverters are available that produce the AC output in three waveforms (illustrated in Figure 5-24). These include **pure sine wave, modified sine wave,** and **square wave.**

Square wave inverters are rarely used today except in the most inexpensive systems. Some legacy PV installations may still have them incorporated as part of the system. This type of inverter is not compatible with modern electronics or complex motors. These inverters can run simple tools with universal motors or lights, but little else.

FIGURE 5-24:
WAVEFORMS UTILIZED WITHIN VARIOUS INVERTERS

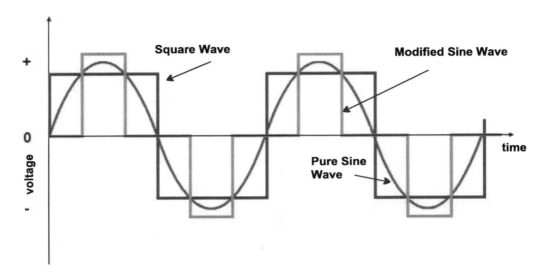

Modified sine wave inverters adjust the AC electrical signal to more closely reflect a sine wave, but not quite. Converting the signal from digital to analog takes energy. Most motors running off a modified sine wave inverter will use about 20% more power than it would use if being powered by a sine wave inverter. Other equipment such as photocopiers, battery chargers, digital clocks, and some fluorescent lights will not work at all in a system that uses a modified sine wave inverter. These units are, however, less expensive than a pure sine wave inverter.

Sine wave inverters cost two to three times as much as a similarly sized modified sine wave inverter, however, they have a major advantage in that all electronic equipment on the market are designed to function with the power output from these units. This matches the wave form of AC electricity produced by utility company generators.

Grid-Tied Inverters

GRID-TIED INVERTER

Grid-tied inverters are used in systems that are connected to the utility grid. These units are designed to match the energy form coming from the utility company.

ANTI-ISLANDING

They also include **anti-islanding** features, which automatically shut down the PV system whenever they detect that there is no power on the grid. This is a safety issue – protecting utility workers who may be repairing the grid. The utility worker may assume there is no electricity flowing (as the power grid is down), however the PV system could still be pumping electricity onto supposedly "dead" lines... and the result could be fatal. An example of a grid-tied inverter specification sheet can be seen in Figure 5-25.

Rapid Shutdown of PV Systems on Buildings

Perhaps the most dramatic change in the 2014 NEC, from the perspective of PV design, was the incorporation of a provision (in Section 690.12) that all

Input Data:	PRIMO 3.8-1	PRIMO 5.0-1	PRIMO 6.0-1	PRIMO 7.6-1	PRIMO 8.2-1
Recommended PV power (kWp)	3.0 - 6.0 kW	4.0 - 7.8 kW	4.8 - 9.3 kW	6.1 - 11.7 kW	6.6 - 12.7 kW
Max. usable input current (MPPT 1 / MPPT 2)	18 A / 18 A				
Operating Voltage Range	80 - 600V				
Max. input voltage	600V				
Nominal input voltage	410V	420V			
Admissable conductor size DC	AWG 14 - AWG 6				
MPP Voltage range	200 - 480V	200 - 400V	240 - 480V	250 - 480V	270 - 480V
Number of MPPT	2				
Output Data:	PRIMO 3.8-1	PRIMO 5.0-1	PRIMO 6.0-1	PRIMO 7.6-1	PRIMO 8.2-1
Max Output Power (208/240V)	3800 W	5000 W	6000 W	7600 W	7900 W/8200 W
Max. Continuous Output Current (208/240V)	18.3A/ 15.8A	24A/ 20.8A	28.8A/ 25.0A	36.5A/ 31.7A	38.0A/ 34.2A
Recommended OCPD/AC breaker size (208/240V)	25A/ 20A	30A/ 30A	40A/ 35A	50A/ 40A	50A/ 45A
Max Efficiency	96.7%		96.9%		97.0%
Admissable conductor size AC	AWG 14 - AWG 6				
Grid connection	208/240V				
Frequency	60 Hz				
Total harmonic distortion	< 5.0%				

FIGURE 5-25: EXAMPLE OF GRID-TIED INVERTER SPECIFICATIONS

(FROM FRONIUS)

rooftop arrays must incorporate a method where they can be effectively shut down at the source.

The reason for this provision comes from the increasing risk faced by first responders when responding to a fire at a structure where a PV system is installed.

Historically, when a first responder disconnects power from the grid (by pulling the meter, for example), they can safely assume that they will not come into contact with any live electrical wires within the building. However, if a PV system is present and the sun is shining, wires within the structure may still be energized, even when the main AC disconnect has been opened and the PV system is no longer operating (power can still be present from the panels to the string inverter where the anti-islanding feature has shut down the system).

In the 2017, NEC Section 690.12(B) defined the term **array boundary** as one foot from the array in all directions. Those direct-current PV conductors that lay outside the array boundary, and those that extended more than three feet inside a building, were required to be controlled to not more than 30 volts within 30 seconds of rapid shutdown initiation. Those direct-current PV conductors that lay inside the array boundary were required to be controlled to not more than 80 volts within 30 seconds of rapid shutdown initiation.

ARRAY BOUNDARY

Rapid shutdown is not required for ground mounted systems or systems installed on non-enclosed structures (such as a carport).

MODULE-LEVEL
POWER ELECTRONICS

PV HAZARD
CONTROL SYSTEM

UL STANDARD 3741

RAPID SHUTDOWN
EQUIPMENT

RAPID SHUTDOWN
SYSTEM

System designers normally comply with the rapid shutdown requirement by attaching some type of **module-level power electronic (MLPE)**, such as power optimizers or micro inverters directly to the solar panel.

In 2020, the NEC introduced the term **PV hazard control system (PVHCS)** and used an informational note bringing attention to **UL Standard 3741** that details how these systems can comply with this new provision. This standard takes a systems approach to determine that all the parts (panels, combiner boxes, MLPE, inverters, racking systems, etc), when combined, are minimizing the risk to those who come in contact with them.

The specific component parts used must be listed (by UL or some other listing agent) that they comply with rapid shutdown provisions (**rapid shutdown equipment**, or PVRSE) or as a system, when designed to work together (**rapid shutdown system**, or PVRSS).

Rapid Shutdown Initiator

The rapid shutdown provisions of the NEC are designed to ensure that power is not present within the building when the main PV disconnect is turned off (disconnected from the grid).

However, one of the main advantages of incorporating batteries into a PV system is to ensure the array remains functional and power is available, even when the grid goes down. Obviously these two goals are in conflict.

RAPID SHUTDOWN
INITIATOR

In grid-tied systems that do not incorporate batteries, the AC disconnect is considered the **rapid shutdown initiator (RSI)**. In other words, when the main AC disconnect is opened (turned off), shutdown takes place at the array.

FIGURE 5-26:
REMOTE BATTERY RAPID
SHUTDOWN INITIATOR

With systems that incorporate batteries, a second RSI is required.

Most AC-coupled and DC-coupled systems incorporate a manual RSI within the battery-based inverter unit. However, since the RSI must be accessible by first responders, this requires the inverter to be mounted outside (normally within 10 feet of the meter). If this is not possible or desirable, an auxiliary RSI (for the battery system) must be installed.

Most battery-based inverters will offer a remote disconnect switch designed to integrate into their unit, such as the RSI switch illustrated in Figure 5-26.

Inverter Standards

In 2003 the **Institute of Electrical and Electronics Engineers (IEEE)** first published Standard 1547 to establish guidelines on how **distributed energy resources (DER)** devices (such as grid-tied inverters) must function when interconnected with the electrical grid. This standard was adopted within the U.S. in 2005. It was originally designed for generators that produce less than 10 MVA.

Other similar standards exist and are used around the globe. Germany and many other European nations have adopted *"Power Generating Plants Connected to the Low-voltage Grid"* Application Rule (**VDE-AR-N 4105**). Australia and New Zealand have adopted **AS4777.2** that defines how inverters must function.

UL 1741 Listed

UL 1741: Standard for Inverters, Converters, Controllers and Interconnection System Equipment for Use with Distributed Energy Resources is intended to supplement and be used in conjunction with IEEE 1547.

An inverter listed as complying with UL 1741 demonstrates that the product is certified for interconnection to the grid, with particular respect to grid voltage and grid frequency requirements. It also addresses provisions such as anti-islanding, ground fault protection, and arc fault protection.

The European equivalent of UL 1741 is the *IEC 62109: Safety of Static Inverters* standard.

Smart Inverters

In late 2017, California became the first U.S. state to require the use of advanced, or **smart inverters** in solar arrays. The changes outlining their use through updates to **Rule 21**, seek to lessen the impact of distributed energy systems on the grid as more and more are integrated into the utility's distribution system. Most jurisdictions now require that inverters must meet these new requirements.

Grid-tied PV arrays not only draw power from the grid, but they also feed excess power onto the grid. Historically, PV systems have been required to disconnect from the grid when voltages or frequencies on the grid fall outside normal operating ranges. But if a large amount of these systems shut down, and then all try to reconnect at the same time, the grid is stressed.

Smart inverters allow PV systems to remain connected to the grid under a wider range of voltage and frequency levels. Once smart inverters are widely deployed and connected to the grid, they have the potential to not only avoid stress, but can actually be used to improve the stability of the power on the

Chapter 5: Parts of the PV System

grid. For example, dynamic volt/VAR operations (also called **dynamic reactive power compensation**) within the smart inverters can allow PV systems to assist in counteracting voltage deviations on the grid.

Dynamic Reactive Power Compensation

Where smart inverters are required, units installed must comply with **UL1741-SA** (as opposed to the previous listing of UL1741). Most manufacturers of inverters have compliant products available for the market. This does mean, however, that many older models of inverters may not be suitable for current installations in many parts of the country.

UL1741-SA

In 2023 the standard was further updated with the release of **UL1741-SB**. The testing requirements for UL1741-SB are more stringent and include a few more testing requirements for inverter manufacturers. It is important to know what inverter standard is required within the jurisdiction where the system will be installed.

UL1741-SB

Functions within compliant smart inverters include:
- **voltage and frequency ride-through** (able to accept a wider range of deviation from the grid),
- **soft start reconnection.** By requiring PV systems to ramp up during reconnection or to reconnect randomly within a time window, the grid should avoid sharp transitions and as a result power quality problems of voltage spikes, harmonics, and oscillations including the possibility that the disruptions caused by the reconnection of large numbers of PV systems actually result in another power outage,
- **fixed power factor** - this allows the inverter to maintain a fixed power factor, adjusting the amount of real and reactive power output in response to changing conditions,
- new **ramp rate** requirements, designed to help avoid sharp transitions and as a result, power quality problems of voltage spikes or dips, harmonics, and oscillations on the grid,
- updated anti-islanding provisions.

Voltage and Frequency Ride-Through

Soft Start Reconnection

Fixed Power Factor

Ramp Rate

Selecting a Grid-Tied Inverter

When selecting a grid-tied inverter, consider:
- the maximum DC voltage the unit can receive from the solar array. In the U.S., this voltage is limited by the NEC and should never exceed 600 Vdc for one to two family residential systems.
- the minimum DC voltage the unit can receive from the solar array and still produce power.
- AC output voltage. Output voltages from grid-tied inverters in the U.S. are typically 240 Vac for residential systems. European systems output at 230 Vac, and three-phase systems at other voltages (typically 208 Vac or 277/480 Vac).
- the power rating, or the watts the inverter can generate. For grid-tied inverters, this power rating can be related to the amount of power the array

is designed to generate. It is not based on the load the home is expected to draw (as is the case with a stand-alone inverter).
- the output frequency (measured in hertz). In the U.S. this should be 60 Hz. In much of the rest of the world it will be 50 Hz.
- conformity with the local utility provider. These units are designed to match the utility power exactly, so discuss the inverter selection with the local utility provider to assure compatibility. The inverter must be marked with a UL1741 listing, stating it has been approved as a **utility inter-tie device**.
- the conversion efficiency of the unit.

Bimodal Inverters

Bimodal inverters (sometimes called hybrid inverters) combine the functionality of stand-alone and grid-tied inverters (Figure 5-27). These are used in systems that require a dependable back up system.

FIGURE 5-27: SOL-ARK 12K BIMODAL INVERTER

Bimodal systems function like a grid-tied system, until the electrical grid goes down. In a grid-tied system, the homeowner would simply be without power while the grid is down. A bimodal inverter senses that the grid is down, then physically disconnects from the grid. It then redirects power from the battery bank to critical loads, now acting like a stand-alone system. The inverter continues to monitor the grid. Once power is restored, it switches back to grid-tied mode.

Module Level Power Electronics (MLPE)

The rapid shutdown provision within the NEC essentially mandated that an electronic device be attached to each solar panel installed on a rooftop. This device senses a loss of signal from the grid and shuts down the distribution of power from within the boundary of the array.

As a result, nearly all string inverter manufacturers have developed their own MLPE device or partnered with other manufacturers to develop MLPE devices that are compatible with their inverter system. These electronic components (Figure 5-28) attach either to the panel itself or to the rail system and are then connected directly into the panel.

But once an electronic device is connected to the panel, other features can be added to the unit that enhance the performance of the system. MPPT control can be placed at the individual panel level, dramatically reducing the effect of shading on the array.

Figure 5-28: Tigo module level power electronic device

Individual panel monitoring can be incorporated into the device, so production and problems can be isolated to specific panels. Also, voltage can be regulated (as in fixed voltage power optimizers) and power can be converted from DC to AC, as in micro inverters. All these options are available in various MLPE devices.

Micro Inverters

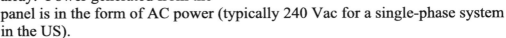

Micro inverters, such as the unit pictured in Figure 5-29, are becoming increasingly popular in PV systems as their price falls, becoming more competitive with string inverter systems. These are small inverters matched to individual solar panels. An individual micro inverter is typically attached to each panel in the array. Power generated from the panel is in the form of AC power (typically 240 Vac for a single-phase system in the US).

Figure 5-29: Micro inverter (from Enphase)

Use of micro inverters greatly simplify the design process of a system (as demonstrated in Figure 5-30), although the up front investment in multiple inverters may be more than that of a single inverter (depending on the system size).

Other advantages may include:
- increased system efficiency due to minimizing voltage drop.
- scalability. The system can easily grow over time. Panels can be easily added as needed or as money becomes available.
- use of mismatched panels. Standard system design requires that all the panels in the array (or at least in each string) be identical. With micro inverters, this is not necessary.

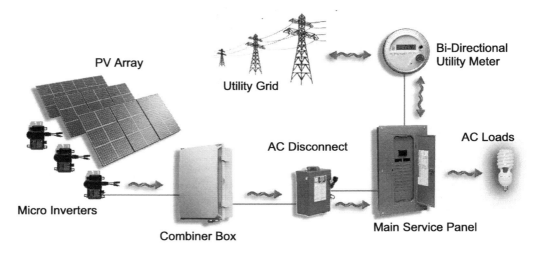

Figure 5-30: PV system incorporating micro inverters

- individualized panel efficiency. Each inverter is matched (through MPPT) to each panel. If a panel becomes defective, dirty, or shaded, the performance of that panel will not affect any other portion of the array.
- aging panel mismatch. As panels age, they become slightly less efficient over time (studies show that due to slight variations during manufacturing and aging differences, two identical panels operating under identical conditions may produce a power output that varies from 0.4 - 2.4%. This can result in losses as high as 12% within a single string).
- the entire system shuts down at the panel when the AC disconnect is in the open position.

Power Optimizers

Another popular MLPE system also exists that lies somewhere between a rapid shutdown MLPE and micro inverters. These DC-to-DC converters (generally referred to as **power optimizers**) attach to each solar panel (like a micro inverter). Like a micro inverter, the power optimizer incorporates MPPT functions as well as anti-islanding functions.

Unlike a micro inverter, however, the device does not convert the DC voltage from the panel to AC, but rather converts it to a fixed DC voltage. The array is then connected to a string inverter, as shown in Figure 5-31, that operates at a set voltage (rather than a range) and does not need to incorporate MPPT or anti-islanding functionality.

Like micro inverters, power optimizers:
- allow the use of mismatched panels,
- increase individualized panel efficiency,
- reduce the effects of aging panel mismatch,
- shut down the system at the panel when the grid is down.

Additional advantages of power optimizers over string inverters include:
- no temperature adjustments are required for string calculations (as the MPPT adjusts for temperature allowing the panel to always generate a fixed voltage),
- longer strings are possible due to the fixed nature of the panel output,

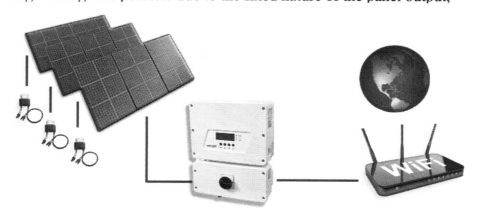

FIGURE 5-31: POWER OPTIMIZER SYSTEM

- improved safety, as high DC voltage is not present in the circuit until the entire system is connected to the grid.

Power optimizer producers cite a number of advantages of their systems over the use of micro inverters. These include:
- Lower cost. Although recent declines in the price of micro inverters have cut into this claim.
- Scalability. As the system size increases, the price per watt decreases.
- Comparability. Power optimizers work with a wide variety of panels and can harvest up to 950 watts per panel.
- Increased efficiency (99.5% peak efficiency, up to 98.8% weighted efficiency),
- No proprietary wiring system required (wire made specifically for the product), which can be quite expensive with micro inverter systems.

Solar Power Generation Monitoring

After the inverter (and most often incorporated into the inverter), a solar power **generation meter** is installed to measure just how much AC electricity the PV system has produced. This meter can be a digital readout, integrated on the Internet (with no physical meter at all)– or even an analog meter exactly like those used by the electric utility company.

These meters are critical for systems hoping to sell renewable energy credits (RECs). The power generation meter will track production and provide accountability to those purchasing the RECs. All solar energy systems that wish to earn SRECs must report system production based upon readings from a **revenue-grade meter** that meets ANSI Standard C12.1-2008.

Most inverter manufacturers offer online monitoring for installers, either free with the unit or by subscription. Inverter-supplied monitoring systems (shown in Figure 5-32) include features that let installers view multiple systems and track performance, diagnose failures, remotely update inverters and commission systems.

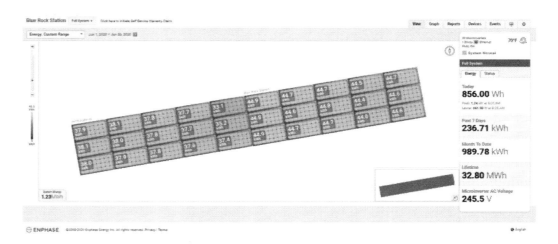

Figure 5-32: Online production & monitoring system for Enphase micro inverter system

End-users can use these apps to monitor their system's performance. Some inverter suppliers even offer monitoring of a system's battery storage.

AC Disconnect

Next in the system comes the **AC disconnect,** as pictured in Figure 5-33. This disconnect allows the entire PV system to be isolated from the standard household electrical wiring system. Once the PV system is disconnected (the AC disconnect turned off), electricians and/or the utility company can more safely work on the remaining electrical system.

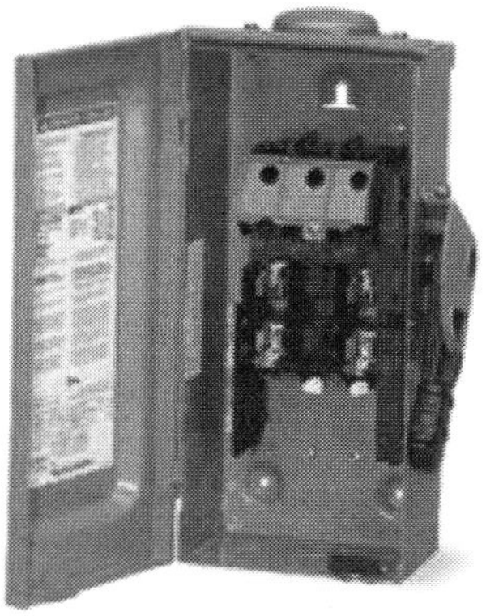

FIGURE 5-33: A TYPICAL AC-RATED DISCONNECT

These disconnects should be located in an accessible location, and should be compatible with the energy output (volts and amps) of the inverter(s).

Issues to consider when selecting an AC disconnect include:
- the ampacity rating of the AC output from the inverter(s).
- the voltage rating of the AC output from the inverter(s).
- the power phase - single phase or three-phase.
- the number of poles. Typically disconnects will have either 1, 2 or 3 poles. For 120 Vac single-phase circuit, only one pole is required to interrupt the line (hot) conductor. For 240 Vac service, two poles will be required to interrupt line 1 (black) and line 2 (red). For three-phase service, three poles will be required.
- the cabinet rating. If placed outdoors, ensure the unit is rated for outdoor use (NEMA 3R or better).
- whether the unit is fused (contains overcurrent protection) or unfused.
- the location of the disconnects. AC and DC disconnects should be located at or near the inverter. Many inverter systems incorporate these disconnects within their wire management assembly. If the disconnects are not incorporated in the unit, then they must be located within sight of the inverter and within 10 ft (3 m) of the equipment.

An AC disconnect must also be located within 10 ft (3 m) of the service connection point (generally the utility meter). They must be visible, externally operable, and lockable. They must also be installed so that the center of the operating handle, at it's highest position, is not more than 2 meters (6 feet, 7 inches) above the floor or working platform.

Current Transformers

Monitoring a PV system often involves obtaining detailed array production data as well as a load-versus-time profile of the home's energy consumption.

FIGURE 5-34: INSTALLED CURRENT TRANSFORMER (CT)

CURRENT TRANSFORMERS

Current transformers (CTs) are installed around the hot conductors in the solar production circuit (normally within the combiner box or service panel). The arrow on the CT must be pointing toward the load, as illustrated in Figure 5-34.

Load consumption data is collected through a second set of CTs and sent to a monitoring system. These consumption monitors are typically placed over the hot conductors between the utility meter and the main disconnect in the service panel.

This information can be used to monitor the flow of power within the system. Through the use of CTs, the system owner can monitor in real time how much power is being produced by the array, how much is flowing to or from the battery bank, and how much energy is flowing to or from the utility grid.

ELECTRICAL SERVICE PANEL

FIGURE 5-35: A TYPICAL US SERVICE PANEL WITH PV SYSTEM CONNECTED

Electrical Service Panel

From the AC disconnect, the system is then wired into the building's **electrical service panel** in smaller residential PV systems. This panel is commonly referred to as the circuit breaker box.

In most residential solar PV installations, the final connection to the grid will take place within the service panel (Figure 5-35).

The existing electrical service panel often presents a number of challenges when installing a PV system. The busbar rating may be too small to handle the added power from the array. Or there may simply be no space to accommodate another double-pole breaker from the PV system.

Smart Electrical Panels

Electrical service panels haven't changed much since circuit breakers began replacing the old fuse boxes. But increasingly, homeowners are moving to **smart electrical panels** that allow them to manage each circuit (usually on an app on their phone).

SMART ELECTRICAL PANEL

Smart panels, pictured in Figure 5-36, function much as a standard electrical service panel, except they have sensors that allow for the monitoring and control of each individual circuit. This allows the user to monitor electrical consumption on a circuit-by-circuit basis, and turn on or off individual circuits remotely.

When integrated with a grid-coupled PV system, during a power outage the user can monitor and adjust backup loads remotely as conditions and preferences change.

Bi-Directional Meter

Typically, when a PV system is hooked up to the grid, the grid itself acts as a virtual battery bank. When the sun is producing ample power to meet all the building's needs, all the electrical devices are being powered by the solar array. If there is more energy than required, the extra electricity is "sold back" to the grid. In effect, the electric meter runs backwards as electricity is pushed onto the grid.

FIGURE 5-36: A SPAN SMART SERVICE PANEL

On cloudy days, or at night, the electrical loads within the building are powered from the grid (as if there were no PV system). Most states in the U.S. (but not all), have passed laws that require **investor-owned utilities** (electrical co-ops and municipal systems are typically exempt from these laws) to **net meter** PV systems (in other words, allow the bi-directional flow of electricity and charge only for the "net" amount of electricity consumed by the home or business).

INVESTOR-OWNED UTILITIES

NET METER

Periodically, the utility will compute how much energy has been used from the grid, and how much energy the system has produced over that same time period. The customer will then be required to pay the difference (if there is one). This is referred to as the **true-up period**, which varies from utility to utility. If the true-up period is monthly, then the customer may find they are paying for energy they needed during low-solar months (such as the winter) and not get credit for excess energy they produced in the summer months. For this reason, an annual true-up period is preferred.

TRUE-UP PERIOD

The utility will typically provide the **bi-directional meter** – although the building owner may be required to pay for it and its installation.

BI-DIRECTIONAL METER

The 2023 NEC (Article 230.85) requires that all residential dwellings must have an **emergency utility disconnect** built into the utility meter mounting equipment, similar to that shown in Figure 5-37. The disconnect must be clearly marked as the "Emergency Disconnect, Service Disconnect."

FIGURE 5-37: EMERGENCY UTILITY DISCONNECT BUILT INTO THE METER BASE

EMERGENCY UTILITY DISCONNECT

Chapter 5: Parts of the PV System

Some Tools of the Trade

Familiarity with a number of tools is required for those working on PV systems.

MULTIMETER

Multimeter: (also referred to as a multi tester, a volt/ohm meter or a Vom). A **multimeter** is an electronic measurement instrument that can be used to test several electrical functions, such as voltage, current and resistance.

A multimeter can either be digital (such as the one shown in Figure 5-38), or analog. These hand-held devices range in price from as little as $6, to more than $5,000 for a top-of-the-line model.

Most multimeters include fuses that are designed to protect the unit from current overload. Testing a current larger than the unit is designed to handle will cause the fuse to blow. Digital multimeters are rated based on their intended function. Make sure the unit used is rated for the task performed.

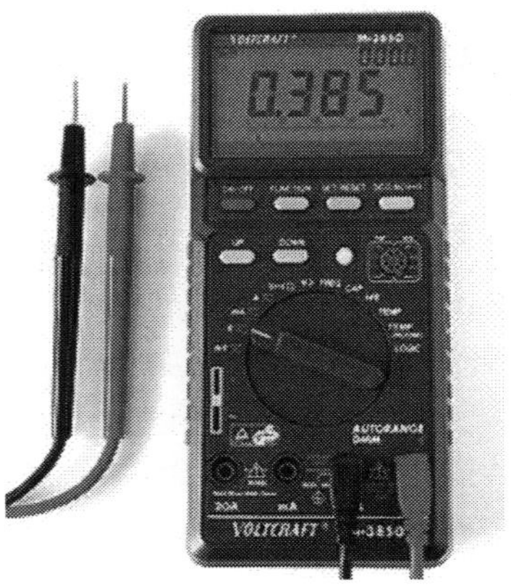

FIGURE 5-38: TYPICAL LOW-COST DIGITAL MULTIMETER

These include:
- **Category I**: used where equipment is not directly connected to the mains (the utility power).
- **Category II**: used on single phase mains final sub-circuits.
- **Category III**: used on permanently installed loads such as distribution panels, motors, and three phase appliance outlets.
- **Category IV**: used on locations where fault current levels can be very high, such as supply service entrances, main panels, supply meters and primary over-voltage protection equipment.

Ammeter

AMMETER

An **ammeter** is a device that measures current as it flows along a circuit.

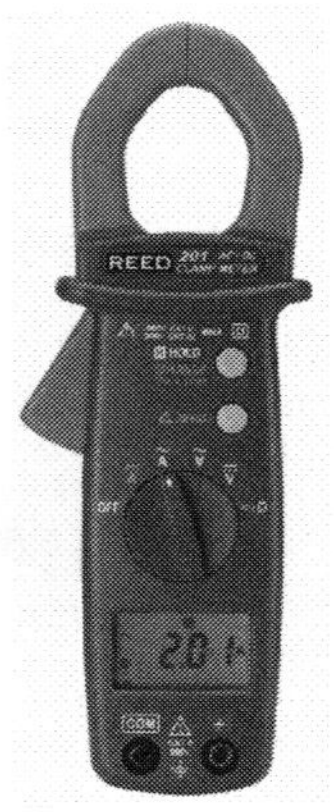

FIGURE 5-39: ZERO-CENTER AMMETER DISPLAY (LEFT) AND CURRENT CLAMP MEASURING TOOL (RIGHT)

This function is often included in most clamp type multimeters, or can take the form of a zero-center display readout or even as a current clamp measuring device (both shown in Figure 5-39).

Crimping Tool

CRIMPING

Crimping is the joining together of two pieces of metal by deforming one or both of them to hold the other in place. The bend or deformity is called the crimp.

Often wire is crimped to the end connector with the use of a special tool. These tools can be quite specific (provided by the manufacturer, designed to work with only one form of connector) or fairly standard. They are designed to crimp and strip wires, allowing connections to standard blade, ring, or spade **connectors** as shown in Figure 5-40.

FIGURE 5-40: ASSORTMENT OF VARIOUS CRIMP CONNECTORS

CONNECTORS

Wire Strippers

Hand held tools such as the one in Figure 5-41 are designed to remove the insulation from around the conductor of a wire. **Wire strippers** come in many designs and styles, but they all perform the same basic function.

WIRE STRIPPERS

Other Typical Installation Tools

- tool belt,
- flat pry bar or roofing bar (also called a ripping bar - Figure 5-42),
- utility knife,
- standard framing hammer,
- tape measure,
- flat head and Phillips screwdrivers,
- open-ended wrenches,
- ratcheting socket driver with deep-well sockets,
- cordless power drill,
- wood/metal drill bits ("impact ready" preferred),
- drill/driver bits (1/4" shank): T-25, T-30 (star drive), 1/4" shank extension,
- nut driver bits (1/4" shank adapter): 1/4", 5/16", 3/8",
- hole saw (1/4" shank),
- standard channel locks (set of 2),
- tin snips,
- pliers and needle nose pliers.

FIGURE 5-41: BASIC WIRE STRIPPER

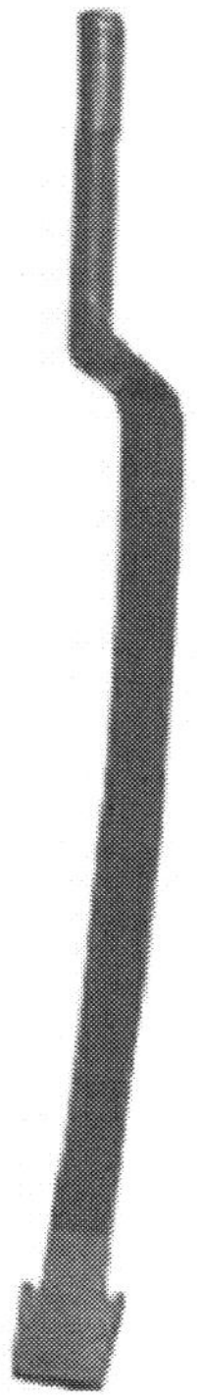

FIGURE 5-42: FLAT PRY BAR OR ROOFING BAR (ALSO CALLED A RIPPING BAR)

Chapter 5 Review Questions

1. Commonly a solar panel used in a stand-alone PV system will be described as a 12-volt or 24-volt panel. When referenced in this manner, we are speaking of the panel's _____.
   A) open-circuit voltage
   B) nominal voltage
   C) short-circuit voltage
   D) maximum power voltage

2. The highest voltage a panel can generate (under standard test conditions), with no load connected to it is referred to as the panel's _____.
   A) open-circuit voltage
   B) nominal voltage
   C) short-circuit voltage
   D) maximum power voltage

3. The ratings (Isc, Imp, Voc, Vmp) noted on most solar panels are the result of laboratory testing using _____.
   A) generally accepted accounting principles (GAAP)
   B) standard test conditions (STC)
   C) the National Electrical Code (NEC)
   D) American National Standards Institute (ANSI)

4. As the temperature of the PV array rises above 25 degrees C, the panel will:
   A) produce more power
   B) produce less power
   C) temperature does not affect the ability of a panel to produce power
   D) automatically disconnect at 35 degrees C

5. Today, most solar panels incorporate:
   A) MC1 connectors
   B) MC2 connectors
   C) MC3 connectors
   D) MC4 connectors

6. Which of the following components within a PV system is typically not considered as part of the balance of systems (BOS)?
   A) PV panel
   B) battery system
   C) disconnects
   D) controller

7. Which of the following is NOT a widely available type of PV solar cell on the market today?
   A) monocrystalline PV
   B) bicrystalline PV
   C) polycrystalline PV
   D) amorphous PV

8. A number of solar panels connected in series is referred to as a(n):
   A) array
   B) bank
   C) string
   D) band

9. The place where multiple strings from an array are connected in parallel is referred to as the:
   A) junction box
   B) MC4 connector
   C) raceway
   D) combiner box

10. Which of the following will help minimize system damage in the event of a lightning strike?
    A) surge protection
    B) fuses and circuit breakers
    C) ground fault detection
    D) all of the above

11. What is the primary difference between a combiner box and a junction box in a PV system?
    A) Combiner boxes are used in roof mounted systems, while junction boxes are used in ground mounted systems.
    B) Combiner boxes incorporate overcurrent protection, while junction boxes do not.
    C) Combiner boxes connect PV source circuits in parallel while junction boxes do not.
    D) Combiner boxes are placed in the PV output circuit while junction boxes are located within the inverter input circuit.

12. Which of the following represents the **LARGEST** diameter wire?
    A) 0 AWG
    B) 2 AWG
    C) 6 AWG
    D) 12 AWG

13. Which of the following wire can carry the greatest current?
    A) 0 AWG
    B) 2 AWG
    C) 6 AWG
    D) 12 AWG

14. When selecting the appropriate wire gauge, which of the following is of greatest concern?
    A) the price
    B) its ampacity rating
    C) its resistance rating
    D) its voltage rating

15. Which of the following wire insulation is rated for outdoor wet conditions and is considered an acceptable alternative to PV wire by the NEC?
    A) THHN
    B) THWN
    C) NEC
    D) USE-2

16. A #4 AWG TW copper conductor has an ampacity rating of (use Table 5-1 from this chapter):
    A) 35 A
    B) 70 A
    C) 85 A
    D) 95 A

17. If this same #4 AWG TW conductor was placed in a conduit with 14 other similar conductors, it would now have an ampacity rating of (use Table 5-3):
    A) 35 A
    B) 70 A
    C) 85 A
    D) 95 A

18. The NEC ampacity ratings for conductors (TABLE 310-16) assumes the wires are operating at:
    A) a temperature of 20° C (68° F).
    B) a temperature of 25° C (77° F).
    C) a temperature of 30° C (86° F).
    D) a temperature of 40° C (104° F).

19. According to the NEC, when connected to a functionally grounded inverter (which is normal in today's market), the color of the **NEGATIVE** wire must be:
    A) black, red, or any other color (other than green or white).
    B) green or green with yellow stripes.
    C) white or gray.
    D) bare copper wire.

20. According to the NEC, when connected to a functionally grounded inverter (which is normal in today's market), the color of the **POSITIVE** wire must be:
    A) black, red, or any other color (other than green or white).
    B) green or green with yellow stripes.
    C) white or gray.
    D) bare copper wire.

21. If wire is placed in conduit that is exposed to sunlight and is located with the bottom of conduit resting on the deck of the roof, then:
    A) holes must be drilled into the side of the conduit to vent off hot air.
    B) a conduit cover must be placed over the run to shade the installation from direct sunlight.
    C) 33° C must be added to the highest measured ambient air temperature of the site and the ampacity of the wire must be derated.
    D) high temperature wire and terminations (90°C (194°F)) must be used to ensure the circuit can handle its rated ampacity.

22. Colored tape may be placed around the wire near each termination point rather than requiring that the entire wire have the appropriate color when:
    A) the wire is run through metal conduit.
    B) the wire is #4 AWG or larger.
    C) the system is solidly grounded rather than functionally grounded.
    D) the wire run is direct buried.

23. The NEC permits a maximum of _____ disconnect(s) for any electrical service.
    A) one
    B) two
    C) three
    D) six

24. The DC disconnect for the inverter within a PV system that utilizes a string inverter must be located:
    A) within sight of it and no more than 10 feet (3 m) from the junction/combiner box.
    B) within sight of it and no more than 10 feet (3 m) from the inverter.
    C) within sight of it and no more than 10 feet (3 m) from the utility meter.
    D) within sight of it and no more than 10 feet (3 m) from the AC disconnect.

25. Which of the following is **NOT** considered an overcurrent protection device?
    A) a module level power electronic (MLPE)
    B) ground-fault protection of equipment (GFPE) breaker
    C) ground-fault circuit interrupter (GFCI)
    D) heavy-duty cartridge fuse

26. When current passing through a breaker exceeds its rating - regardless of the direction from which the power originates (line and load side) – it is considered:
    A) back-fed.
    B) damaged.
    C) a circuit interrupter.
    D) an ampere interrupt device.

27. The charge controller handles the various stages of charging within deep cycle batteries. These stages typically include:
    A) bulk, absorption, floating.
    B) full, tapering, leveling.
    C) equalization, absorption, floating.
    D) equalization, leveling, absorption.

28. Older charge controllers used a thermostat type control known as a series controller. Today most charge controllers use:
    A) parallel controllers.
    B) inverters.
    C) MC4 connectors.
    D) pulse width modulation (PWM).

29. Most charge controllers sold today that are designed to work with lead-acid batteries are:
    A) series controllers.
    B) MPPT charge Controllers.
    C) load diverter controllers.
    D) pulse width modulation (PWM) controllers.

30. Lithium-ion battery systems, unlike lead-acid battery systems, generally must incorporate:
    A) a battery management system (BMS).
    B) a DC-to-DC voltage converter.
    C) maximum power point tracking (MPPT).
    D) a module level power electronic (MLPE).

31. What technology used within inverters is generally used today to most closely mimic utility power, and avoids problems with sensitive electronic devices?
    A) square wave
    B) modified sine wave
    C) pure sine wave
    D) net-metering

32. When selecting a stand-alone inverter, all the following specifications are important **EXCEPT**:
    A) system nominal voltage rating.
    B) power rating.
    C) conversion efficiency.
    D) anti-islanding feature.

33. An inverter that can function both as a grid-tied inverter as well as a stand-alone inverter is referred to as a(n) _____ inverter.
    A) bimodal
    B) micro
    C) UL 1741 listed
    D) inter-tie

34. Inverters that comply with Rule 21 and are listed as UL1741-SA compliant are often referred to as:
    A) smart inverters.
    B) bimodal inverters.
    C) rapid shutdown inverters.
    D) AC coupled inverters.

35. When sizing a GRID TIED inverter, the size of the inverter is directly related to:
    A) the daily load demand.
    B) the maximum power draw.
    C) the size of the battery bank.
    D) the size of the array.

36. When sizing a STAND ALONE inverter, the size of the inverter is directly related to:
    A) the daily load demand.
    B) the maximum power draw.
    C) the size of the battery bank.
    D) the size of the array.

37. Which of the following inverter systems comply with the NEC 2014 "rapid shutdown" provision?
    A) bimodal inverters with MPPT
    B) multi-modal inverters with PWM
    C) fixed voltage inverters with power optimizers
    D) none of the above

38. Changes to the rapid shutdown provisions that went into effect within the 2017 NEC include all the following **EXCEPT**:
    A) when the PV system disconnect is turned off, conductors within one foot of the array boundary (rather than 10 ft) must be energized to no more than 30 volts within 30 seconds.
    B) inverters must comply with Rule 21 and be listed as UL1741-SA compliant.
    C) within the array boundary, there is no conductor with a voltage higher than 80 Vdc when rapid shutdown is initiated.
    D) a label must be attached to the rapid shutdown disconnect indicating the system is in compliance with the new 2017 code requirement.

39. In a simple grid-tied PV system, the AC disconnect shuts off all power from the array within the building. In AC and DC coupled systems, opening the AC disconnect would simply cause the system to operate in stand-alone mode. In order to turn off all power within the building, a second disconnect is required for coupled systems known as a:
    A) ground-fault interrupter.
    B) anti-islanding disconnect.
    C) rapid shutdown initiator.
    D) back-fed breaker.

40. Which of the following is **NOT** an advantage of a system that includes micro inverters?
    A) Can use mismatched panels.
    B) The array can easily be expanded over time.
    C) The panels will not need to be replaced as often.
    D) The effect of shading is reduced.

41. Which of the following is **NOT** considered a module level power electronic (MLPE)?
    A) micro inverter
    B) power optimizer
    C) rapid shutdown device
    D) load controller

42. All PV systems that wish to participate in a state's SREC program must incorporate a:
    A) revenue-grade meter.
    B) rapid shutdown initiator.
    C) utility inter-tie.
    D) load controller.

43. An AC disconnect must be:
    A) located in an accessible location.
    B) externally operable, and lockable.
    C) located within 10 ft (3 m) of the service connection point.
    D) all of the above.

44. Production from the array and consumption from the grid are measured by:
    A) current transformers.
    B) module level power electronics.
    C) bi-directional utility meters.
    D) all of the above.

45. Under a utility's net metering policy, they will determine a period of time at which any credits are removed from the PV owner's account (not carried forward forever). This is known as the:
    A) true-up period.
    B) commissioning period.
    C) interconnection period.
    D) distribution period.

46. The 2023 NEC requires that all residential dwellings must have a/an _____ built into the utility meter mounting equipment.
    A) revenue-grade unit
    B) current transformer
    C) emergency utility disconnect
    D) rapid shutdown initiator

Lab Exercises : Chapter 5

5-1. Using Wire Ampacity Tables

Lab Exercise 5-1:

Supplies Required:
- Copy of NEC Table 310.16

Tools Required:
- calculator

The purpose of this exercise is to practice locating ampacity limits for various wire sizes in the NEC.

a. Locate the correct ampacity limit reading from the NEC Table 310.16 for a #12 AWG USE-2 copper conductor (use table on page 111).

b. What would be the ampacity limit if UF wire (#12 AWG) were used rather than USE-2?

c. What would be the ampacity limit if using THWN wire #4 AWG?

d. Assuming the temperature rating of either termination point of the cable run is unknown, determine the ampacity limit for the run if using #12 AWG USE-2 copper wire.

5-2. Derating Ampacity for Heat and Conduit Fill

Lab Exercise 5-2:

Supplies Required:
- Copy of NEC Table 310.16
- Copy of NEC Table 310.15(B)(1)
- Copy of NEC Table 310.15(C)(1)

Tools Required:
- Calculator

The purpose of this exercise is to practice adjusting wire ampacity limits for temperature and conduit fill.

if it is co-located in a conduit containing 12 similar wires.

a. Determine how many amps of current a #12 AWG USE-2 conductor is capable of carrying safely.

b. Assume the conduit will be located on a rooftop, the bottom of the conduit will be located directly on the surface of the roof deck. Assume the hottest ambient air temperature the site will receive is 35° C (95° F). Using NEC TABLE 310-15 (B) (1) , adjust the ampacity limit for this cable run.

c. Now assume there are four conductors in the conduit. Using NEC TABLE 310.15(C)(1) further adjust the ampacity limit for each wire based on conduit fill.

d. What happens to the ampacity limit in task "e" if the conduit is mounted 6 inches off the roof deck, rather than on the roof deck?

e. Perform the same tasks as outlined in steps "a-c", but replace the #12 AWG wires with #4 THW wires (all other environmental factors remain the same).

Chapter 6

Conducting a Site Survey

Chapter Objectives:

- Determine available sunlight for a specific site.
- Understand how panel orientation affects power generation.
- Identify the various mounting options and note the advantages/ disadvantages of each type.
- Determine the various environmental concerns when locating a PV system on the site.
- Describe shading issues and the tools used to measure its impact on a specific site.
- Discuss the limitations and advantages of remote site assessments.
- Identify site-specific issues that might affect the location of the system.
- Determine the best location for the BOS electronics.
- Explore issues and options when connecting the PV array to the existing AC electrical system.
- Understand the economics of installing a residential scale PV system.

It will be necessary to take a look at the specific site to see if it is appropriate for a photovoltaic system. And if so, determine what system is best for the particular application and location.

General Site Information & System Goals

Before beginning the design process of a PV system, it is important to understand a number of site-specific details.

For example:
- How much of the existing electrical load is to be serviced by the array?
- What are the budget limitations?
- Is there grid power available? (This will help determine what type of system to design).
- Are there any deed restrictions, city ordinances, utility restrictions, etc. that might effect the design of the system?
- How much sunlight is available at the location?

Clearly the amount of sunlight in Phoenix, Arizona is different than what is available in Boston, Massachusetts. And even at a specific location the amount of sunlight will vary throughout the year (typically more in the summer and less in the winter) and where the array is placed (shading and orientation).

Available Sunlight

There are many websites that can provide detailed reports of the available **solar irradiance** for a specific site.

SOLAR IRRADIANCE

The available solar irradiance, on a clear day, at sea level, when the temperature is 25°C (77°F) - is 1000 watts per square meter. These conditions (in the solar world) are referred to as standard test conditions. But the universe rarely provides such ideal conditions.

The amount of sunlight striking any square meter of land at a specific location will vary from moment to moment. Usually, it will contain less energy than 1000 watts, sometimes more.

PEAK SUN HOURS

So the accumulated total sunlight striking a square meter of earth over a 24-hour period, divided by 1000 watts (the ideal energy received in an hour) will provide the number of **peak sun hours** for a specific location.

Expressed mathematically, it is:

Peak Sun Hours = Total Daily Energy from the Sun (watts) / 1000 (watts)/hr

For example, if the total amount of energy from the sun hitting a square meter area (angled directly at the sun) in Nebraska during the month of October is measured, the answer for an average day will be about 4,760 watts. This would then be expressed as 4.76 hours of "ideal" sunlight received each day for that month.

Peak Sun Hours = 4,760 watts / 1000 watts/hr = 4.76 hrs

INSOLATION

The terms peak sun hours, irradiance and **insolation** are sometimes used interchangeably. Insolation is short for *incoming solar radiation*.

What are the average peak sun hours for the entire year at a particular location? The answer to that question might not be completely straightforward. The amount of energy hitting a solar panel will depend on its angle in relation to the horizon and the direction it faces.

Solar Panel Orientation

The homeowner may already have some firm ideas regarding where the system should be located. These may or may not be practical. Regardless of where the array is placed, the panels should be oriented as much as possible towards the south (if located in the northern hemisphere).

AZIMUTH
SOLAR NOON
TRUE SOUTH

As the sun tracks across the sky over time, it essentially "moves" in two directions. During each day, it "rises" in the east and "sets" in the west. This movement (measured in degrees) is referred to as the **azimuth** (as depicted in Figure 6-1). In the northern hemisphere, at **solar noon**, the sun will be at **true south** (or solar south).

It should be noted that solar noon has no relationship to a specific time of day. It is simply that precise moment that the sun is aligned with true south on the horizon.

True or solar south should not be confused with **magnetic south**. When using a compass to locate north, the compass is not actually pointing to the north pole, but towards a mass of magnetized liquid iron beneath the surface of the earth.

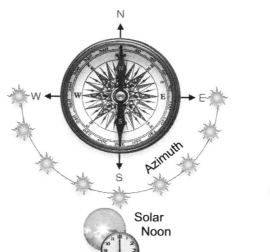

FIGURE 6-1: AS THE SUN TRACKS FROM EAST TO WEST EACH DAY, ITS AZIMUTH IS MEASURED IN RELATION TO TRUE SOUTH

(MAGNETIC SOUTH)

This magnetic mass is always in motion, Currently, the magnetic mass sits below the Arctic Ocean north of Alaska and is moving about 30-40 miles (50-60 kilometers) each year toward Siberia.

The difference between geographic north and magnetic north will have a varying affect on the orientation of a solar array in different locations. Just east of the Mississippi River, for example, the two line up fairly well and magnetic north and geographical north are nearly the same (known as the **agonic line**). But at the extreme northern coast of the U.S. (Maine and Washington) they can differ by as much as 20 degrees. These variations can be graphically depicted in a **magnetic declination map** (like that shown in Figure 6-2).

(AGONIC LINE)

(MAGNETIC DECLINATION MAP)

This map demonstrates that for a solar array located near Memphis, Tennessee – true south and magnetic south will be roughly the same. But if the array is

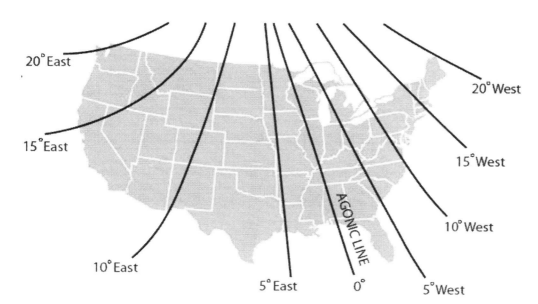

FIGURE 6-2: MAGNETIC DECLINATION MAP OF THE UNITED STATES

(U.S. GEOLOGICAL SURVEY)

FIGURE 6-3:
ADJUSTING FOR SOLAR NORTH USING A COMPASS

(U.S. GEOLOGICAL SURVEY)

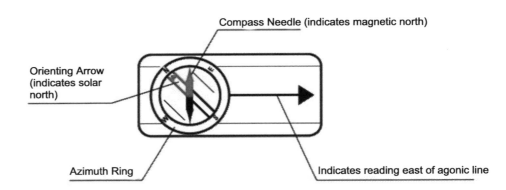

located in San Francisco, California – then true south is going to be about 14 degrees different (magnetic north is skewed to the east of true north) than magnetic south.

To determine how to orient an array at a specific location, take a compass, align it so the needle is pointing north (logically the opposite end will point south). If the adjustment is in degrees west (this would be the case in installing an array in New York for example), rotate the compass clockwise (so the orienting arrow that normally aligns with north moves towards the west). As Figure 6-3 demonstrates, rotate the compass the specified number of degrees adjustment for the location and the compass needle will indicate true north.

The initial orientation of the array can be set to take advantage of local needs. For example, a utility might orient their array with a bias towards the west (rather than an azimuth angle of true south) so that the array generates more power in the afternoon, when electrical demand is at its peak.

Any variation in orientation of the array from solar south will result in a loss of power generated by the array, as illustrated in Figure 6-4. Unless there are

FIGURE 6-4: LOSS IN ARRAY OUTPUT WHEN ORIENTED OFF SOLAR SOUTH

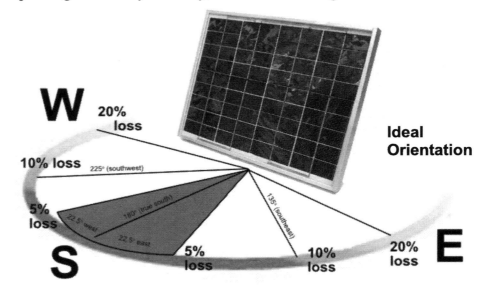

unusual circumstances, the array should always be facing within 45 degrees of true south (true north in the southern hemisphere).

Angle of the Array

Over the course of a year, the sun also tracks higher in the sky during the summer and lower in the sky during the winter. This movement (also measured in degrees) is referred to as its **altitude** (or **solar elevation**).

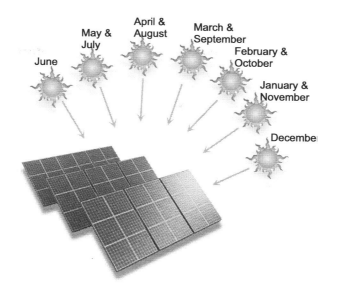

FIGURE 6-5: THE IDEAL ANGLE OF A SOLAR PANEL WILL VARY DEPENDING ON THE MONTH OF THE YEAR

ALTITUDE

SOLAR ELEVATION

ZENITH

When the sun is at the horizon, its altitude would be zero degrees. When the sun is directly overhead, its altitude will be 90°. The highest elevation the sun reaches on a given day is referred to as its **zenith**.

As indicated in Figure 6-5, the angle of the solar panel should be adjusted for altitude in order to capture the maximum energy from the sun. The more directly the panel faces the sun, the more energy the solar module will capture. The difference between an imaginary line that points to the sun and an imaginary line that points straight out (at a 90° angle) of a PV panel is known as the **incidence angle**. Solar panels are most efficient when pointing directly at the sun, so installers want to minimize this angle at all times, ideally lowering it to zero.

INCIDENCE ANGLE

If the energy contained within the direct rays of the sun is measured at 1000 watts per square meter (W/m^2), then tilting that same square meter surface will "spread out" that available energy over a larger surface (as shown in Figure 6-6).

Orienting the panel as close as possible to the direct angle of the rays of the sun will increase the amount of energy the panel receives.

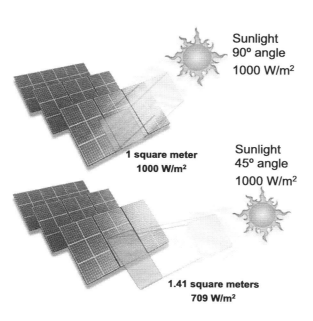

FIGURE 6-6: THE AMOUNT OF ENERGY IN SUNLIGHT STRIKING A SQUARE METER OF SURFACE AREA FROM DIRECTLY OVERHEAD WILL ACTUALLY SPREAD OVER A 1.41 SQUARE METER SURFACE IF THE ANGLE IS INCREASED TO 45 DEGREES.

SUN TRACKING SYSTEM

DUAL-AXIS TRACKING SYSTEM

SINGLE-AXIS TRACKING SYSTEM

There are devices, or **sun tracking systems**, that automatically adjust the angle of the solar array to face the sun as it tracks in the sky. If the system adjusts for both altitude and azimuth, it is referred to as a **dual-axis tracking system**. If it only adjusts for one of these dimensions, it is a **single-axis tracking system**.

One of the advantages of installing a solar array is its inherent lack of required maintenance. A tracking system is a mechanical device likely to require periodic maintenance or be subject to wear and tear. The cost of such systems may often exceed the cost of simply adding more solar panels to increase the array's output.

FIXED-MOUNT

If the array is attached to a racking system that cannot be adjusted, this is considered a **fixed-mount** system. With such a system, the array is only directly facing the sun twice each year.

Recent studies have shown that for smaller PV systems (less than 10 kW), the average installed cost of systems with tracking was 19% higher than fixed-mount systems. Costs increase for larger systems by about 15% when tracking systems are incorporated.

However, manufacturers of tracking systems claim that they will increase power output over a fixed-mount array by as much as 30-40%. So despite the maintenance concerns, tracking systems may be a practical consideration in some instances.

But at the current time, fixed-mount systems comprise the vast majority of installed systems. If the array is to be mounted and never moved, and it is the desire of the installer to get the best possible amount of sunlight averaged over the entire year, and each month that location receives the same amount of sunlight (which never happens in reality), then the angle should be set at the site's latitude. At the spring and the autumn equinoxes, the rays of the sun will point directly (at a 90° angle) at the array.

array angle = latitude

LATITUDE

For example, if an array were to be placed in Washington, DC (**latitude** 39° N), the optimum year-round angle would be 39°.

SUMMER BIAS

If local conditions are such that there is a great deal of sunlight during the summer months, and a relatively small amount of sunlight during the winter months (as is often typical in mid-latitude regions), then the array might be oriented with a lower angle (pointing more directly overhead during the summer). This altitude adjustment is referred to as a **summer bias**.

To find the ideal summer bias, take the latitude and subtract 15°.

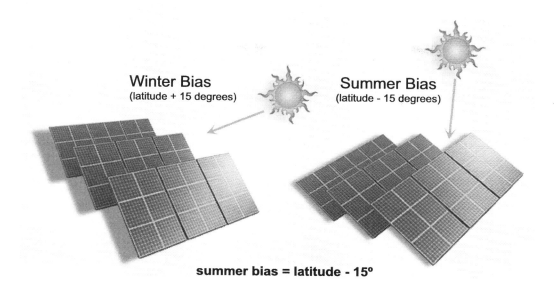

FIGURE 6-7: WINTER AND SUMMER BIAS FOR SOLAR PANEL ORIENTATION

summer bias = latitude - 15°

An array in Washington DC with a summer bias would be at a 24° angle.

To adjust this same array for a **winter bias**, add 15° to the altitude of the year-round position of the array. In this example, an angle biased towards generating more power in the winter months would be 39° + 15° = 54°, as indicated in Figure 6-7.

winter bias = latitude + 15°

WINTER BIAS

Calculating the ideal angle for a fixed-mount array at a specific location may not be as simple as selecting the year-round, summer or winter bias settings.

The monthly variation of solar insolation can make for some fairly complicated calculations. Fortunately there are software programs available, such as the **PV Watts** program available from NREL, output from which is illustrated in Figure 6-8. By running various altitude angle settings through the program, it is possible to determine the optimum angle for a site, given the variations of available sunlight from month-to-month.

PV WATTS

An example of such a sampling can be seen in Table 6-1. If peak annual production is the primary objective for this location, then the array's altitude angle would be set between 30 - 33 degrees.

Some smaller ground-mounted racking systems have pre-drilled holes in the mounting brackets that allow the home owner to make periodic adjustments manually. This fairly low-tech feature can greatly increase the efficiency of the system – allowing for maximum production during the winter as well as during the summer. Racks of this type are referred to as an **adjustable frame** mounting system.

ADJUSTABLE FRAME

FIGURE 6-8: ENERGY GENERATING CHART FOR COLUMBUS, OHIO WITH ARRAY ANGLE SET TO 28 DEGREES ORIENTED TO SOLAR SOUTH

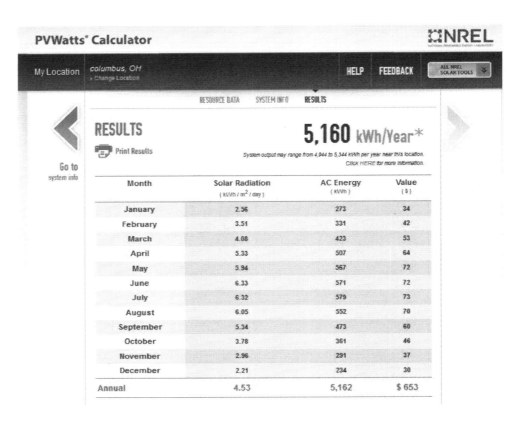

Measuring the Angle of the Roof

When installing a system on a typical residential home, the existing angle and orientation of the roof will nearly always determine the altitude and azimuth of the array.

In the past, fairly complicated calculations were required to determine the angle of the roof. Today, most smart phones can be outfitted with a free app that gives the precise angle when placed against the roof decking, as demonstrated in Figure 6-9 (showing a 40 degree angle).

The pitch of a roof (in construction terms) is generally referred to as a ratio between the **rise and run.**

(RISE AND RUN)

TABLE 6-1: PEAK HOURS OF SOLAR RADIATION FOR COLUMBUS, OHIO AT VARIOUS DEGREES OF ALTITUDE

Array Angle	Avg. Daily Peak Solar (hrs)	Array Angle	Avg. Daily Peak Solar (hrs)
65°	3.77	34°	4.29
60°	3.91	33°	4.3
55°	4.03	32°	4.3
50°	4.13	30°	4.3
45°	4.21	28°	4.29
40°	4.26	25°	4.28
35°	4.29	20°	4.24

For instance, if the pitch rises 4 inches (101 mm) vertically (the rise) for every 12 inches (305 mm) of horizontal distance (the run), the roof is said to have a 4:12 pitch. If it rises 12 inches (305 mm) vertically for every 12 inches (305 mm) horizontally, it has a 12:12 pitch.

Table 6-2 provides a handy conversion between this form of measurement and degrees of pitch.

FIGURE 6-9: MEASURING THE ANGLE OF A ROOF USING A SMART PHONE.

Sun Charts

It is sometimes helpful when designing a PV system to somehow capture an image of how the sun tracks at a specific location. But this movement actually takes place in several dimensions (east to west, up and down, and over time). Fortunately there are ways to graphically represent these movements. Such a representation is referred to as a **sun chart**.

Using the sun chart in Figure 6-10 to optimize the available solar resource, a sun tracking system would orient its azimuth setting to directly face the sun at about 90°E at about 8 am (solar time) on June 21st. The altitude setting is found at this location to be somewhere between 30° and 45° (probably about 40°).

SUN CHART

Locating the System on the Site

The homeowner may already have some firm ideas regarding where the array should be located. This may or may not be the ideal spot for this particular site. But they have probably given little or no thought to where the electronic components and disconnect switches should be located. The placement of all the components of a PV system must be considered during a **site survey**.

SITE SURVEY

With a solar PV system, the greatest space requirements will be for the solar array itself. During the site inspection, it is important to determine where the

Array Angle	Roof Pitch	Array Angle	Roof Pitch
4.8°	1 rise / 12 run	30.3°	7 rise / 12 run
9.5°	2 rise / 12 run	33.7°	8 rise / 12 run
14°	3 rise / 12 run	36.9°	9 rise / 12 run
18.4°	4 rise / 12 run	39.8°	10 rise / 12 run
22.6°	5 rise / 12 run	42.5°	11 rise / 12 run
26.6°	6 rise / 12 run	45°	12 rise / 12 run

TABLE 6-2: ARRAY ANGLE AT VARIOUS ROOF PITCHES.

FIGURE 6-10: EXAMPLE OF A SUN CHART FOR 48°N SHOWING THE ALTITUDE AND AZIMUTH OVER THE COURSE OF A YEAR

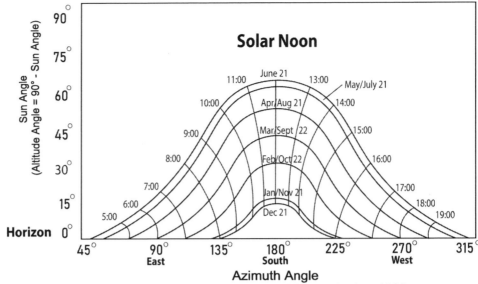

array should be placed, as well as where all the remaining balance of system (BOS) should be located.

Mounting Options

There are several mounting options available when placing a PV array, with advantages and disadvantages for each option.

Roof Mounted:

Many PV systems are mounted on top of the building's roof. This is perhaps the best option where space is limited (no room for a ground-based system) or security is an issue (gets the array away from possible damage or theft).

For aesthetic purposes, it is best to mount the rooftop array parallel to the roof surface, centered and squared with the edges of the roof and the roof lines (as demonstrated in Figure 6-11). In high wind areas, it is often a good idea to set the array back from the eave at least 10 inches (254 mm) to minimize the effect of lift from the wind.

Several concerns should be addressed when considering placing an array on a rooftop.

FIGURE 6-11: EXAMPLE OF A PROPERLY MOUNTED ROOF MOUNTED PV ARRAY

These include:
- azimuth of the roof. Is the roof facing the right direction (south)?
- available space. Is there enough available space on the

roof to hold all the panels required? Given the current efficiency of crystalline panels, they will generate between 15-20 W per square foot. A safe "rule of thumb" is to estimate 10 W for every available square foot of roof space (100 W per square meter). This accounts for unusable space on the roof. So a roof with 365 sq ft (3.65 sq m) of usable space can generally accommodate a 3,650 W array.

- pitch of the roof. Is the angle of the roof appropriate for the desired altitude of the panels? If the pitch is too great (60 degrees, for example) it may give the array a winter bias when a summer bias is desired.
- structural condition. Is the condition of the roof suitable to support the weight and stress of a PV system? Check the roof for wear and damage. A structural engineer may need to be called in to assess the building if it appears that it may not support the added weight or construction activity.
- The weight of a system is not just the weight of the panels and racking system themselves (referred to as **dead weight**). Typically the dead weight of a roof-mounted system (racks and panels) is about 3-5 pounds per square foot.
- The **live weight** of the system may also include the weight of accumulated snow, as well as the weight from the **lift force** and **drag force** of wind as it passes over the panels (acting a bit like the wing of an airplane).
- Wind forces might add as much as 50 pounds per square foot of force on the structure. Make sure the mounting hardware, screws, panels and roof structure can resist the forces that the system will impose on the building. This may be especially important in areas periodically exposed to very high winds.
- age of the roofing materials. Determine if the shingles or other roofing materials will need to be replaced within the next few years. If so, the PV system will need to be removed to accomplish this. It may be best to mount the system elsewhere, or wait until a new roof has been installed before installing a rooftop PV system.
- panel ventilation. It is important to select a mounting system that allows at least 3-6 inches (76-152 mm) of air gap between the bottom of the panel and the surface of the roof. This allows for the **passive cooling** of the panels, and avoids any condensation, corrosion, or accumulation of debris (such as leaves).
- minimize roof penetrations. Roof penetrations are also a

FIGURE 6-12: A BALLASTED ARRAY MOUNTED ON A FLAT

BALLASTED MOUNTING SYSTEM

problem. Most commercial mounting systems try to minimize the number of penetrations, to guard against leaking. But anytime a roof is penetrated, problems are bound to happen. On flat roofs (very common in commercial structures), penetration of the roof can be avoided by installing a **ballasted mounting system** such as the example shown in Figure 6-12. These systems normally rely on weights to hold them in place, rather than lag screws.

Ground Mounted:

GROUND MOUNTED SYSTEM

A number of array mounting systems are designed to be mounted on the ground, such as the one illustrated in Figure 6-13. These do require space relatively close to the building, and may be subject to theft and vandalism concerns.

Foundation work will be required for such systems, as well as determining if any buried utilities will be affected during construction.

INTER ROW SHADING

Often larger arrays will be oriented in rows. It is important to ensure that one row does not cast a shadow upon the row behind it (a situation known as **inter row shading**). A good rule to follow is to allow a distance of three times the height of the top of the array between the rows (at 40° latitude). Less offset is required closer to the equator (twice height at 30°), and more offset is required in locations closer to the poles (4 times height at 45°). For example, assuming the array in Figure 6-14 is located at 40° latitude, if the top of the array is 6 ft (1.8 m) off the ground, then the row behind should be no closer than 18 ft (5.5 m).

LANDSCAPE

PORTRAIT

Another consideration when placing solar panels is to determine whether they are set in a **landscape** orientation, or a **portrait** orientation, both illustrated in Figure 6-15. Typically this will be determined by the available space, underlying supports, and the racking system selected.

FIGURE 6-13: A RELATIVELY SMALL GROUND-MOUNTED ARRAY

FIGURE 6-14: AVOID INTER ROW SHADING BY OFFSETTING THE BACK ROW BY THREE UNITS FOR EVERY ONE UNIT IN HEIGHT (IN THE MID LATITUDES)

Pole Mounted

Smaller systems, or those that incorporate a tracking system, may be mounted on poles (shown in Figure 6-16). These systems generally use galvanized steel pipe buried in concrete – the size and depth of which depends upon the size and weight of the system.

Concerns with Ground Mounted Systems:

Several concerns that must be addressed when selecting to use ground mounted PV systems include:
- zoning and land use restrictions,
- terrain (hills, slopes and valleys),
- soil type,
- nearby ground cover (fire regulations require that brush be cleared within 10 ft (3 m) of the perimeter of ground mounted systems),
- water and drainage issues,
- foundation requirements,
- security (fences, for example, to prevent damage and/or vandalism),
- and vehicle access.

Portrait Orientation

Landscape Orientation

FIGURE 6-15: LANDSCAPE AND PORTRAIT ORIENTATION OF PV PANELS

FIGURE 6-16: POLE-MOUNTED PV ARRAY

(BUILDING INTEGRATED PHOTOVOLTAIC (BIPV))

(SOLAR SHINGLES)

(ELECTRIC VEHICLES (EVs))

FIGURE 6-17: FLEXIBLE SOLAR SHINGLES INTEGRATED INTO THE ROOF

(PHOTO FROM DOW CHEMICAL)

FIGURE 6-18: SOLAR PANELS UTILIZED TO PROVIDE COVERED PARKING

Building Integrated Photovoltaics (BIPV): Architects and designers are increasingly integrating photovoltaics into the design of the building or structure itself. More times than not, these **building integrated photovoltaic (BIPV)** systems utilize thin-film panels, that can be modified (such as **solar shingles,** as illustrated in Figure 6-17) to become part of the building itself.

Another growing application is the use of PV panels to provide shaded parking, such as is illustrated in Figure 6-18. This will become even more widespread as the fleet of **electric vehicles (EVs)** expands – providing not only a covered parking area, but easy access to solar-powered EV recharging stations.

Shading Issues

A shadow cast upon just a portion of a solar panel can reduce the output from that panel dramatically, as illustrated in Figure 6-19. So it is important to ensure that the array is placed to avoid the shade of nearby trees, buildings, telephone poles, or any other obstruction that may potentially block the sun. PV arrays should be placed so that there is no shading for a minimum of six hours during the middle of the day.

It is also important to remember that shadows cast by nearby objects will change over the course of the day and over the course of a year. Aside from the ever-changing angle of the sun in relation to the location of the array, also keep in mind that plants tend to grow and bare branches will be covered with leaves in the spring.

Figure 6-19: Shadow cast from one array to the next reduces the output of a panel by 90%

Solar Pathfinder

To avoid the complexity of calculating the angles of all nearby potential obstructions, most PV installers use a tool such as a **Solar Pathfinder®** pictured in Figure 6-20. This relatively inexpensive tool (starting at $200-$300 dollars) combines a sun chart for the location's latitude with a reflective dome that provides a visual representation of nearby potential sources of shade.

Place the unit level to the ground (there is a level bubble to assist with that task). Point it towards true south (a compass is also integrated into the unit to assist with this). The clear plastic dome will then act as a lens, bending the reflected image of all objects within 180 degrees, imposing them over the sun chart inserted below the lens.

Tools other than the Solar Pathfinder, such as the SunEye® by Solmetric (Figure 6-21) as well as a number of apps for smart phones are also available to assist with shading analysis.

When using a Solar Pathfinder, the reflection of a nearby tree or building can be seen superimposed over the selected sun chart. Where the reflection falls will indicate that the location will be shaded by that object on that day and time (as indicated on the sun chart). In this way, the unit projects an image of shading issues for that location throughout the year.

With the basic model, white grease pencils are provided to assist in tracing the outline of the obstructions onto the sun chart - or easier still - simply take a photo of the unit as illustrated in Figure 6-22). More expensive models incorporate GPS devices, true south locators, and cameras to capture the image for later analysis.

Figure 6-20: Solar Pathfinder with stand and carrying case

(Photo from Solar Pathfinder)

Figure 6-21: SunEye 210 Shade Tool

The sun chart also incorporates small white numbers inside the half-hour divisions. These are values that have been calculated to account for the relative energy contained in sunlight during each half-hour time period throughout the day. Hours near solar noon have a higher relative energy than those nearer to dawn or dusk. All numbers added horizontally will add up to 100 (or 100%). The sum of all the numbers that are unshaded will provide an approximate percentage of time during that month that the site will be free from shading.

For example, if the numbers in the unshaded region for a given month add up to 72, then 72% of the available energy in the sun during that month will not be impacted by nearby obstructions. If shading is from deciduous trees, assign only half the value as shaded during months when leaves will be off the trees.

Remote Site Assessments

Soft costs comprise more than 50% of the cost of a solar installation. The site assessment is a major part of these costs - not just for the project under construction, but an installer may have to visit and assess dozens of sites before making a sale.

Increasingly, designer/installers are turning to **remote site assessment** tools that allow them to assess a site remotely, without ever leaving the office. NREL has estimated that remote assessment solar design software can save about $0.17/W per 5-kW system (or about $850 per install).

There are a number of remote assessment systems available on the market - and each functions in slightly different ways. However the process for conducting a remote site assessment is fairly uniform.

Figure 6-22: Trace the shading line on the sun chart or take a photograph

(Photo from Solar Pathfinder)

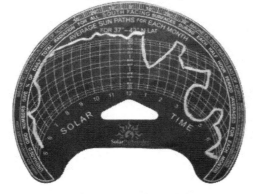

Site Tracing on a Monthly Sunpath Diagram

After entering an address of the location in question, the system will bring up an image of the property (generally from Google Earth's database). The designer can use a tool to define the area where the array will likely be placed (either on the roof or on the ground), as illustrated in Figure 6-23.

Once the array boundaries have been defined, system information such as the panel selected, the orientation and angle of the array, the height of the building, and setback limits can be entered. The system will then populate the defined area with panels and estimate energy output for the array.

The designer can modify the system, changing orientation, altitude, setbacks, row spacing, ground cover ratio, or even change the selected panel if desired. Obstructions such as air conditioning units, access doorways, vent pipes or trees can be identified and some systems will even plot the shading from those objects and restrict panels from being placed within the shadow.

Nearly all remote assessment systems integrate data from PV Watts and generate an estimate of annual production based on the information entered and the data from PV Watts.

Potential shading issues may be clear from the image or some systems incorporate a 3-D feature that allows the designer to see shading in three dimensions.

These systems often utilize **lidar** data that creates a three-dimensional image of the property, allowing the designer to determine not only orientation but the angle of the roof, height, and shading effect from nearby trees and other obstructions.

LIDAR

Lidar, which stands for Light Detection and Ranging, is a remote sensing method that uses light in the form of a pulsed laser to measure variable

FIGURE 6-23: REMOTE ASSESSMENT DESIGN TOOL

distances on the Earth's surface. This light pulse data can generate precise, three-dimensional information about the shape and surface characteristics of the site.

One concern with the remote assessment process is that the data used (Google Earth maps, lidar images, etc) may be out of date, making the assessment inaccurate and unreliable. An on-site assessment will be necessary at some point prior to construction to finalize the design.

Total Solar Resource

For any given site, based on its location on the earth and weather conditions, there is a certain anticipated amount of power available, known as that location's **total solar resource**.

However, the actual orientation and angle of an array rarely match the ideal for a given location. The slope and orientation of the roof will often dictate how a system is installed. And shading might also contribute to limiting the amount of energy the array can gather. A number of factors will reduce the actual output of the array from its ideal orientation and angle, as indicated in Figure 6-24.

So once the orientation (azimuth) and the angle (altitude) of the array have been determined and an analysis of shading issues at the site has been completed, the resulting energy that can be produced by the system can be determined. This will not be the total energy available from sunlight at the site, but rather the amount the array can gather given the actual orientation and accounting for any shading. This may be expressed as a percentage (actual/ideal), often called the **total solar resource fraction (TSRF)**.

Tools such as PV Watts will calculate the energy available based on the angle and orientation. The actual energy the system can gather based on the angle and orientation (actual/ideal) is known as the **tilt and orientation factor (TOF)**.

Tools such as the Solar Pathfinder or the SunEye can assist in calculating how much will be lost due to shading. The amount lost due to shading is known as the **shading factor (SF)**, or shade impact factor. Combined, the TOF and the SF are used to determine the TSRF.

Some grants and incentive programs that assist in the funding of solar projects may require a minimum TSRF in order for a project to qualify. The state of Oregon's residential energy tax credit program, for example, requires systems have a 75% TSRF or better to qualify.

For example, assume it has been determined that at a given location the optimum orientation of the array would be for it to face solar south (180°

azimuth) at an angle of 33° altitude. With no shading, this orientation would provide 100% of the total solar resource available to the array.

However the roof where the array will be placed actually faces south-east (120° azimuth) with a roof pitch of 50°. Using the PV Watts tool at the ideal orientation, the site will receive 5.92 hours of insolation per day. At the actual orientation, it will only receive 5.12 hours of insolation.

5.12 hrs actual insolation /5.92 hrs theoretical max = .865 or 86.5%

So the tilt and orientation factor (TOF) for this site is 86.5%.
A shading analysis is conducted and it is found that 10% of the array will be shaded throughout the year, so the shading factor (SF) is .90 or 90%.

The total solar resource fraction (TSRF) is then calculated by:

TSRF = .895 (TOF) x .90 (SF) = .8055 = 80.55%

In other words, the array, given its orientation and shading issues, will receive only about 80% of the available energy from sunlight that it would have received if the orientation had been ideal and no shading affected the array.

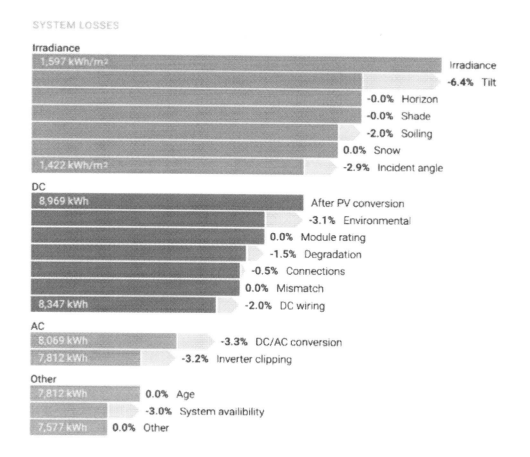

FIGURE 6-24: A NUMBER OF SYSTEM LOSSES CAN REDUCE THE TOTAL SOLAR RESOURCE

Other Issues Impacting the Location of the Array

Assuming the perfect location for the array has been found, sketch out the system (with locations and measurements) for future reference. That task completed, there are still a few other issues to consider during the site inspection.

These include:
- aesthetics. How will the array look when integrated onto the site? Are there ways to design the system to assist with it fitting into the character of the property and/or neighborhood?
- legal issues. Some sub-divisions or municipalities have restrictions regarding the placement of solar arrays. Check to make sure there are no ordinances or other legal restrictions that limit or prohibit the placement of a solar array in the location selected.
- insurance issues. Is adequate insurance in place to protect all parties during the installation, and is insurance in place to provide for the added value of the property after the installation is complete?
- **future proof.** It is hoped that this system will provide power to the homeowner for 20 or 30 years. So check for trees or bushes that might grow and cause a problem in the future. Try to take these into account if possible. Also ask about any future building plans that might impact the array or if the homeowner has any plans that may affect how much energy the site will need in the future (adding an electric vehicle, for example).
- a site **hazard assessment**. During installation, panels and ladders, trucks and people will be loaded and unloaded. Determine if there are any hazards (low electric lines, nearby swimming pool, vicious dogs, whatever) that may impact construction.
- determine staging/lifting/access areas. Make sure there is room to bring the materials to where they need to be. And determine how panels will be raised up on the roof (should that be their destination).
- determine where the cables will run. Calculate their distances, and see if there are any obstacles (driveways, other utilities, gardens, etc) that must be taken into account.
- unusual conditions. Find out if there are any other unusual conditions that come into play at this location (such as excessive wind, snow, earthquakes, flooding, etc) that will affect the design of the system.
- authorities having jurisdiction. Confirm the electrical utility and other authorities that have jurisdiction at this location. Make sure what you are planning does not violate any of their local rules and obtain whatever permissions are required to install the system at that site.

Determine the Location of the Balance of Systems (BOS)

The balance of systems (BOS) may include the battery bank, charge controller, disconnects, racks, wire, conduit, meters and/or remote sensing equipment. Basically everything except the solar panels and the inverter(s).

Often these are sub-categorized into two groups, the **electrical BOS** and the **structural BOS.** The electrical BOS includes components such as charge controllers, cables, monitoring electronics, meters, etc. The structural BOS includes the racks, other support structures, conduit, shelving, mounting brackets, etc.

> ELECTRICAL BOS

> STRUCTURAL BOS

These items should be located as close as possible to the array (to minimize cabling runs) but also meet the following criteria:
- electronics should be located in a protected, waterproof and weather protected location.
- they should NOT be located in direct sunlight (to avoid heat and UV issues).
- they should be insulated against extreme temperatures (many of the electronic components will list the range of ambient temperatures they are designed to operate within). Some of these components may be rated only for indoor installation (check the manufacturer's specifications).
- battery banks should be housed in well ventilated locations (lead acid batteries give off hydrogen gas when charging), and a temperature controlled environment (extreme heat and cold will also impact the functioning of the batteries).
- the batteries should also be protected from any combustible sources. They should NOT be co-located where open flames are present (from a gas furnace, for example).
- the BOS should be protected from children, pets, rodents, vandals and any potential physical hazard (such as flying debris from lawn mowers).
- there may be **clearance** requirements for the equipment (above, below, beside - and working clearances in front of the units). These are typically 36 inches deep x 30 inches wide x 6 feet 6 inches high (1 m x 760 mm x 2 m).
- consider **height requirements,** the NEC requires that disconnects (the center of the grip of the handle or the circuit breaker) not be located more than 6 feet 7 inches (2 meters) above the floor or working platform. Outdoor disconnects and equipment should be located high enough off the ground to ensure they are not impacted by drifting snow.
- AC and DC disconnects must be located near any installed string inverter. It may also be necessary to install additional disconnects, such as an AC disconnect located within 10 ft (3 m) of the utilities service connection.

> CLEARANCE

> HEIGHT REQUIREMENTS

Locate the **service entrance** used by the existing utility provider (it is likely required that a system disconnect be placed at this location). The utility meter is usually located near the existing service entrance.

> SERVICE ENTRANCE

It may be practical to locate inverters and other electronic components of the BOS (such as charge controllers) near the existing service panel or breaker box (typically indoors). Ensure there is ample room for the equipment and that the area is protected from excessive heat and moisture.

Enclosures

Cabinets and **enclosures** are rated by the **National Electrical Manufacturers Association (NEMA)**. They are rated based on certain performance criteria, as outlined in Table 6-3.

Solar installers are particularly concerned with three ratings for enclosures used in PV systems.

- *NEMA Type 1* – These enclosures are for indoor use only and provide a limited degree of protection against access to hazardous parts, as well as to equipment inside (against falling objects, dirt, etc).
- *NEMA Type 3R* – These enclosures are for indoor or outdoor use. They provide a limited degree of protection against access to hazardous parts, as well as to equipment inside (against falling objects, dirt, etc). But they also provide protection for the equipment from the harmful effects of water (rain, sleet, snow) and will remain undamaged by the external formation of ice on the enclosure.
- *NEMA Type 12* – These enclosures are for indoor use only but provide protection against water, unlike NEMA Type 1 enclosures.

Battery Location

NFPA 855 (Standard for the Installation of Energy Storage Systems) is a relatively new National Fire Protection Association Standard developed to define the design, construction, installation, commissioning, operation, maintenance, and decommissioning of stationary energy storage systems.

The organization adopted NFPA 855 in 2019. It set out to standardize and codify installations of lithium-ion batteries, as well as other new technologies. The standards also include lead-acid and nickel-cadmium (Ni-CD) batteries.

The 2021 versions of International Fire Code (IFC), International Residential Code (IRC), and NFPA 1 base their ESS fire code requirements on this document.

This standard restricts the placement of ESS at a site. For example, within residential homes (1-2 family dwellings), the standard states that *"ESS shall not be installed in a living area of dwelling units or in sleeping units other than within utility closets and storage and utility spaces."*

NFPA 855 further outlines that storage systems in a residential setting can only be installed in:

- attached garages separated from the dwelling unit (maximum size allowed in this location is 80 kWh)
- detached garages or other non-occupied buildings (maximum 80 kWh)
- outdoors either attached to the exterior wall or mounted on the ground, located at least 3 feet from any doors or windows (maximum 80 kWh)

TABLE 6-3: NEMA ENCLOSURE RATING SYSTEM

NEMA Rating	Description
Type 1	Indoor. Protects against dust, light, and indirect splashing but is not dust-tight.
Type 2	Indoor. Drip-tight. Similar to Type 1 but with addition of drip shields; used where condensation may be severe.
Type 3, 3R	Indoor/outdoor. Weather-resistant. Protects against falling dirt and windblown dust, against weather hazards such as rain, sleet and snow, and is undamaged by the formation of ice.
Type 4, 4X*	Indoor/outdoor. Watertight.
Type 5	Indoor. Dust tight, drip and splash protection.
Type 6	Indoor/outdoor. Submersible.
Type 7	Indoor. For use in areas with specific hazardous conditions.
Type 8	Indoor/outdoor. For use in areas with specific hazardous conditions.
Type 9	Indoor/outdoor. For use in areas with specific hazardous conditions.
Type 10	Meets the requirements of the Mine Safety and Health Administration
Type 11	General-purpose. Protects against the corrosive effects of liquids and gases. Meets drip and corrosion-resistance tests.
Type 12	Indoor. Provides some protection against dust, falling dirt, and dripping non-corrosive liquids
Type 13	Similar to 12 but also meets oil exclusion tests.
	*X indicates additional corrosion protection; commonly used near salt water.

- or in an enclosed utility closet or room or storage room (maximum size allowed within the building is 40 kWh).

Each ESS unit can be up to 20 kWh in size, and adjacent units must maintain at least 3 feet of separation to mitigate the risk of fire propagation between units.

Interestingly the NFPA 855 does allow the home owner to use an electric vehicle parked outside or in the garage as a battery backup system for the home - but only on a temporary basis. It states that the use of an EV for battery backup should not exceed 30 days.

Existing Electrical Equipment

When doing a site evaluation, it will be necessary to locate existing electrical equipment, such as the service meter, the main distribution panel, any sub panels, grounding electrode, any generators, or other electrical equipment that will impact the PV installation.

For larger commercial systems, also locate and obtain the ratings of the utility transformer, service disconnecting means, service entrance conductors, supply breaker ratings, and even load breaker ratings.

FIGURE 6-25:
PROPERLY INSTALLED SERVICE PANEL WITH PLENTY OF ROOM FOR EXPANSION

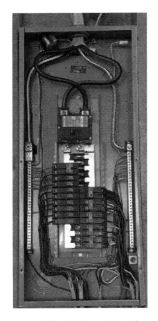

(SHORT-CIRCUIT CURRENT RATING)

(SHORT-CIRCUIT INTERRUPTING RATING)

(LOAD SIDE CONNECTION)

(BUSBAR)

The Service Panel

Locate the building's main service panel, similar to an example illustrated in Figure 6-25, and any subpanels that may be present.

Document the make and model, and verify the busbar rating. Remember that different manufacturers require different breakers (this is not a one-size-fits-all industry). Also verify the level of service entering the building from the utility (for example 100 amp or 200 amp service). These amp ratings will be important in determining if a load side connection is possible.

The busbar rating on the service panel is generally referred to as the **short-circuit current rating**. A short-circuit current rating (SCCR) is the maximum current a device or system can safely withstand for a specified time (such as 0.05 seconds), or until a specified fuse or circuit breaker opens and clears the circuit.

In service panels, the term short-circuit current rating and **short-circuit interrupting rating (SCIR)** are often used interchangeably. Utilities normally require the SCIR rating of the service panel on interconnection applications.

Also determine if there is enough space in the service panel for another two-pole breaker required for the PV system interconnection. If the service panel is inadequate for the connection, it may be necessary to upgrade the panel or find another location at which to connect the PV system.

How a PV system is interconnected to the grid is primarily dictated by the system size (based on AC output current), the requirements of the local utility company, the size and location of existing electrical equipment, and possibly rebate and incentive program requirements.

Load Side Connection

Many residential and small commercial systems are net-metered, and are connected to the building through a **load side connection** (typically a back-fed circuit breaker in an AC service panel) of existing service equipment.

While this is the easiest method of connecting a PV system to the grid, larger systems will run up against a sizing limitation set by the NEC. The NEC requires that the sum of all input breakers amperage may not exceed 120% of the panel's **busbar** rating.

A common residential panel (100 amp rating) will only allow a maximum PV input breaker of 20 A. Since the NEC requires there be a 25% safety margin when sizing wires, 125% of the inverter's output current cannot exceed 20 A (in this example). Assuming the inverter's output is 240 Vac, then the maximum array size in this situation would be:

20 A (allowable 120% of busbar rating) x 240 V (inverter output) / 125 % (NEC safety margin) = 3,840 W (maximum inverter output)

A 200-amp rated service panel with a 200-amp main disconnect will allow twice the inverter output (which corresponds with the array size) of 7,680 watts.

40 A x 240 V /125 % = 7,680 W (maximum inverter output)

The 120% rule applies to the breaker size, not to the current of the load feeding that breaker.

The breaker from the array (inverter) should be located as far away from the main breaker as possible, so that current feeding the loads flow from opposite ends of the busbar.

If a utility disconnect box is present at the meter, the array can be connected into one of the double-pole breaker slots in a manner similar to how it would be connected within the service panel. If connected in this manner, it may eliminate the need to penetrate the shell of the building to make the connection.

Supply Side Connection

For larger systems, or systems where there is simply no space to add a solar array to the existing service panel, a **supply side connection** may be necessary.

A supply side connection, or sometimes known as a **line side connection,** is interconnected directly to the utility on the utility side of a building's main service disconnect (the supply side), normally somewhere between the main breaker and the utility meter.

SUPPLY SIDE CONNECTION

LINE SIDE CONNECTION

A supply side connection requires coordination with the utility company, as the service to the building may need to be "shut off" on the utility's side of the system before the connection can be made. In many cases the utility also requires that their electricians be involved in any supply side connection.

When making a supply side connection, the connection must be made on the customer's side of the **service point.** This is the **demarcation point** between the utility and the customer. On the utility's side of the demarcation point, they are responsible for the equipment and wires. On the customer's side, the building owner is responsible.

SERVICE POINT

DEMARCATION POINT

It is often assumed that the service point is at the meter. But in some cases, the service point is at the power pole, or at the top of the mast above the meter, or at an underground distribution transformer.

Conduit Pathways and Existing Grounds

During the site inspection, try to get a general idea of where the conduits and/or cable trays, as well as the grounding conductors will be placed.

When planning for conduit pathways, consider:
- equipment placement,
- hazards or obstacles,
- local regulations,
- climate,
- aesthetics.

Locating and identifying the grounding components during the site survey is also important. The location of the existing grounding electrode conductor (GEC) and grounding electrode (often a ground rod) will be critical in determining the path of the PV system grounding conductor(s) to interconnect the two systems.

Generators

Many commercial businesses (such as data centers) have **generators** in place that provide emergency power when the grid goes down. Increasingly, residential customers are installing smaller generator backup systems as well.

Many AC and DC-coupled systems anticipate that a generator may be incorporated into the system and there are connection ports available in the smart switch. Simply ensure that the generator is compatible as specified in the PV system guidelines.

Occasionally a homeowner will wish to add a generator to a home that has an existing grid-tied solar array in place. It is possible to do so, but trying to run both the generator and the PV array while the grid is down can lead to problems.

If the output from the generator is stable and of high quality, the inverter will monitor the generator's output voltage, frequency and waveform, just as it does with power from the grid. If the AC waveform from the generator is grid quality, the inverter will attempt to synchronize with the generator.

If the AC input to the inverter does not dip, sag or surge when the PV system generates power, then the inverter will remain online. But if the generator's output falls out of normal operating ranges, the inverter will shut off.

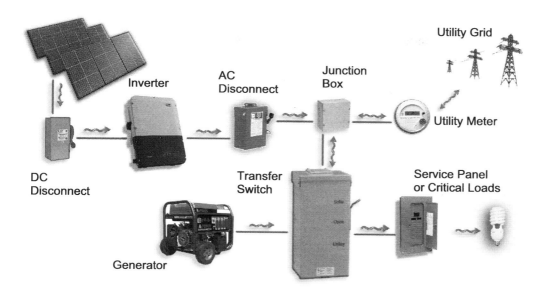

Figure 6-26: Connecting a grid-tied PV system with a generator

But assume the inverter remains operational. If more power is available from the PV system than is required to service the building's loads, then it will seek to go somewhere (normally onto the grid when the grid is connected).

Since the grid is down (which has caused the generator to kick on), the excess power from the array, with no place to go, will cause a rise in the system voltage. This will in turn cause the inverter to shut down, as the system is now out of operational limits. So unless the generated power from the array is always less than the load demands of the business or home, a PV system connected in this fashion has no chance of working.

So normally it is best to connect a generator to a grid-tied PV system ahead of the generator transfer switch and the subpanel feeding the critical loads, as illustrated in Figure 6-26. Connected this way, the generator feeds the backup loads, just as if there was no PV system present.

Connected in this manner, when the grid goes down, the PV system goes down as well. The generator's transfer switch disconnects the building from the grid and from the PV system as well.

Site Hazard Assessment

OSHA requires that site managers perform a safety assessment on every project. Installing a PV system can be risky. There are inherent risks associated with lifting and placing solar panels onto rooftops. Installers also work with electrical systems that present a risk from shocks or arcs.

Installers often work under extreme weather conditions, facing injury from either extreme heat or cold.

During the site inspection, it should be determined:
- What type of ladders or scaffolding will be needed?
- Is fall protection required, and where will it be secured?
- What barriers will be required and where will they be placed (around skylights, access hatches, roof edges, etc.)?
- Where will materials be staged during the installation?
- Where will vehicles and equipment be parked on the site?
- Is any cleanup required prior to construction to ensure a safe and uncluttered work environment?
- Is the building structure adequate for the installation and personnel?
- Is there adequate access to the site during the installation?
- Are there any hazards imposed from traffic (both people and vehicles) during installation?
- Are there any potential slip and fall hazards present?
- Are there any dangerous substances stored on the site?
- and many other site-specific concerns.

Economics of Solar

According to the EnergySage Market Report (Sept 2023) the median price of a residential solar system installed in the US in 2023 was $2.90/W and the average residential system size installed was 10.4 kW.

Of the total cost of installing a system, only $0.98/W, or about 34% of the cost is for the hardware used (panels, inverters, wire, conduit, etc.). The remaining $1.92 is for soft costs, such as labor, marketing, overhead, permits, etc. A complete breakdown is illustrated in Figure 6-27.

(SUPPLY CHAIN COSTS)

Any costs incurred between the manufacturer and the installer are referred to as **supply chain costs**. These can include: distributor profits and overhead, transportation costs, storage fees, credit card processing fees, etc. The supply chain can make up a significant portion of the overall project cost.

It is estimated that the installation cost for systems installed during the initial construction of a home (versus a retrofit on an existing building), are only about half this figure, due to savings on installation (install labor costs will be cut in half due to the efficiency of building at the time of initial construction), and permitting, inspection and interconnection savings (piggybacked on the existing project).

System Type Economic Comparison

Various system configurations (simple grid-tied with string inverter, micro inverter system, power optimizer system, AC or DC coupled, or stand-alone) have very different cost structures. Adding a battery bank into the system can add significant costs to a PV system.

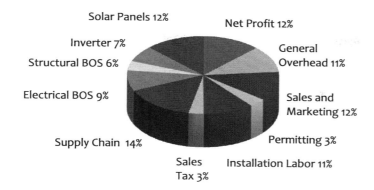

Figure 6-27: Cost breakdown per Watt of an installed Residential PV system

(Data From NREL)

In order to get a sense of the differences in cost between the various systems, this exercise compares the various configurations for a household with a given load demand. Prices quoted are obtained from a single distributor during December, 2023. No effort was made to shop around for the best price - products were selected based on current popularity.

The scenario:
The selected home consumes the U.S. average about 890 kWh of electricity per month. It was determined that the home will need an 8 kW roof mounted array to fully meet their electrical demand.

The homeowner is installing the system, so there is no overhead or installation costs factored into the total. Also assume the BOS costs for wire, conduit, racking, remain constant for all configurations at $0.48/W.

Grid-Tied with String Inverter Configuration:

Panel Selected:
Canadian Solar 395W Mono PERC BOB Tier 1 Solar Panel
    8,000 W array / 395 W panel = 20.25 = 20 panels
    Price = $0.38 per watt or $152.05 per panel x 20 = $ 3,041

Inverter: Fronius 7.6 kW Grid Tie Inverter
    Price = $2,050
MLPE: Optimized Rapid Shutdown Unit
    Price = 20 x $43.50 = $870
BOS Costs: $0.48/W x 8,000 W = $3,840

Total Initial Cost: $9,801

All the items in the installation with the exception of the inverter have at least a 20-year warranty. Adjusted for the life expectancy of the inverter brings the cost to:
- Inverter (10 years warranty) = $2,050 x 2 = $4,100

String Inverter cost over 20-Year period: $11,851 (or $1.48/W)

Chapter 6: Conducting a Site Survey

Grid-Tied with Micro Inverters Configuration

Canadian Solar 395W Mono PERC BOB Tier 1 Solar Panel
    8,000 W array / 395 W panel = 20.25 = 20 panels
    Price = $0.38 per watt or $152.05 per panel x 20 = $ 3,041
Inverter: Enphase IQ 8+ micro inverters
    Price = $185 x 20 = $3,700
Combiner Box: Enphase Energy IQ Combiner 5 with IQ Gateway
    Price = $799
BOS Costs: $0.48/W x 8,000 W = $3,840

Total Initial Cost: $11,380

All the items in the installation have at least a 20-year warranty. Adjusted for the life expectancy of the inverter brings the cost to:

Micro Inverter cost over 20-Year period: $11,380 (or $1.42/W)

Grid-Tied with Power Optimizer Configuration

Panel Selected:
Canadian Solar 395W Mono PERC BOB Tier 1 Solar Panel
    8,000 W array / 395 W panel = 20.25 = 20 panels
    Price = $0.38 per watt or $152.05 per panel x 20 = $ 3,041
Inverter: SolarEdge SE7600H-US HD Wave Grid Tie Inverter
    Price = $1,760
MLPE: SolarEdge 440W Optimizer
    Price = 20 x $115 = $2,300
BOS Costs: $0.48/W x 8,000 W = $3,840

Total Initial Cost: $10,941

All the items in the installation with the exception of the inverter have at least a 20-year warranty. An extended warranty is available for purchase:
- Inverter (12 years warranty extended to 20 years) = $230

Power Optimizer System cost over 20-Year period: $11,171 (or $1.39/W)

As will be apparent, the addition of battery backup to the system adds a great deal of cost. While the 20-year hardware cost in this example for a simple grid-tied system averages around $1.44/W, adding a 10 kWh ESS will add about $1.00 per watt to the hardware costs.

Note: For the battery-based systems we will simply compare initial costs.

DC Coupled Configuration with 10 kWh Battery

Panel Selected:
Canadian Solar 395W 108 Half-Cell Mono PERC BOB Tier 1 Solar Panel
    8,000 W array / 395 W panel = 20.25 = 20 panels

    Price = $0.38 per watt or $152.05 per panel x 20 = $ 3,041

Inverter:
SolarEdge Energy Hub SE7600H-US 7.6kW Single Phase Hybrid Inverter
    Price = $2,825

MLPE: SolarEdge 440W Optimizer
    Price = 20 x $115 = $2,300

Smart Switch: SolarEdge Home Backup Interface
    Price = $2,299

Energy Storage System: SolarEdge Home Battery 10kWh
    Price = $5,999

BOS Costs: $0.48/W x 8,000 W = $3,840

DC Coupled System Initial Cost: $20,304 (or $2.54/W)

AC Coupled Configuration with 10 kWh Battery

Panel Selected:
Canadian Solar 395W Mono PERC BOB Tier 1 Solar Panel
    8,000 W array / 395 W panel = 20.25 = 20 panels
    Price = $0.38 per watt or $152.05 per panel x 20 = $ 3,041

Inverter: Enphase IQ 8+ micro inverters
    Price = $185 x 20 = $3,700

Combiner Box: Enphase Energy IQ Combiner 5 with IQ Gateway
    Price = $799

Smart Switch: Enphase IQ System Controller 2
    Price = $2,225

Energy Storage System: Enphase IQ Battery 10kwh Encharge-10T
    Price = $5,699

BOS Costs: $0.48/W x 8,000 W = $3,840

AC Coupled System Initial Cost: $19,304 (or $2.41/W)

Stand-Alone Configuration

Panel Selected:
Canadian Solar 395W Mono PERC BOB Tier 1 Solar Panel
    8,000 W array / 395 W panel = 20.25 = 20 panels
    Price = $0.38 per watt or $152.05 per panel x 20 = $ 3,041

Inverter: Sol-Ark 8K Hybrid Inverter Pre-Wired System
    Price = $4,999

MLPE: Optimized Rapid Shutdown Unit
    Price = 20 x $43.50 = $870

Energy Storage System: HomeGrid Stack'd 9.6kWh - 2 Battery Modules
    Price = $6,330

BOS Costs: $0.48/W x 8,000 W = $3,840

Stand-Alone System Initial Cost: $19,080 (or $2.39/W)

PV Systems Compared to Utility Rates

The average installed cost of a U.S. residential PV system varies by state, ranging between $2.35 (Arizona) to $3.79 (Indiana) per watt. If it is assumed the cost is $3/W, then an 8 kW grid-tied system would cost about $24,000 to install. In an area with four hours of peak sunlight each day, a system of this size could be expected to produce about 890 kWh of electricity each month.

So how does this compare to the cost of electricity purchased from the utility company?

FIGURE 6-28: TYPICAL RESIDENTIAL ELECTRIC BILL

(GENERATION CHARGE)

Electric Account Summary	
Amount Due	$0.00
Last Payment Received	$0.00
Balance Forward	$0.00
Current Charges/Credits	
Electric Supply Services	$145.89
Delivery Services	$133.04
Total Current Charges	$278.93
Total Amount Due	**$278.93**

Total Charges for Electricity

Supplier (Eversource) (Basic Svc Fixed)		
Generation Service Charge	1356 kWh X .10759	$145.89
Subtotal Supplier Services		$145.89

Delivery (Rate A1-RESIDENTIAL)		
Customer Charge		$6.43
Distribution Charge	1356 kWh X .05585	$75.73
Transition Charge	1356 kWh X -.00243	-$3.30
Transmission Charge	1356 kWh X .02307	$31.28
Renewable Energy Charge	1356 kWh X .00050	$0.68

The costs listed on a typical residential electric bill fall into a number of broad categories. The **generation charge** or supply charge is the cost associated with producing the energy used. If most of the energy used by the home is being supplied by the solar array, then these charges will go down (as energy is not being generated by the utility).

(TRANSMISSION CHARGE)

(DISTRIBUTION CHARGE)

The **transmission charges** and the **distribution charges** are costs associated with bringing the electric power from the power plant to the home. The transmission is the portion of the grid from the power plant to the local substations. These lines may or may not be owned by the local utility.

The distribution systems are the lines that carry the power from the local substations to the home or business. The delivery charges are generally based on the amount of energy used, so they normally decrease as the amount of power purchased from the utility decreases.

(CUSTOMER CHARGE)

There are also a number of additional charges, such as an access charge or **customer charge** - a fee assessed each month just for hooking up to the grid. Or the renewable energy charge noted on the bill shown in Figure 6-28, which is assessed to help pay for a state's clean energy measures.

(NON-BYPASSABLE CHARGE)

Many of these charges may be based on the amount of energy purchased from the utility. But some will be assessed regardless of how much power is imported from the grid. These charges will not be reduced by installing a solar PV system. These are known as **non-bypassable charges**.

As of September, 2023, according to the **Energy Information Administration (EIA)**, the average cost of residential electricity in the U.S. was about 16.29 cents per kWh. This is the net price of electricity and does not include many of the additional costs tacked onto the average electric bill. Prices vary widely from state to state, the highest rates paid in Hawaii (41.52 cents/kWh) and lowest rates found in Washington (11.38 cents/kWh).

ENERGY INFORMATION ADMINISTRATION (EIA)

Production of 890 kWh per month will, at the rate published by the EIA, offset a monthly electric bill of $144.98 (890 kWh x $0.1629) or about $1,740 per year.

Assuming the price of electricity remains constant over the years (not a very safe assumption), even these very crude figures (not factoring RECs, tax incentives, cost of money, etc.) show that the system will pay for itself ($24,000 / $1,740 per year = 13.79 years) in about 14 years (the **payback period**).

PAYBACK PERIOD

Typically the major components of the system will be guaranteed for 25 years.

By taking advantage of the current 30% federal tax credit on the cost of the installation of PV systems, this payback period falls to less than 10 years.

24,000 x 70% = $16,800 after tax incentives /$1,740 electric savings = 9.65 years

In markets such as Hawaii, where electric rates are closer to $0.4152 per kWh, the payback period for a system such as this declines to under four years.

890 kWh/mth x $0.4152 x 12 months = $4,434 per year
$16,800 after tax cost / $4,434 annual electric savings = 3.79 years

Chapter 6 Review Questions

1. Which of the following would be useful in determining if a specific site has enough sunlight available to support a PV system?
   A) solar insolation map
   B) a solar window
   C) a declination chart
   D) a topographical relief diagram

2. A two-dimensional diagram representing the tracking of the sun over an entire year for a given latitude is referred to as:
   A) an insolation map.
   B) an I-V curve.
   C) a sun chart.
   D) all of the above.

3. Peak sun hours is a measurement of a location's:
   A) declination .
   B) azimuth.
   C) latitude.
   D) insolation.

4. When adjusting the orientation of a solar panel from east to west, this adjustment affects its:
   A) azimuth.
   B) altitude.
   C) declination.
   D) resistance.

5. When adjusting the orientation of a solar panel up and down in relation to the horizon, this adjustment affects its:
   A) azimuth.
   B) altitude.
   C) declination.
   D) resistance.

6. If no bias is introduced into the altitude of a fixed solar panel, the angle will likely be based on:
   A) solar south.
   B) the azimuth.
   C) longitude.
   D) latitude.

7. The winter bias of an array is:
   A) solar south less 15 degrees.
   B) the angle of declination on Dec 22nd.
   C) the assumption that arrays produce less energy in the winter.
   D) latitude plus 15 degrees.

8. Using the declination map in Figure 6-2, estimate the adjustment required to a magnetic compass while siting an array in Denver, Colorado.
   A) no adjustment is required
   B) adjust the array 11 degrees east of magnetic north
   C) adjust the array 7 degrees east of magnetic south
   D) adjust the array 7 degrees west of magnetic south

9. An array facing solar west will likely produce about _____ energy than would be produced by the same array in the same location facing solar south.
   A) 20% less
   B) 40% less
   C) 10% more
   D) 12.53 % less

10. Using the sun chart in Figure 6-10, determine the optimal angle of the array to capture the most energy from the sun on March 30th at 10 am.
    A) Azimuth: solar south (180°), Altitude: latitude (48°)
    B) Azimuth: solar south (180°), Altitude: 65°
    C) Azimuth: solar south (180°), Altitude: 25°
    D) Azimuth: southeast (135°), Altitude: 45°

11. Which type of mounting system will increase the amount of time the panel directly faces the sun?
    A) rooftop
    B) ground
    C) pole mounted
    D) sun-tracking

12. Most PV arrays utilize a _____ cooling system.
    A) sun tracking
    B) solar thermal
    C) passive air
    D) hydraulic assisted

13. When evaluating a roof for a possible PV array, which is not a consideration?
    A) age of the shingles
    B) structural condition of the rafters
    C) pitch of the roof
    D) its demarcation point

14. When placing an array on a roof, which of the following should be considered?
    A) the live weight
    B) the dead weight
    C) condition of the rafters
    D) all of the above

15. Which of the following is an example of a dead weight of the PV system?
    A) The weight of an installer walking on the roof.
    B) Pressure from wind flowing over an array.
    C) The shipping weight of the racking and panels.
    D) The weight of the shingles and decking on the roof.

16. Which is preferred, landscape or portrait orientation?
    A) Landscape is preferred on ground mounted systems, while portrait is preferred on rooftop installations.
    B) Landscape is preferred on rooftops built before 1970 to more evenly distribute the weight across more support rafters.
    C) The preferred method is determined by local building officials.
    D) This is largely determined by the racking system design.

17. A potential client has a home with a north/south-facing roof. The portion of the roof facing south measures 30 feet x 40 feet. There are no unusual features such as excessive dormers, chimneys, or valleys. What size array could this roof support using a quick "rule of thumb" estimate for crystalline solar panels?
    A) 3 kW
    B) 4 kW
    C) 7 kW
    D) 12 kW

18. When placing several rows of ground-mounted panels, in order to avoid inter-row shading:
    A) mount the panels using a landscape orientation.
    B) make sure each row is set at least 9 feet apart.
    C) adjust the azimuth to not less than 15 degrees off true south.
    D) none of the above.

19. A software program supplied by NREL is commonly used to determine available sunlight on a site. This program is called:
    A) PV Watts.
    B) PV Sun.
    C) Solar Pathfinder.
    D) SolarApp+ .

20. A tool commonly used to determine if a site will be shaded is called:
    A) PV Watts.
    B) PV Sun.
    C) Solar Pathfinder.
    D) SolarApp+ .

21. Determining restrictions on the height of buildings adjacent to the property where an array is mounted might be an example of:
    A) hazard assessment.
    B) building integrated PV.
    C) commissioning the system.
    D) future proofing.

22. One limitation of conducting a remote site assessment includes:
    A) lidar imagery is inherently unreliable.
    B) assessment tools are too expensive to be practical.
    C) on-site visits save time and money.
    D) imaging systems often rely on outdated information.

23. The total theoretical amount of solar energy available at any given site is known as that site's:
    A) total solar resource fraction (TSRF).
    B) total solar resource (TSR).
    C) tilt and orientation factor (TOF).
    D) shading factor (SF).

24. The total actual amount of solar energy collected as compared to the theoretical amount of solar energy available at any given site is known as that site's:
    A) total solar resource fraction (TSRF).
    B) total solar resource (TSR).
    C) tilt and orientation factor (TOF).
    D) shading factor (SF).

25. A solar array is sited at a location that has an optimal solar resource of 4.68 hours of insolation each day. Given the orientation of the array, it actually receives 3.92 hours of insolation each day. The array will also experience 5% shading over the course of the year. What is the total solar resource fraction (TSRF) for this installation?
    A) 3.72 hours
    B) 4.45 hours
    C) 79.57 %
    D) 83.76 %

26. When placing the BOS of a PV system, ensure that:
    A) they are facing as close to solar south as possible.
    B) all cabinets and cases are rated for interior conditions.
    C) they do not contain any overcurrent protection.
    D) none of the above.

27. A cabinet rated for outdoor wet conditions would be:
    A) NEMA Type 1.
    B) NEMA Type 2.
    C) NEMA Type 3.
    D) NEMA Type 5.

28. Which of the following is typically NOT a consideration when determining where to place disconnects?
    A) height requirements
    B) clearance requirements
    C) NEMA rating
    D) tilt and orientation factor

29. NFPA 855 further allows storage systems (batteries) installed in a residential setting to be placed:
    A) in attached garages separated from the dwelling unit (maximum 80 kWh).
    B) in detached garages or other non-occupied buildings (maximum 80 kWh).
    C) outdoors either attached to the exterior wall or mounted on the ground, located at least 3 feet from any doors or windows (maximum 80 kWh).
    D) all of the above.

30. By code, what is the maximum amount of power that can be generated by an inverter which is load-side connected to a 100 amp rated service panel box connected to 100 amp service from the utility?
    A) 2.4 kW
    B) 3.84 kW
    C) 4.8 kW
    D) The panel has already reached its rated capacity, so no PV array can be connected.

31. You have decided to connect your solar array to the grid using a double-pole breaker located within the main service panel (breaker panel). This is known as a:
    A) line side connection.
    B) load side connection.
    C) supply side connection.
    D) busbar connection.

32. A client wishes to increase the power output of their 5-year-old solar array. Which of the following options would likely be the best option (barring unusual circumstances)?
    A) Add additional panels to the system.
    B) Add an active tracking system.
    C) Replace all the solar panels with more efficient models.
    D) Add additional batteries to the system.

33. If the total cost of a grid-tied residential solar system is estimated to be about $3 per watt. Based on industry averages, the hard costs (panels, inverters, wire, racks, equipment, etc) for this installation will be about:
    A) $1.00 per watt.
    B) $1.50 per watt.
    C) $2.00 per watt.
    D) $2.50 per watt.

34. Which of the following is not a typical charge found on a residential electric bill?
    A) demand charge
    B) access charge
    C) generation charge
    D) distribution charge

Lab Exercises : Chapter 6

6-1. Locate Solar South

Lab Exercise 6-1:

Supplies Required:
- Copy of declination map

Tools Required:
- Magnetic compass

The purpose of this exercise is to practice locating solar or true south at your location.

a. Using a declination chart, determine how many degrees east or west your site varies from magnetic south.
b. Go outside to a likely solar array site and using the compass, locate magnetic south.
c. Adjust the reading by the variation indicated by the declination map to locate solar south.

6-2. Determining the Altitude Setting of a Fixed-Mount Array

Lab Exercise 6-2:

Supplies Required:
- PV Watts online tool

Tools Required:
- Calculator

The purpose of this exercise is to determine the optimum angle for the altitude adjustment of a fixed-mount array.

a. Determine the latitude for your location.
b. Calculate the proper angle (for altitude) to which your array should be adjusted at your location to give the array a summer bias and a winter bias.
c. If a single setting must be selected, what is the optimum angle for your array in your location if your net metering agreement is "trued up" on an annual basis?
d. How would your net metering agreement be impacted if it was "trued up" independently each month?
e. How much more power would your system generate (percentage) if the array were manually set to winter bias on September 21st and summer bias on March 21st of each year?

6-3. Research Shading using a Solar Pathfinder

Lab Exercise 6-3:

Supplies Required:
- None

Tools Required:
- Solar Pathfinder

The purpose of this exercise is to become familiar with the use of a Solar Pathfinder and utilize it to determine how a site will be subjected to shading.

a. Take the Solar Pathfinder unit outside to a potential site for a ground-mounted array. Assemble and level the unit. Make sure the sunchart inserted is reflective of your latitude.
b. Orient the unit to true south (adjust for declination).
c. Observe to see if any shading concerns are visible in the dome of the unit.
d. Take the unit to the "other end" of the array (estimate about three feet per panel) and check for shading issues at this new location.
e. Repeat the process in a location where you feel there will be shading issues (so you can see the obstacles reflected in the dome).
f. Repeat the process in an area you believe is free from shading issues.

Chapter 7

Design/Install the Array and Inverter

Once the site inspection is complete, typically the next step in the design process is to size the array and select the panels and inverter.

Chapter Objectives:

- Calculate the array size required to service a given load.
- Understand factors that comprise the derate factor used in the design.
- Explore the criteria used in selecting a solar module.
- Determine the maximum and minimum number of panels in each string of a given array.
- Note the various specifications that are important in selecting the right inverter.
- Understand the difference between kVA and kW.
- Determine the proper size of an inverter based on system type and load demand.

The size of the system may be constrained by a number of factors. These include:
- the space available for the array,
- the budget for the installation,
- the load demand of the building (many jurisdictions restrict the amount of energy that can be reduced to a percentage of the load).

Determining the Size of the Array

The size of the array is expressed in terms of its total peak-watts of generating capacity. In other words, how much power the array will generate under ideal solar conditions (standard test conditions).

The amount of power the home or business will need over the course of a month should have already been determined, either through a load analysis, or more likely, by checking the customer's most recent utility bills. Once the total demand is known, the homeowner will need to determine what percentage of the power they would like the array to offset.

Budget also will have a major role in determining the size of the array. But if the budget will allow it, it is often best to oversize the array, making it bigger rather than smaller. Electrical use tends to increase over time. Or if money is tight, perhaps design the system so that it can easily be expanded in the future as more money becomes available.

When designing a grid-tied system, it is important to consider the rules and requirements of the utility provider. Many utilities place severe penalties on systems that produce more than they consume.

Some states have placed net metering requirements on investor-owned electric utilities. This typically means that the utility must pay for excess electricity generated by their customers through PV systems.

But if a customer generates more power than they consume (generally determined over the course of a year), they my be recognized as a utility provider, like any other power plant. The homeowner then falls outside the scope of the net metering law.

A number of states now require that solar energy systems must be installed on all new homes. For example, the **California solar mandate** is a building code that requires all new construction homes have a solar photovoltaic (PV) system as an electricity source. This code, which went into effect on January 1, 2020, applies to both single-family homes and multi-family homes.

CALIFORNIA SOLAR MANDATE

The PV array needs to be large enough to meet the annual electricity usage of the building. As there is no history of electricity usage in a new building, builders use an estimate for each property that's based on the building's floor space and the climate zone in which it's located.

Future-Proofing Designs

As the various sizes and ratings of the components in the system are calculated, bear in mind that situations change over time.

Will the home or business expand at some future date? Is there a reason to expect that additional electricity will be needed in the future (might they purchase an electric vehicle in a few years, for example)? Might battery backup be added to the system in the future?

It is a lot easier to place larger wires and larger components today, than redesign and perhaps have to replace parts of the system at a later date.

It is, of course, a balancing act. Customers do not want to pay for things that will never be used. But if the system is expanded in the future, they will find it very cost effective to install larger than currently required wiring and disconnects rated for more amps than needed, which may cost a little more during the initial installation, but will save a lot of money in the future.

Array Size Example

In this example, assume the site assessment has been conducted and it has been decided to install a grid-tied rooftop array designed to offset as much of their utility bill as possible, without exceeding their average monthly use. The customer is confident that electrical use will not increase over the life of the system.

Load Analysis

A review of the customer's electric bills for the past year shows they use about 1,220 kilowatt-hours (kWh) of power each month on average. This has remained fairly consistent from month-to-month and from year-to-year.

The utility supplier offers net-metering, trued-up over the entire year (months when the array over produces are credited to those months when production falls below consumption).

Calculate the Array's Generating Capacity

Assume this site is located in Columbus, Ohio. Columbus sits at 39.96 degrees north latitude – or close enough to call it "40 degrees."

A search of PV Watts shows that given the angle and orientation of the roof, the average annual insolation figures (hours of peak sunlight) for this location is about 4.1 hours per day.

The household needs 1,220 kWh of power each month, and there are (on average) 30.5 days in a month. So the average daily load for this site is:

　　1,220 kWh (per month) / 30.5 days (per month) = 40 kWh (per day)

Each day (as determined from PV Watts), there are 4.1 hours of peak sunlight available at this site. So the next step is to determine how large the array must be, given the available insolation:

　　40 kWh / 4.1 hrs (insolation) = 9.76 kW

This calculation reveals that the array must be sized to 9.76 kW (or 9,756 watts) given the insolation available at the site, in order to offset the entire electric bill with power from the array. But this assumes the array is 100% efficient.

Unfortunately, PV systems are not 100% efficient. A certain amount of energy produced is lost as the DC signal travels from the panels, through the wires and connections (voltage drop), and is converted to AC power within the inverter. In fact, up to 10% of the power generated can be lost as the DC energy is converted to AC by an inexpensive string inverter.

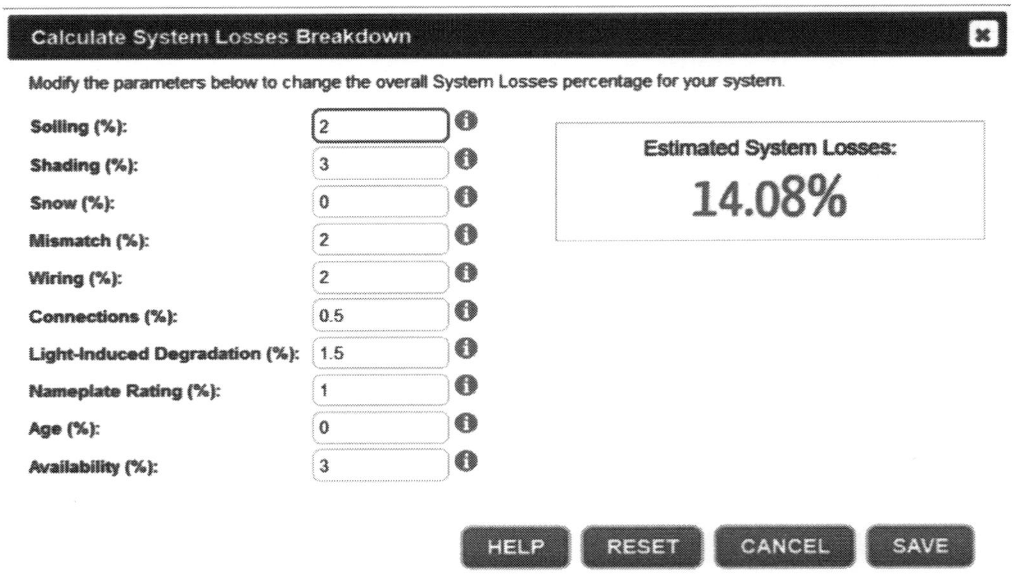

FIGURE 7-1: DERATE FACTORS FOR SOLAR ARRAY

(FROM NREL)

Calculate System Losses Breakdown	
Modify the parameters below to change the overall System Losses percentage for your system.	
Soiling (%): 2	
Shading (%): 3	Estimated System Losses:
Snow (%): 0	**14.08%**
Mismatch (%): 2	
Wiring (%): 2	
Connections (%): 0.5	
Light-Induced Degradation (%): 1.5	
Nameplate Rating (%): 1	
Age (%): 0	
Availability (%): 3	

[HELP] [RESET] [CANCEL] [SAVE]

Shading, dust, aging panel mismatch and even air temperature (remember, panels are less efficient the higher the air temperature) will also effect how the system performs.

DERATE FACTOR

So it is necessary to adjust for these inefficiencies by applying a **derate factor** when sizing the array.

NREL offers a calculator, illustrated in Figure 7-1. In this example, the defaults are set to a derate of 14.08%, or about 86% efficient.

Every system is different, and over the years systems have become more efficient, so derate factors change over time. But for purposes of this example, we will assume the system is 87% efficient (or a derate factor of 0.87).

In our example, it has been determined that the system will be 87% efficient, or have a derate factor of 0.87. The 100% efficient array size previously determined will need to be adjusted for anticipated inefficiencies:

9,756 W (array size) / 0.87 (derate factor) = 11,214 W (adjusted array size)

So roughly speaking, this site will require an 11.2 kW array to meet the current electrical needs of this household.

Derate Factor Components

A number of factors combine to determine the derate factor used in sizing a PV system.

These derate factors include:
- **Module Nameplate DC Rating**: This accounts for variations in the efficiency of PV panels during manufacture. Not all panels are identical. Typically the maker of the panel will indicate that each panel will operate at the rated levels, +/- 5% (or some other fudge factor). Some panels will work better than the rating, some less well.
- Inverter: Some power is lost as the signal is converted from DC to AC within the inverter. This efficiency factor will be provided in the inverter specifications.
- **Module Mismatch**: Whenever two panels are not operating at exactly the same voltage-current, then neither will operate at peak efficiency when connected with the other panel. Panels connected in series with no MLPE connected always operate at the lowest individual panel production values.
- Diodes and Connections: Each diode and connection inserts a bit of resistance into the circuit. This results in a loss of voltage.
- DC Wiring: This accounts for the voltage drop experienced in the wiring system from the panels to the inverter.
- AC Wiring: This accounts for the voltage drop experienced in the wiring system from the inverter to the utility meter.
- **Soiling**: Any dirt, pollen, snow or film from pollution on the front surface of the panel will reduce its performance. This number can be significant in some dirty environments that receive infrequent rainfall.
- **System Availability**: This accounts for the amount of time over the year when the system might be down for maintenance or utility outages. While this might be useful in calculating annual power production, it is unlikely to be helpful during the design of the system.
- Shading: A well placed and well oriented array should have no problem with shading. But if shading cannot be avoided, it should be accounted for in the derating of the system.
- **LID (Light Induced Degradation)** is a loss of performances within silicon crystalline modules that occurs in the very first hours of exposure to the sun. The LID loss is related to the quality of the silicon wafer and may be of the order of 1% to 3%.
- Sun-Tracking: Systems that employ a mechanical single or dual-axis tracking system will find that they do not always maintain the optimum angle at all times. Inefficiencies in these systems are accounted for in this factor.
- **Age:** As panels age, they will experience a loss of performance (typically about 0.1% to 0.5% per year). New panels should not be affected by this derate factor, but aging or used panels will need to take this loss of performance into account.
- **Round-trip battery efficiency**. For stand-alone systems, the round-trip battery efficiency must be added into the overall system derate factor.

(MODULE NAMEPLATE DC RATING)

(MODULE MISMATCH)

(SOILING)

(SYSTEM AVAILABILITY)

(LIGHT INDUCED DEGRADATION (LID))

(AGE DERATE)

(ROUND-TRIP BATTERY EFFICIENCY)

If designing a stand-alone system, the inefficiencies inherent in battery banks will also need to be taken into account. Add an additional 15% of inefficiency to the system to account for internal loss of power within a lead-acid battery bank. As a result, a larger array is needed to supply the same amount of power to the loads.

Note: Adjustment for Aging Solar Panels:
Many within the PV industry will recommend that the minimum string size be adjusted to account for the degrading performance of a solar panel as it ages. They suggest factoring a 0.5% loss of power for each year over the 20-year projected life of the system. This would mean that the array might produce 10% less power on the last year of this cycle.

However, sizing the array 10% larger in anticipation of production falloff over 20 years might result in overproduction during the initial years of service, resulting in conformity issues with utility and net metering regulations.

Selecting a Solar Module

Figuring out the number of required panels at this point is straightforward. Simply take the total generating capacity required (in this case 5,600 W) and divide it by the number of watts generated by each panel.
But before this can be accomplished, a panel must be selected.

After working on several systems, designers usually settle in on a PV panel and/or manufacturer that provides the quality, reliability, service and warranty that they desire. But how do they come to this determination?

Several things to look for when selecting a panel provider include:
- buy a reputable brand. Bear in mind, even large brand names may come and go. But larger, more stable companies are likely to be around long enough to support their products and honor their warranties.
- review the warranty offer (many panels are warranted for up to 25 years or even longer).
- make sure the panel has passed independent testing requirements. Depending on the jurisdiction and/or type, solar panels may be rated under various national and international standards, such as IEC 61215 (crystalline), IEC 61646 (thin film), IEC 61730, EN 61215 and EN 61730, EN 61646 and EN 61730, and UL 61730 (for crystalline panels in the US).
- ask around. Find out what experiences other installers have had with various panels and panel manufacturers.
- the **efficiency** of a panel (how much of the energy of the sun is actually transformed into electricity) may vary widely (for crystalline panels generally between 15-20%). Unless space is an issue, when panels are purchased in terms of dollars per watt, efficiency is not a critical issue.

(EFFICIENCY)

Tier 1 versus Tier 2 and Tier 3 Solar Panels
Solar panels are divided into three broad quality categories; tier 1, tier 2 and tier 3.

A **tier 1** designation is awarded to about 2% of panel manufacturers. It indicates that they have:
- produced panels for more than 5 years;
- are either publicly traded on a major stock exchange or have a strong and stable balance sheet;
- have a fully automated production process with a high degree of vertical integration;
- and have invested significantly in marketing their brand.

A **tier 2** manufacturer is a somewhat younger company that does not generally have the same research & development, as well as marketing budget of their larger competitors. They usually buy the wafers from a tier 1 manufacturer. Not all of the manufacturing processes are automated in a tier 2 company, which means there is a higher risk of faults during production.

A **tier 3** manufacturer is usually a very young company. They often employ many of the same manufacturing techniques of the tier 1 group. However, the cells may be soldered by hand which again means that there is a higher fault risk than with an automated process.

Typically tier 1 solar panels cost 10-30% more than tier 2 and tier 3 solar panels.

The tier designation was originally determined by Bloomberg New Energy Finance and was designed to describe the economic health of the manufacturing company - not the quality of the panel produced.

A quick online check of solar panels offered for sale will demonstrate that they come in a wide variety of sizes (both physical, but more importantly, in terms of **peak watts** generated).

Typically most solar panels installed in a residential grid-tied project (in 2023) range from 350 watts to 500 watts of peak generating capacity. There is a trend in recent years towards larger panels, as it reduces the installation cost per watt of generating capacity (it takes just as much time to carry and install a 450 W panel as a 200 W panel).

Larger panels (up to 800 W) are available, but primarily used in commercial or utility-scale systems. The extra-large size (and added weight) is typically not well suited for most residential rooftop installations and challenging to install.

FIGURE 7-2: PANEL SPECIFICATIONS HYPERION 400 W BI-FACIAL SOLAR PANEL

HYPERION SOLAR 400-WATT SOLAR PANEL SPECIFICATIONS

Specification	HY-DH108P8
Peak Power Watts (PMAX)	400W
Power Tolerance (PMAX)	0 ~ +5W
Maximum Power Voltage (VMPP)	31.01V
Maximum Power Current (IMPP)	12.90A
Open Circuit Voltage (VOC)	37.07V
Short Circuit Current (ISC)	13.79A
Module Efficiency	20.5%
Solar Cells	Monocrystalline
Cell Count	108
Temperature Coefficient of PMAX	-0.36%/°C
Temperature Coefficient of VOC	-0.304%/°C
Temperature Coefficient of ISC	0.05%/°C
Operational Temperature	-40~+85°C
Maximum System Voltage	1500VDC
Max Series Fuse Rating	30A
Glass	0.08" (2.0mm)
Frame	30mm anodized aluminum alloy (black)
J-Box	IP 68 rated
Connectors	Staubli MC4
Cable Length	
Dimensions	67.76" x 44.60" x 1.18" (1721mm x 1134mm x 30mm)
Weight	49.8 lb (22.6 kg)
Warranty	12 Year Product, 30 Year Performance

The panel selected will depend on the manufacturer's available options, price (one size may be on sale), product availability, and system configuration.

POWER TOLERANCE

Note the listed **power tolerance** of the panel. This is generally expressed as a % +/-. In other words, a 400 watt rated panel with a power tolerance of +/- 5% is actually rated at between 380 W and 420 W under ideal conditions. Generally the tighter the range (smaller the percentage of variation), the better. In this example the power tolerance is expressed as +/- 5 W, so the expected operating range at STC will be 395 W - 405 W.

Assume that a Hyperion solar panel has been selected (just to pick one at random), rated to generate 400 W per hour of peak sunlight. The rated generating capacity is obtained from the panel's specifications, shown in Figure 7-2.

11,214 W (required) / 400 W (peak generating capacity) = 28.035 panels

Since a partial panel cannot be installed, this result may be rounded up to 29 panels, or rounded down to 28 if there is concern that production not exceed consumption at the site.

Calculating String Lengths

If the system is using micro inverters or power optimizers, then calculations are complete. This system would use 29 (or 28) panels with a similar number of micro inverters or power optimizers (assuming one optimizer per panel). But if a string inverter is to be used, then it will be necessary to calculate the maximum and the minimum number of panels that can be connected in series to the selected inverter.

Determine the Maximum Number of Panels in each String

A traditional solar array is comprised of a number of strings (panels hooked together in series) which are then connected together in parallel (in the combiner box) or run directly to the inverter to form the array. The number of panels connected together in a string will determine the DC voltage of the system.

In the U.S., the National Electrical Code (NEC) restricts the voltage of residential PV systems. Residential systems are currently limited to operate at under 600 Vdc. As a result, 600 Vdc is the highest voltage allowed by the NEC for strings in a residential PV system. It is theoretically possible to design a higher voltage array, but this is not allowed under the current code.

The NEC specifies a maximum voltage limit for multi-family homes (3 or more units) or commercial properties of up to 1,000 Vdc. Ground mounted systems can be designed up to 1,500 Vdc.

As a result of these limits, no commercially available residential string inverter sold for the U.S. Market will accept voltages higher than 600 Vdc. They may, however, limit voltages to below that maximum. It is not uncommon for residential string inverters to have a **maximum input voltage** limit of 550 Vdc, for example.

> MAXIMUM INPUT VOLTAGE

For larger systems, it may be desirable to install a commercial string inverter that has a maximum input voltage limit of 1,000 Vdc on a residential building. The system, however, must still limit the voltage to under 600 Vdc, even though the inverter is capable of handling higher voltages.

The lowest voltage the array can generate and still provide AC power to the loads in a grid-tied PV system is determined by the **minimum input voltage** rating of the inverter selected. For a typical grid-tied string inverter, this will range from 80-200 Vdc. A string inverter, for example, might have a total operating DC input range of 80-600 Vdc.

> MINIMUM INPUT VOLTAGE

The design of the grid-tied system should incorporate voltages high enough to avoid voltage drop issues (which will also save some money on the cost of wiring, as smaller wire can be used with higher voltages), but must also remain within the upper and lower voltage limits of the inverter.

In this example, restrict the design of the voltage from the array to the absolute maximum allowed (600 Vdc).

With this in mind, it is necessary to determine the absolute maximum voltage the array might ever generate under any environmental conditions to which it is exposed.

Under standard test conditions, the absolute maximum voltage a single solar panel can generate is known as its open circuit voltage (Voc). The Voc of the Hyperion 400 watt solar panel is 37.07 Vdc.

This voltage is generated under standard test conditions (which assumes the operating temperature is 25° C). The panel will generate less voltage as the temperature increases. Conversely, a solar panel will generate MORE voltage as the temperature DECREASES. So the colder the air temperature becomes, the more power the panel will generate.

A quick look on the Internet shows that the coldest temperature ever recorded for Columbus, Ohio is -30° C (-22° F).

(TEMPERATURE COEFFICIENT)

In looking at the panel specifications, note that the **temperature coefficient** when measured at Voc for this panel is -0.304%/°C. This means that for every degree Celsius the cell temperature rises over 25°C, the panel will produce 0.304 % less voltage (when measured against its Voc). Conversely, for every degree Celsius DECREASE in temperature, the panel will produce that much MORE voltage than it will at 25°C.

(Δ DELTA)

There is a potential that the ambient temperature in Columbus might drop to -30° C (it has happened at least one time in the past). STC tests at 25°C. So there is a potential of a 55°C (25°C - -30° C) difference, or Δ **delta**, between the operating temperature on the coldest day experienced and the temperature at which the module was tested.

> Δ 55°C (the difference between standard test conditions and the possible lowest temperature) x 0.304% / °C (the temperature adjustment factor for this panel) = 16.72%

So on the coldest day ever experienced, this panel may produce 16.72% MORE voltage than rated under STC. The Voc must be adjusted for possible colder temperatures (in this case, 37.07 Voc x 1.1672 = 43.27 Vdc).

Ambient Temperature (°C)	Temperature Correction Factor
-1 to -5°C	1.12
-6 to -10°C	1.14
-11 to -15°C	1.16
-16 to -20°C	1.18
-21 to -25°C	1.2
-26 to -30°C	1.21
-31 to -35°C	1.23
-36 to -40°C	1.25

TABLE 7-1: AMBIENT AIR TEMPERATURE CORRECTION FACTORS

(FROM NEC TABLE 690.7(A))

It is theoretically possible for the panel to generate up to 43.27 Vdc should there be perfect sunlight on the coldest day ever recorded for this location. With this possibility in mind, use 43.27 Vdc as the "worst case" output voltage number from the panel.

To determine the maximum number of panels in each string, divide 600 Vdc (the maximum voltage determined the array should generate) by the maximum voltage output possible from a single panel (in this case, 43.27 Vdc on the coldest day ever recorded).

$$600 \text{ Vdc} / 43.27 \text{ Vdc} = 13.87 \text{ panels}$$

This means that the maximum number of panels that can be designed into a string (and still remain below 600 volts on the coldest of days at this site) is 13.87 panels. Since a portion of a panel cannot be installed, this number must be rounded DOWN - in order to ensure the voltage output remains below 600 volts. So the strings of this array can be no longer than 13 panels.

There is an alternative way of determining the maximum panels allowed in each string. By using temperature correction factors found in the NEC Table 690.7 (A) (a portion reproduced in Table 7-1), it is possible to arrive at a similar, although less precise result.

For example, if the coldest temperature ever recorded for a site was -30° C, then multiply the Voc by the correction factor found in the appropriate row in the table (1.21 in this example).

$$37.07 \text{ Voc} \times 1.21 = 44.85 \text{ Vdc}$$
$$600 \text{ Vdc} / 44.85 = 13.38 \text{ panels, rounded down to 13 panels}$$

Determine the Minimum Number of Panels in Each String

The maximum number of panels that can be in each string has been determined, but the inverter also is rated to require a minimum voltage in order for it to function.

Assume this minimum is 150 Vdc input (this rating is obtained from the inverter specifications). There must be at least that amount of voltage reaching the inverter from the array in order for the system to generate usable power.

In this calculation, refer to the Vmp rating (the voltage the panel generates at its maximum power point) rather than the Voc. This is because minimum voltage should be based on normal operating conditions, rather than under open circuit conditions that will skew the voltage higher than normal.

The NEC does not address minimum string lengths as this is a performance issue and not a safety issue.

Just as a solar panel will generate more voltage as the temperature drops, it will generate less voltage as the temperature increases. So now it is necessary to find the hottest temperature this site will experience, and adjust accordingly.

The highest temperature ever experienced at this location (Columbus, Ohio in this example) was 41° C (106° F). However, working on a roof can be much hotter than the ambient air temperature. It is recommended by **ASHRAE (American Society of Heating, Refrigerating and Air-Conditioning Engineers)** that the ambient air temperature be adjusted as follows:

- 35° C for arrays mounted less than 6 inches above the deck of the roof.
- 30° C for arrays mounted more than 6 inches above the deck of the roof.
- 25° C for arrays mounted on poles.

Assuming this array is mounted 8 inches above the roof deck, add 30° C to 41° C and find the array might potentially experience operating (cell) temperatures as high as 71° C. Standard test conditions measure at 25° C, so the difference (delta) between the two is Δ46° C.

The temperature coefficient at Vmp (note that the coefficient used is at Vmp rather than Voc, since it is necessary to now test the system under normal operations rather than "worst case"). The Vmp temperature coefficient for this panel is -0.36 % / °C. Often this coefficient will be noted in the panel specifications as the temperature coefficient at the maximum power point.

So in this case, the panel will produce 0.36% LESS voltage for each degree the temperature increases over 25° C (the standard test condition).

$$\Delta 46°C \times -0.36\%/°C = -16.56\%$$

So this panel will produce 16.56% less power on the hottest day ever recorded on this site. 16.56% of the voltage output will need to be deducted from the Vmp (100% - 16.56% = 83.44%) of the panel to adjust for possible higher temperatures.

$$31.01 \text{ Vmp} \times 83.44\% = 31.01 \text{ Vmp} \times .8344 = 25.87 \text{ Vdc}$$

In order to determine the minimum number of panels required in each string, take the minimum operating voltage of the inverter (in this example 150 Vdc) and divide it by the minimum voltage a single panel might generate on the hottest day experienced at this location (25.87 Vdc).

$$150 \text{ Vdc} / 25.87 \text{ Vdc} = 5.79 \text{ panels}$$

In this case, since the result is the minimum number of panels required – round UP. So the string must be at least six (6) panels in order to safely generate the minimum power required by the inverter on the hottest day of the year.

These calculations show the strings can range from 6 to 13 panels in length in order to stay within the minimum voltage requirement and the maximum voltage restriction.

Determine how Many Strings are Required

It has been determined how many panels are required in the array to fully generate the required power (in this case, roughly 28 panels). It has also been determined how many panels can be in each string (between 6- 13) to stay within the voltage requirements of the inverter. With this information in hand, it is possible to determine which configuration makes the most sense for the array.

The 28 panels required in the array cannot be neatly divided by any of the string size options. Two strings of 14 panels would be ideal, but the system is limited to a maximum of 13 panels in each string. Two strings of 13 will only require 26 panels (slightly under sizing the system).

Assuming the client wishes to service the entire load (and perhaps over generate a bit) then three strings of 10 panels may be the best option, as indicated in the one-line drawing in Figure 7-3.

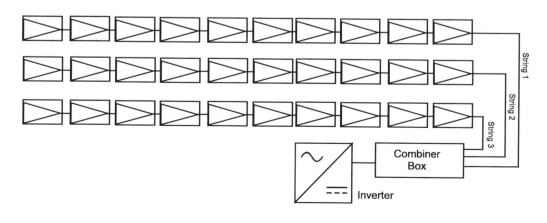

FIGURE 7-3: ONE LINE DRAWING OF AN ARRAY CONSISTING OF THREE STRINGS OF 10 PANELS

Changing the size of the system (larger or smaller), changing the panel selection, or changing the inverter selection (one that allows a lower minimum operating range, for example), might result in a configuration that more closely matches the required capacity of the system. This "give and take" exercise is all part of the design process.

If sized in a configuration of three strings of 10 panels, this array will provide a bit more power than has been calculated to service the entire anticipated load. Now the array will generate 400 W x 30 panels = 12 kW rather than the original 11.2 kW design.

The operating voltage of the inverter at STC coming from the array will be 310 Vdc (10 panels x 31.01 Vmpp = 310.1 Vdc).

Just as a string inverter has a maximum voltage and a minimum voltage, it also has a voltage range where it operates most efficiently. This ideal operating range is known as the **MPPT input voltage range.** If the inverter selected had a MPPT input voltage range of between 240 Vdc - 480 Vdc, then this system would be operating well within that range under STC.

In this example it was also determined that there are three strings in the system. Assuming each string will be terminated at the inverter (so that each string's maximum power point can be controlled independently) it is important that the selected inverter can accommodate at least three strings.

String Calculations for a Micro Inverter Systems

Calculating the strings for a system that incorporates micro inverters designed for single panels is about as straightforward as it gets. Once it has been determined how many panels are required to power the load, then one micro inverter attaches to each panel.

A micro inverter that connects to a single module is referred to as a "1-in-1" micro inverter. This is far and away the most common type of micro inverter used in residential installations. However, manufacturers do make "2-in-1" models (that hook one micro inverter to two panels) and even **"4-in-1" micro inverters** like the model shown in Figure 7-4.

When using micro inverters, each micro inverter is its own string. Either a string of one, two or four - depending on the type of micro inverter selected.

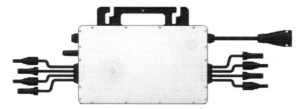

FIGURE 7-4: 4-IN-1 MICRO INVERTER

(FROM HOYMILES)

In many instances such a configuration where four modules connect to a single micro inverter may result in significant cost savings (assuming the cost of the 4-in-1 unit is less than four times the cost of a 1-in-1 micro inverter).

Branch Circuit Calculations for a Micro Inverter System

Micro inverters are connected together in parallel using a proprietary cable obtained from the manufacturer. When connecting in parallel, the ampacity increases rather than the voltage. So the gauge of the wire used in the system will limit the number of micro inverters that can be connected together in a **branch circuit.**

> BRANCH CIRCUIT

The Enphase IQ 7 micro inverter, for example, will produce 1.0 amp (240 VA continuous output / 240 Vac) when operating in a single phase system. When installed in a three phase 208 Vac system, each unit will produce 1.15 A (240 VA / 208 Vac).

The proprietary trunk cable connecting these units together has an ampacity rating of 20 amps. Incorporating the 1.25 NEC safety margin, each circuit can carry a maximum of 16 amps (20 A / 1.25 = 16 A)

So no more than 16 Enphase IQ 7 units can be connected together in parallel in each branch circuit when producing 240 Vac single phase power. When operating in three-phase, no more than 13 IQ 7 micro inverters can be connected in parallel (16 A / 1.15 A each = 13.9 rounded down to 13).

Different micro inverters operate at different ampacity, and the proprietary cable may vary, so check the specification sheet for the inverter selected.

String Calculations for Power Optimizer Systems

A single power optimizer can be attached to a single module or to multiple panels. It is common for installers to attach two panels in series, attached to a single optimizer (make sure the optimizer is rated for the combined power output of the two panels).

This practice can save a significant amount of time and money when installing larger systems. However, the advantages of the optimizer (MPPT functions, monitoring, etc) will now be shared over two panels. Environmental conditions that might affect one panel (shading, for example) will also affect the second panel to which it is attached.

Special connectors are available to allow three and four panels to be connected in parallel and then connected to a single optimizer.

With the increasing availability and use of high voltage panels (such as 96-cell high-efficiency panels), special care must be taken to avoid violating the 80-volt limit within the array boundary rule imposed in the rapid shutdown provisions of the NEC.

If multiple panels are connected in series to a single MLPE, such as a power optimizer, the combined voltage may exceed the 80-volt limitation. For example, a Panasonic 325 W, 96-cell panel has a Voc rating of 69.6 Vdc. Connecting two together in series would clearly violate 80-volt limitation.

Even panels with a Voc rating less than 40 Vdc might violate this limitation when connected in series if temperatures fall below 25°C (standard test conditions).

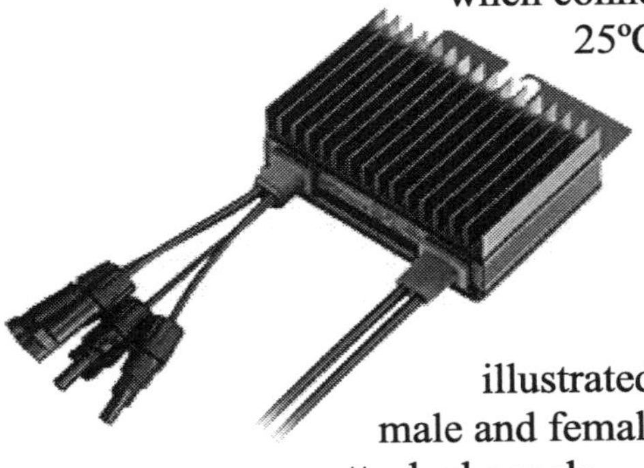

FIGURE 7-5: DUAL INPUT POWER OPTIMIZER

(FROM SOLAREDGE)

DUAL INPUT

Some MLPE manufacturers have addressed this issue by allowing two panels to be connected to the unit in parallel, keeping the voltage within limits. MLPEs, such as the one illustrated in Figure 7-5, incorporate two sets of male and female connection points (**dual input**) for attached panels.

Most power optimizer system manufacturers perform the string calculations for their various models and provide them within the product manual. In Figure 7-6, for example, it is demonstrated that when using a SolarEdge single phase system with a P440 optimizer, the minimum number of optimizers in each string is 8, and the maximum is 25.

Selecting the Inverter

In a grid-tied system, the size of the inverter selected is determined by the size of the PV array (and not the load demand placed upon the system). In a system designed to provide 100% of the load demand, these numbers (load and array size) are directly related (but not necessarily the same).

Often the PV array is only designed to provide a portion of the load requirements. Excess load demand is serviced by the grid, electricity that never passes through the inverter. So it is the array generating capacity, not the load requirements that are critical when selecting an inverter.

FIGURE 7-6: STRING CALCULATION CHART FOR POWER OPTIMIZERS

(FROM SOLAREDGE)

If it has been determined that the array might generate 8,600 watts during ideal operating conditions – then an inverter must be selected that is capable of handling an array this size.

PV System Design Using a SolarEdge Inverter[5]		SolarEdge Home Wave Inverter Single Phase	SolarEdge Home Short String Inverter Three Phase	Three Phase for 230/400V Grid	Three Phase for 277/480V Grid	
Minimum String Length (Power Optimizers)	S440, S500	8	9	16	18	
	S500B, S650B	6	8	14		
Maximum String Length (Power Optimizers)		25	20	50		
Maximum Continuous Power per String		5700	5625	11,250	12,750	W
Maximum Allowed Connected Power per String[6] (In multiple string designs, the maximum is permitted only when the difference in connected power between strings is 2,000W or less)		6800[7]	See[6]	13,500	15,000	W
Parallel Strings of Different Lengths or Orientations		Yes				

Technical data	Sunny Boy 6.0-US		Sunny Boy 7.0-US		Sunny Boy 7.7-US	
	208 V	240 V	208 V	240 V	208 V	240 V
Input (DC)						
Max. PV power	9600 Wp		11200 Wp		12320 Wp	
Max. DC Voltage	600 V					
Rated MPP Voltage range	220 – 480 V		245 –480 V		270 –480 V	
MPP operating voltage range	100 – 550 V					
Min. DC voltage / start voltage	100 V/ 125 V					
Max. operating input current per MPPT	10 A					
Max. short circuit current per MPPT	18 A					
Number of MPP Tracker / string per MPP Tracker	3 / 1					
Output (AC)						
AC nominal power	5200 W	6000 W	6660 W	7000 W	6660 W	7680 W
Max. AC apparent power	5200 VA	6000 VA	6660 VA	7000 VA	6660 VA	7680 VA
Nominal voltage / adjustable	208 V/ •	240 V/ •	208 V/ •	240 V/ •	208 V/ •	240 V/ •
AC voltage range	183 – 229 V	211 – 264 V	183 – 229 V	211 – 264 V	183 – 229 V	211 – 264 V
AC grid frequency	60 Hz/ 50 Hz					
Max. output current	25.0 A	25.0 A	32.0 A	29.2 A	32.0 A	32.0 A
Power factor (cos φ) / harmonics	1/ <4 %					
Output phases/ line connections	1 / 2					
Efficiency						
Max. efficiency	97.3 %	97.7 %	97.3 %	97.9 %	97.3 %	97.5 %
CEC efficiency	96.5 %	97.0 %	96.5 %	97.0 %	96.5 %	97.0 %

FIGURE 7-7: PARTIAL SPECIFICATIONS FOR SUNNY BOY STRING INVERTERS

(FROM SMA)

Assume a search of available options has been conducted and it has been decided that the SMA Sunny Boy string inverter model line (specifications detailed in Figure 7-7) is selected.

This inverter comes in a variety of sizes, but based on the design, the 7.7 US model appears to be the best choice (producing 7,680 W from a 8,600 W array). But note that the 7.7 US inverter can actually connect to an array as large as 12,320 W. In fact, even the 6.0 US model (with an AC output of only 6,000 W) can connect to an array as large as 9,600 W.

So, which inverter would be the right choice for the system?

Matching Inverter Output with Panel Output

Inverters come in a variety of sizes based on the maximum output they are capable of producing (measured in watts or volt-amps). Typical sizes for residential systems include 3,800 watts, 5,000 watts, 6,000 watts, 7,600 watts, and more. Many models are compatible with both single and three-phase power systems. Newer models are also compliant with smart inverter requirements.

It might seem logical that the array output must match the inverter's output. This is not, in fact, the case. To better understand this concept, it might be easier to visualize a single micro inverter connected to a single panel (but the same holds true for an array connected to a string inverter).

The Enphase IQ8+-72-2-US, for example, can accept up to 440 Wdc (can be connected to panels ranging in size from 235 Wdc -440 Wdc). However, the output from the inverter will be limited to a maximum continuous output of 290 Wac (Figure 7-8).

If a 375 Wdc panel were connected to a 290 Wac micro inverter, the **DC-to-AC ratio** of the system would be 1.3 (375 Wdc/ 290 Wac). This ratio is referred to as the **inverter load ratio (ILR)**.

> DC-TO-AC RATIO
>
> INVERTER LOAD RATIO (ILR)

FIGURE 7-8: PARTIAL SPECIFICATIONS FOR ENPHASE IQ8 SERIES MICRO INVERTERS

(FROM ENPHASE)

IQ8 Series Microinverters

INPUT DATA (DC)		IQ8-60-2-US	IQ8PLUS-72-2-US	IQ8M-72-2-US	IQ8A-72-2-US	IQ8H-240-72-2-US	IQ8H-208-72-2-US
Commonly used module pairings	W	235 – 350	235 – 440	260 – 460	295 – 500	320 – 540+	295 – 500+
Module compatibility		60-cell/120 half-cell	\multicolumn{5}{l}{60-cell/120 half-cell, 66-cell/132 half-cell and 72-cell/144 half-cell}		
MPPT voltage range	V	27 – 37	29 – 45	33 – 45	36 – 45	38 – 45	38 – 45
Operating range	V	25 – 48			25 – 58		
Min/max start voltage	V	30 / 48			30 / 58		
Max input DC voltage	V	50			60		
Max DC current (module Isc)	A			15			
Overvoltage class DC port				II			
DC port backfeed current	mA			0			
PV array configuration		1x1 Ungrounded array; No additional DC side protection required; AC side protection requires max 20A per branch circuit					
OUTPUT DATA (AC)		IQ8-60-2-US	IQ8PLUS-72-2-US	IQ8M-72-2-US	IQ8A-72-2-US	IQ8H-240-72-2-US	IQ8H-208-72-2-US
Peak output power	VA	245	300	330	366	384	366
Max continuous output power	VA	240	290	325	349	380	360
Nominal (L-L) voltage/range	V			240 / 211 – 264			208 / 183 – 250
Max continuous output current	A	1.0	1.21	1.35	1.45	1.58	1.73
Nominal frequency	Hz			60			
Extended frequency range	Hz			50 – 68			
AC short circuit fault current over 3 cycles	Arms			2			4.4
Max units per 20 A (L-L) branch circuit		16	13	11	11	10	9
Total harmonic distortion				<5%			
Overvoltage class AC port				III			
AC port backfeed current	mA			30			
Power factor setting				1.0			
Grid-tied power factor (adjustable)				0.85 leading – 0.85 lagging			
Peak efficiency	%	97.5	97.6	97.6	97.6	97.6	97.4
CEC weighted efficiency	%	97	97	97	97.5	97	97
Night-time power consumption	mW			60			

CLIPPED

When the DC power feeding an inverter is more than the inverter's AC output, the resulting power is **clipped** and lost, as Figure 7-9 illustrates.

This situation can exist for string inverters as well as micro inverters. An 8,600 Wac string inverter supplied by a 7,680 Wdc array would have a 1.12 inverter load ratio (8,600 Wdc/ 7,680 Wac).

While this might seem wasteful, many will argue that supplying an inverter with panels rated at a higher output is actually more cost efficient, especially as the cost of panels has declined rapidly in comparison to the cost of inverters.

FIGURE 7-9: ADDED PRODUCTION EXCEEDS LOST PRODUCTION WHEN ATTACHING A 250 W MICRO INVERTER TO A 280 W PANEL

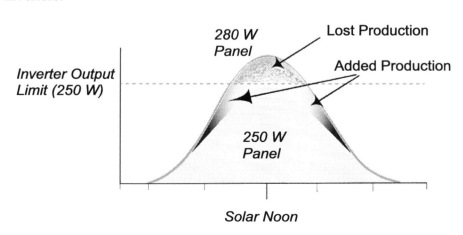

It is rare that a panel generates at its maximum output level. So comparing the two production curves, a greater amount of power is produced over a longer period of time with the larger panel or array. In other words, more power is captured when the panel is producing at less than capacity than is lost when it is producing at its maximum rated capacity.

While it is common for systems to be designed within a 1.12 - 1.30 range inverter-load ratio, the economics increasingly push for higher ratios, as much as a 1.6 ratio. There is a limit, however, that will be indicated by the maximum DC input designed into the inverter.

Inverter Specifications

In addition to maximum array size, minimum and maximum input voltages, MPPT operating ranges, and string connections, an inverter will also specify a **maximum operating input current** (32 amps for the 7.7 US model). Make sure the wiring system and overcurrent protection (fuses and circuit breakers) in the DC portion of the system conform to this limitation.

> MAXIMUM OPERATING INPUT CURRENT

By their very nature, a grid-tied inverter must be capable of matching the electrical signal obtained from the grid (it will be both receiving power from and sending power to the existing grid).

In the United States, most homes and small commercial buildings with less than 400 amps of service receive power in the form of single-phase, 240 Vac, at a nominal frequency of 60 Hz. In Europe they operate at 230 Vac, with a nominal frequency of 50 Hz.

Many commercial properties use three phase 208 Vac power. And larger systems use three phase 480/277 Vac. Many inverters are capable of accommodating both 240 Vac and 208 Vac power. Make sure the inverter selected is compatible with the power fed to the site where it is to be installed.

When selecting an inverter, ensure that it has obtained the UL 1741SA listing (IEC 61727 in Europe), which indicates it has been approved as a utility inter-tie device. Talk with the local utility prior to selecting the inverter, since they may have specific brands or conformance requirements with which the inverter must comply as part of their interconnection agreement.

The inverter specifications will also list the **maximum output current**. This number will be used in sizing wires that follow the inverter in the system.

> MAXIMUM OUTPUT CURRENT

Also, note the efficiency of the inverter in converting DC current to AC. Modern inverters have achieved efficiencies of 95-99 %. This number will be required in calculating the derate value for the system.

- kVA (KILO-VOLT-AMPS)

- APPARENT POWER (VA)

- REAL POWER (W)

- REACTIVE POWER (VAR)

- POWER FACTOR

kVA versus kW

The power output from a generating device, such as an inverter or a generator or even the grid itself, is normally expressed as **kVA (kilo-volt-amperes)**.

The power equation states that watts (W) = amps (I) x volts (V). So, it is common to assume that kVA would be exactly equal to kW. But this is not the case.

It may be best to think of these terms in this manner. kVA is the amount of power available for use, referred to as **apparent power**. This is the amount of power supplied by the inverter or the grid, measured in volt-amperes (VA).

kW is the amount of power that is actually used by the loads, referred to as **real power** (or often referred to as active power) measured in watts (W). In a 100% efficient system, these two would be equal, or kVA = kW.

But depending on the loads, a certain amount of power required to run the load cannot be used to actually power the loads. This wasted power is called **reactive power (var)**.

The difference between the power that is used by the loads (W) and the power that is required to run the load (VA) is referred to as the **power factor**.

For example, if a building requires 100 kVA to power the loads, but the loads only use 80 kW of actual power to run, then that building would be operating at a power factor of 80 kW / 100 kVA = 0.80. In this case the grid would be supplying power to this building at a power factor of 0.80.

Since grid-tied inverters must match the power on the grid, then the inverter may be producing at a power factor of 0.80 as well. So the output in kVA will not exactly match the kW used by the system.

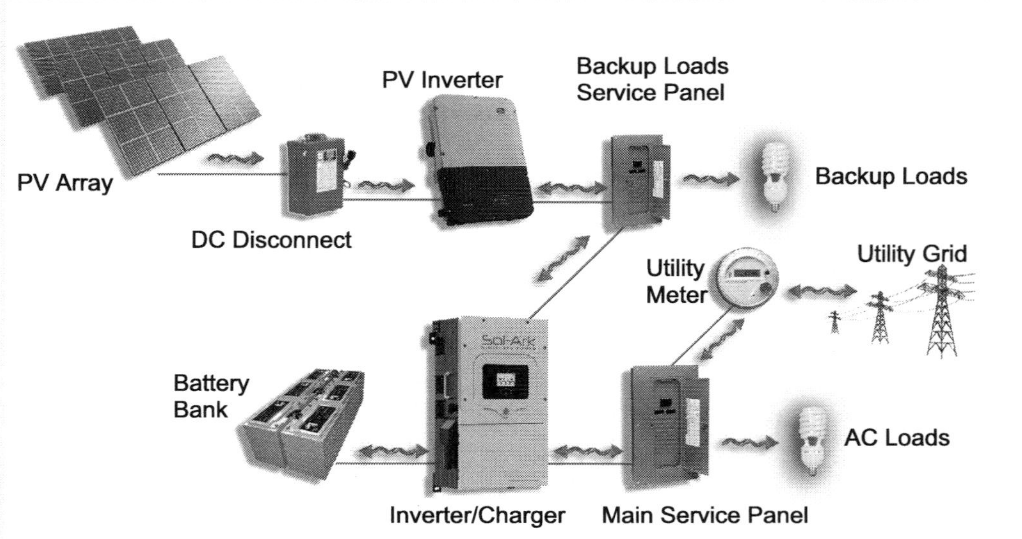

FIGURE 7-10: DIAGRAM OF AN AC-COUPLED MULTIMODE PV SYSTEM

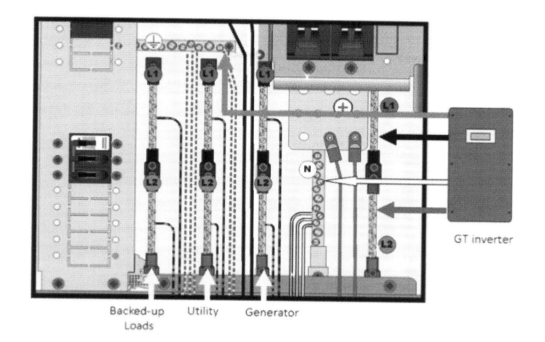

FIGURE 7-11: DIAGRAM OF CONNECTIONS WITHIN AN OUTBACK RADIAN INVERTER CONNECTING TO AN EXISTING GRID TIED INVERTER

(FROM OUTBACK POWER TECHNOLOGIES)

Sizing an AC Coupled Inverter

When selecting an inverter for a DC-coupled system, the process is similar to one used when selecting a string inverter - the difference being that batteries will be added into the system. When modifying an existing grid tied array to an AC-coupled system, an additional inverter is added into an existing system (as shown in Figure 7-10), and care must be taken to ensure that it is the right size and has the proper features.

Not all **battery-based** or **hybrid inverters** are compatible with the existing PV array inverter. So make sure the two inverters are compatible. While systems vary, the connections are made (typically) in a manner similar to those illustrated in Figure 7-11.

When selecting an inverter for an AC-coupled system, the following must be considered:
- solar PV array size (kW)
- inverter power output - continuous and surge rating (kVA)
- battery bank maximum charge rating (A)
- battery compatibility - system voltage and battery type

BATTERY-BASED INVERTER

HYBRID INVERTER

Size of the Battery-Based Inverter

When the grid is operating, all power from the grid-tied inverter flows through the battery-based inverter on its way to service the loads in the main service panel.

In order to avoid a situation where the power flowing from the grid-tied inverter overwhelms the circuits and wiring within the battery-based inverter, it should have a capacity of at least 125% of the existing grid-tied inverter.

Chapter 7: Design/Install the Array and Inverter

For example, a system with an existing 4 kW grid-tied inverter will require a 5 kW battery-based inverter (4 kW x 125% = 5 kW).

Battery-Based Inverter Power Output Rating

The output rating (in kVA) must also be enough to power the critical load maximum power draw (including continuous loads as well as surge demand). Size this in a manner similar to how a stand alone inverter would be sized - based on the critical load demand, rather than the entire home's load demand.

Match PV Array Amps to Maximum Battery Charge Amps

The battery-based inverter also behaves as a charger, taking power from the array and/or the grid and converting the AC current to DC to charge the battery bank. When the grid is operational, the battery-based inverter will use AC power (normally from the grid) to keep the batteries fully charged.

But once the grid signal is lost, all the power generated from the PV array is now available to charge the battery bank (as needed). With this in mind, care must be taken to ensure that the amount of amps flowing from the array does not exceed the **maximum battery charge** current of the battery bank.

(MAXIMUM BATTERY CHARGE)

For example, an Outback 12V EnergyCell 200RE battery has a maximum charge current rating of 53.4 A. A system configured to 48 nominal volts (four of these connected in series) will still have a charge rating of 53.4 A (as voltage rises in series connections, but amps remain the same).

At STC, a 6 kW array's output (assuming a .90 derate factor), after being converted to 48 Vdc power in the battery-based inverter would have an available current of about 112.5 A (6,000 W x .90 / 48 V). The current from the array could overwhelm the battery bank, causing potential damage to the system.

Ensure that the maximum charging current rating of the battery bank is equal to, or greater than, the potential charge from the array or that it is being managed by a charge controller.

Ensure the Battery-based Inverter is Compatible with Battery Bank

Just as every battery-based inverter is not compatible with every grid-tied inverter, they are also not compatible with every battery bank configuration. All battery-based inverters are designed to be used with a specific nominal DC battery bank voltage, the most common being 48 Vdc. Ensure the voltage of the battery bank is compatible with the selected inverter.

Also, some systems will not work with lithium ion batteries - while other systems will only work with lithium ion batteries. Again, ensure the battery bank will work with the selected AC-coupled inverter.

Grounded or Ungrounded Inverters

In the past it was common for the conducting wires to be attached to the DC grounding system at one point, typically within the inverter (shown in Figure 7-12). In the United States, this connection is usually made through the negative (grounded) conductor. The negative wire is connected to the grounding system at only one point, often through the ground fault protection device (GFPD).

These systems incorporate a transformer within the inverter, which creates a physical separation between the DC side of the system and the AC side of the system. There is **dielectric barrier**—such as an air gap, electrical insulation, or both between the AC and DC sides of the inverter. The two sides of the system are **isolated** from each other. A ground fault cannot pass from the DC side of the system to the AC side because of this barrier. Therefore both sides must have their own pathway to ground.

More and more frequently installers are opting to use **transformerless inverters**. This type of PV system has long been the norm in Europe.

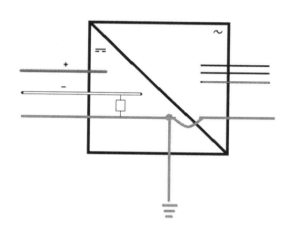

Figure 7-12: Electrical diagram of a grounded inverter

These systems are often referred to as **ungrounded systems,** as there is no physical separation, so no separate DC ground is required. But this term is misleading, as they are certainly grounded (just through the existing AC grounding system). For this reason, these inverters are increasingly referred to as **non-isolated inverters**.

Advantages of ungrounded systems may include:
- enhanced ability to detect ground faults.
- increased safety, as the shock potential is between the positive and negative conductors and not between the current carrying conductors and ground.
- transformerless inverters typically weigh less, are more compact, and more efficient than traditional inverters.
- the more efficient inverters generate less heat, making active cooling components (fans) unnecessary.

Functionally Versus Solidly Grounded Systems

With the inclusion of ground fault protection devices (GFPD) on the DC side of the system, as well as the growing popularity of non-isolated transformerless inverters, the PV industry has generally moved beyond the need for a separate DC system ground.

The NEC defines a solidly grounded system as "connected to ground without inserting any resistor or impedance device." PV systems that incorporate a GFPD on the DC side and/or a non-isolated transformerless inverter do not meet this definition, but are regarded as functionally grounded. In fact, essentially all new PV installations fail to meet the definition of a solidly grounded system.

In the past, systems that incorporated an inverter with a transformer placed overcurrent protection on only one conductor (typically the positive, or ungrounded conductor) and opened only one leg of the DC circuit in the DC disconnect (again, typically the positive conductor).

In systems that incorporated non-isolated transformerless inverters, it is required that overcurrent protection be placed on BOTH legs (both the positive and negative conductors). Also BOTH legs of the circuit must be switched (opened and closed) in the DC disconnect. In other words, a double-pole disconnect is required.

(SOLIDLY GROUNDED)

It should be noted that white wire is now only used for **solidly grounded** systems (as the negative wire would be the grounded conductor). With functionally grounded and non-isolated systems, both DC wires are considered ungrounded. In practice, most installers use a red conductor for the positive leg and a black conductor for the negative leg.

Chapter 7 Review Questions

1. The average U.S. home (in 2021) consumed 890 kWh of electricity per month. Assuming 4.2 hours of insolation at the site and a .90 derate factor for the system, what size array will be required to provide all the power necessary for this home over the course of a year?
   A) 7.72 kW
   B) 6.95 kW
   C) 6.26 kW
   D) 211.90 W

2. A homeowner in Phoenix, AZ has a daily load requirement of 9.6 kWh. She has opted to use micro inverters that will produce 250 W at STC. She has set the altitude at 25 degrees, giving her 6.54 hours of insolation. Assuming she has selected a panel that will produce at least 250 W, and assuming her system is 95% efficient - how many solar panels are required to fully meet her load demand?
   A) 6
   B) 7
   C) 9
   D) 11

3. Which of the following would NOT be a factor to be considered when derating a PV system?
   A) inverter inefficiency
   B) total solar resource
   C) soiling
   D) voltage drop

4. To account for the inefficiencies of a PV system as the result of voltage drop, loss due to connections, inverter inefficiencies, internal battery load resistance, etc, you must _____ the estimated power output of your array.
   A) reduce
   B) future-proof
   C) insolate
   D) derate

5. When selecting a crystalline solar module in the US, always ensure it complies with:
   A) UL 61730
   B) UL 1741-SB
   C) NFPA 855
   D) NFPA 70

6. A Tier 1 solar panel indicates:
   A) that the panel is very high quality.
   B) that the panel is very efficient.
   C) that the panel is very expensive.
   D) that the company that manufactures the panel is economically healthy.

7. The range above or below the rated power of a solar panel that it is expected to operate within under standard test conditions is known as the panel's:
   A) maximum power point.
   B) power tolerance.
   C) temperature coefficient at Vmp.
   D) tier 1 rating.

8. The temperature of the solar panels will be higher than the ambient air temperature. This is important when determining the minimum number of panels in a string. What temperature correction is required when mounting a solar array on a flush-to-roof racking system that is offset less than 6 inches from the deck of the roof?
   A) Add 35° C to the hottest ever recorded ambient air temperature.
   B) Add 30° C to the hottest ever recorded ambient air temperature.
   C) Add 25° C to the hottest ever recorded ambient air temperature.
   D) Add 20° C to the hottest ever recorded ambient air temperature.

Questions 9-10:
It has been determined that a home requires a 6.5 kW array. The homeowner has purchased 26 Solar World 250 W panels from her brother-in-law. At her location, the temperature extremes are -37°C (coldest) and 43°C (hottest). Her string inverter has an operating input range of 100 - 550 Vdc. The solar panels have the following characteristics:

Voc = 37.8 V   Vmp = 31.1 V
Temp. Coeff (Voc) = -.30%/°C   Temp. Coeff (Pmax) = -.45%/°C

9. What is the maximum number of panels in each string?
   A) 8
   B) 10
   C) 12
   D) 14

10. What is the minimum number of panels in each string if the array is ground mounted?
    A) 4
    B) 5
    C) 6
    D) 7

11. You are about to install a micro inverter system using Enphase IQ8+ micro inverters in a residential home that has single-phase 240 V electrical service. The maximum continuous output of each inverter 290 W. The proprietary cable used to hook these inverters together uses #12 AWG wire with an ampacity rating of 20 A. What is the maximum number of micro inverters that can be connected together in each branch circuit under the NEC guidelines?
    A) 12
    B) 13
    C) 15
    D) 16

12. For arrays subject to rapid shutdown requirements, the NEC states that the voltage within the array boundary (the array itself) should not exceed:
    A) 50 V
    B) 80 V
    C) 120 V
    D) 600 V

13. To determine the maximum panels allowable in a string of a grid-tied array, you must know all of the following **EXCEPT**:
    A) the azimuth of the array.
    B) the coldest temperature ever experienced on the site.
    C) the voltage input specifications of the inverter.
    D) the temperature coefficient of the panels.

14. A system has been designed with an 8.5 kW solar array and a 7.6 kW inverter. This system is said to have a 1.19:
    A) inverter load ratio (ILR).
    B) total solar resource fraction (TSRF).
    C) DC-to-DC voltage conversion.
    D) voltage drop.

15. When the DC power feeding an inverter is more than the inverter's rated capacity, the resulting power loss is referred to as:
    A) clipping.
    B) resistance.
    C) ripple.
    D) voltage drop.

16. The output from an inverter is measured in:
    A) kilo-amps (kA).
    B) kilowatts (kW).
    C) kilo-volt-amperes (kVA).
    D) reactive power (VAR).

17. Today most new solar PV inverters are:
    A) functionally grounded.
    B) solidly grounded.
    C) dielectricly grounded.
    D) grounded and isolated.

18. The voltage at which a grid tied inverter operates most efficiently is referred to as its:
    A) total solar resource.
    B) MPPT voltage input range.
    C) PWM maximization point.
    D) maximum power point.

19. Which of the following is NOT an advantage of micro inverters over string inverters?
    A) No string calculations are required.
    B) They are easily accessible for repairs.
    C) They are more efficient than string inverters.
    D) They reduce aging panel mismatch.

20. Which of the following equipment will likely NOT comply with the 2014 NEC requirement for "rapid shutdown"?
    A) power optimizer
    B) micro inverter
    C) UL 1741 listed string inverter
    D) remote disconnect combiner box

21. All grid-tied inverters must:
    A) be rated at no more than 48 volts DC.
    B) be UL 1741 listed.
    C) incorporate a modified sine wave output wave form.
    D) be rated at 125% of the array's output as per the NEC.

Lab Exercises : Chapter 7

7-1. Calculate Array Size

The purpose of this exercise is to practice calculating array sizes using various "real world" conditions.

Calculate the array size for a grid-tied system located in Denver, Colorado designed to provide a residential home currently using 1,200 kWh of electricity each month with 100% of its electric needs. Use the average daily peak sun hours with the angle set at latitude.

a. Using PV Watts, determine the average daily peak sunlight hours (insolation) with angle set at latitude and azimuth at true south.
b. Calculate daily load, divide by insolation number obtained from PVWatts.
c. Derate the system (figure .85 derate factor).
d. Select a panel.
e. Calculate how many panels are required in this array.
f. What size would the array need to be if it was located in Seattle, WA using the same panels and derate factor?

Lab Exercise 7-1:

Supplies Required:
- None

Tools Required:
- Internet access
- Calculator

7-2. Determine the Configuration of the Array

The purpose of this exercise is to determine the string sizes of an array.

Using the arrays calculated in Exercise 7-1 for Denver, CO and Seattle, WA, determine the appropriate configuration. The maximum voltage determined for each system is 550 Vdc.

a. Select a grid-tied inverter that you intend to use with the system.
b. Based on the lowest temperature for each location, calculate the maximum voltage the selected panel might produce.
c. Determine the maximum number of panels in each string.
d. Based on the highest temperature for each location, calculate the minimum voltage the panel might produce.
e. Determine the minimum number of panels in each string.
f. Determine the optimal array configuration, given that the array cannot over-produce energy for this location.

Lab Exercise 7-2:

Supplies Required:
- None

Tools Required:
- Internet access
- Calculator

Lab Exercise 7-3:

Supplies Required:
- NEC Table 310.15 (b) 16
- NEC Table 310.15(B)(2)(a)
- NEC Table 310.15(B)(2)(c)

Tools Required:
- Internet access
- Calculator

7-3. Determine the Wire Gauge

The purpose of this exercise is to size the wire gauge within the PV system's PV source and PV output circuits.

For the Denver array sized in exercise 7-1 and 7-2, determine the wire gauge size for the PV source and PV output circuits. If multiple strings are required, they will be combined in a combiner box located under the array.

Assume that USE-2 wire will be used in both circuits. Assume the PV source circuit will be located 5 inches above the deck of the roof. The PV output circuit will be in conduit located directly on the roof deck and exterior wall (exposed to sunlight).

a. Determine the Isc for each string (using the selected panel), correct ampacity as directed by the NEC.
b. Determine the minimum wire gauge size required for the PV source circuit.
c. Determine the ampacity rating of the PV output circuit.
d. Adjust ampacity of PV output circuit for temperature and conduit fill.
e. Determine the minimum wire gauge size of the PV output circuit assuming the termination ratings at the combiner box and the DC disconnect are unknown.

Chapter Objectives:

- Explore the evolution and types of connectors used in a PV system.
- Identify the various circuits located within a PV system.
- Calculate the ampacity requirements of the various circuits.
- Size overcurrent protection and determine where and when it is needed.
- Comprehend the various types of conduit and raceways and when and where they are used.
- Understand how conduit fill, expansion, location and support are integrated into a solar installation.
- Learn the proper way of bending metal conduit.
- Discuss how wires are pulled through installed conduit.
- Explore the factors used in selection and installation of an appropriate AC disconnect.
- Identify the various options and restrictions when connecting the system to the utility.
- Obtain a basic introduction to the bonding and grounding system.

Chapter 8
Wiring the System

Now that the equipment has been selected, the array sized and sited - it is necessary to connect everything together.

The PV Wiring System

The size of the wire required is dependent upon the amount of current (not voltage) that flows through it. Larger current requires larger wire.

Since watts = volts x amps, as the voltage increases, the amps (current) required decrease to provide the same amount of power (watts). As a general rule, higher voltage systems require smaller wire (lower amps) within the system.

Connectors

In the early days of solar panel installation, junction boxes were attached to the back of each panel and installers connected strings of panels together in series with either UV-resistant cable or cable contained within flexible conduit.

Today most panels come "hard wired" with **pigtail** cables (already attached within the junction box as shown in Figure 8-1), terminated with MC4 or similar connectors (one male and one female). Series connections between panels are made by clicking these connectors together.

PIGTAIL

The strings are then terminated in a combiner box or junction box. Often overcurrent protection is incorporated into the string at this point (provided for the string).

FIGURE 8-1: SHIPMENT OF PANELS, CONNECTORIZED WITH MC4 PIGTAILS

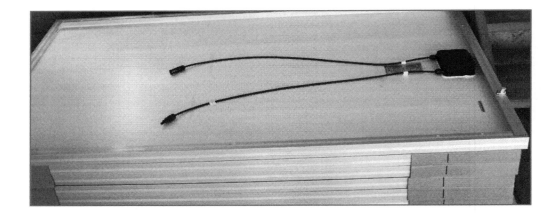

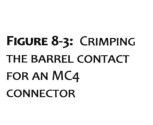

PV JUMPER

Field Terminations of MC4 Connectors

Often a PV technician must manufacture **PV jumpers** in the field to extend the end of the string of pigtails to the combiner or junction box. While various MC4-compatible connectors may look alike, there is no universal PV connector standard. Connectors from different manufacturers are made out of different materials; electrical contacts are made out of dissimilar metals; there are no standard product dimensions or tolerances.

For this reason, many issues (such as excessive voltage drop, poor connections, even fires) are the result of mismatched connectors. Language in NEC 690.33(C) states: *"Where mating connectors are not of the identical type and brand, they shall be listed and identified for intermatability, as described in the manufacturer's instructions."*

FIGURE 8-2: MC4 FIELD CONNECTION KIT

In an ideal world, all wires used in connecting together the array would arrive at exactly the right length and in exactly the right quantity. Rarely does any installation take place in an "ideal world." So it will likely be necessary

FIGURE 8-3: CRIMPING THE BARREL CONTACT FOR AN MC4 CONNECTOR

to terminate MC4 connectors in the field.

Making a field connection of this type requires an MC4 connection kit such as the one depicted in Figure 8-2.

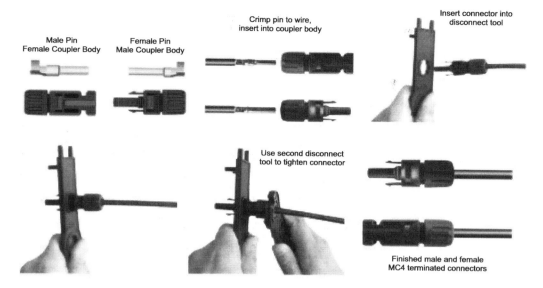

FIGURE 8-4:
ASSEMBLING AND TIGHTENING THE MC4 CONNECTOR

First measure the distance required for the MC4 jumper cable and cut a length of PV wire to match. Then strip about ¼ inch of insulation from the prepared length of PV cable.

Next, crimp the pin to the wire using an MC4 crimp tool, as shown in Figure 8-3.

Then slide the wire with the crimped female pin contact onto the male MC4 coupler body until you hear and feel it "click" into place. Do the same with the male crimped pin contact, sliding it into the female MC4 coupler body until it clicks into place.

Use the two MC4 disconnect tools, as illustrated in Figure 8-4, to tighten the MC4 connectors. Once tightened, the connectors cannot be taken apart.

Combiner Boxes

The purpose of a combiner box is to take the circuit coming from each string in the array and combine them in parallel, resulting in one wire or circuit leaving the array. The combiner box must be designed to accommodate the number of strings of panels feeding from the array, as well as the voltage and the current coming from each string.

In an array that has been configured into two strings of 11 panels, the operating voltage at STC of each string will be 11 times the Vmp (as voltage increases when connected in series). By definition, a string of solar panels are modules hooked together in series.

When any source is connected in series, the voltage rises, but the current remains the same. So the maximum current for each string will be the panel's short-circuit current (Isc) - or the maximum theoretical current generated by each individual panel, regardless of how many panels are connected in series.

If the Isc of a single panel is 9.43 amps, then the maximum current generated by these panels connected in series (regardless of the number of panels), will be 9.43 amps.

While the Isc rating is the most amps a panel may generate under standard test conditions, it is not necessarily the most current the panel may generate under real world conditions. Lower temperatures or more sunlight than are measured at STC will result in a higher current flowing from the panel. With this in mind, the NEC requires that the Isc be multiplied by 1.25 to account for variations in sunlight, temperature, etc. when sizing wires leaving the array.

Also, just as in all cases when sizing wire, the NEC requires that a safety margin of 20% be added to the anticipated maximum current estimates (this is accomplished by multiplying the anticipated current by a factor of 1.25). So, if the Isc of a string is 9.43 amps, then the wire must be sized to handle a current of:

$$9.43 \text{ amps (Isc)} \times 1.25 \text{ (solar variability)} \times 1.25 \text{ (safety margin)} = 14.73 \text{ amps}$$

This sizing requirement also applies to any overcurrent protection (fuses and circuit breakers) as well as the wire in the circuit.

Solar Variability Correction

If the panel is directly connected to a MLPE (module level power electronic) device (which is common on rooftop installations), then the MLPE will control the current output of the panel, adjusting for any variation in available sunlight. In that case, no additional 1.25 **solar variability correction** is required. However, the NEC still requires the standard 1.25 safety margin correction to be incorporated.

Most combiner boxes designed to work with residential grid-tied systems have a voltage rating up to 600 Vdc. Combiners designed for stand-alone systems may be rated at a lower voltage (typically around 150 Vdc).

Assume that after reviewing the various options, it has been decided that a PV system will incorporate a Midnight Solar combiner box MNPV6 (HV), depicted in Figure 8-5.

A review of the spec sheet for this combiner will show that it can accommodate up to four strings, is rated up to 600 Vdc and can accept fuses up to 20 amps.

PV Combiners- MNPV6

Model	Max VDC	Max # of Input Circ.	PV Source Circuits			PV Output Circits			Approved Mounting Orientation	Enclosure Type/ Material	Listing
			Max OCPD Rating Amps	OCPD	Wire Range AWG	Max # of Output Circ.	Max Cont. Current Amps	Wire Range AWG			
MNPV3 (LV)	150	3	20	CB 150V	14-6	1	60	14-1/0	90 to 14°	3R/Alum	UL1741
MNPV3 (HV)	600	3	20	FUSE	14-6	1	60	14-1/0	90 to 14°	3R/Alum	UL1741
MNPV6 (LV)	150	6	20	CB 150V	14-6	2	120	14-1/0	90 to 14°	3R/Alum	UL1741
MNPV6 (HV)	600	4	20	FUSE	14-6	2	80	14-1/0	90 to 14°	3R/Alum	UL1741
MNPV12 (LV)	150	12	30	CB 150V	14-6	2	200	14-2/0	90 to 14°	3R/Alum	UL1741
MNPV12 (HV)	600	10	30	FUSE	14-6	2	200	14-2/0	90 to 14°	3R/Alum	UL1741
MNPV12-250	300	6	50	CB 300V	14-6	2	168	14-2/0	90 to 14°	3R/Alum	UL1741
MNPV16 (HV)	600	16	15	FUSE	14-6	1	240	250MCM	90 to 14°	3R/Alum	UL1741
MNPV16-250	300	12	20	CB 600V	14-6	1	240	14-2/0	90 to 14°	3R/Alum	UL1741

Figure 8-5: The Midnight Solar Combiner Box specifications

(FROM MIDNIGHT SOLAR)

The spec sheet also indicates that this combiner box will accommodate wires ranging in size from 14 AWG (the smallest it will handle) up to 6 AWG (the largest it will accommodate).

Output from this combiner can be up to 2 circuits each with a maximum ampacity of 80 A. Since the ampacity would be larger (combined) by the parallel connection, output wires leaving the box must be larger in size - from 14 AWG to 1/0 AWG.

Junction Box

It is increasingly common in string inverter systems to terminate the panel strings in a junction box, rather than a combiner box. The only difference between the two (typically) is that the strings are not combined in parallel, but simply continue on as separate circuits to the DC disconnect located at or near the inverter.

While this design may require more wire, the inverter can control the output of multiple strings with multiple MPPT devices. The increased production of the array usually compensates for the added wire costs.

Array Circuits

In the past, the wire that runs from the solar panels to the combiner box or the junction box was referred to as the **PV source circuit.**

The circuit that ran from the combiner box to the DC disconnect located at or near the inverter was referred to as the **PV output circuit**.

The circuit connecting the DC disconnect to the inverter was referred to as the **inverter input circuit.** And the circuit leaving the inverter to make a connection to the existing AC power system was known as the **inverter output circuit** (referred to as the renewable output circuit in the Canadian Electrical Code).

- PV SOURCE CIRCUIT
- PV OUTPUT CIRCUIT
- INVERTER INPUT CIRCUIT
- INVERTER OUTPUT CIRCUIT

Figure 8-6: Circuits in a system that incorporate power optimizers

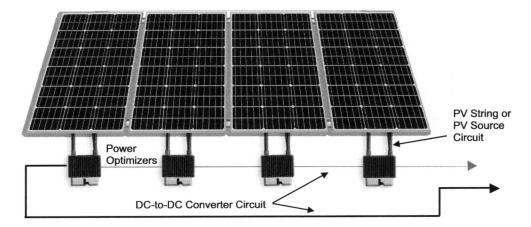

Conductors from different circuits cannot be contained in the same conduit or raceway.

These distinctions made sense when strings of panels were connected to a combiner box, the combined circuit terminated in a DC disconnect, and then the DC disconnect connected to a string inverter. However, today most residential arrays incorporate either power optimizers or micro inverters. So this terminology has become a bit dated.

In the 2023 NEC, the terminology describing circuits within a PV array was updated. For systems that do not incorporate MLPE devices, the wires connecting the panels to the DC combiner box or junction box are still referred to either as a PV source circuit or a **PV string circuit**.

(PV String Circuit)

The circuit that leaves the DC combiner box or junction box and terminates at the DC disconnect located at or near the inverter is now referred to as the PV source circuit (simply a continuation of the same circuit). The term PV output circuit has been eliminated from the code.

For systems that incorporate a power optimizer (a DC-to-DC converter), the wires connecting the panel to the optimizer are still referred to as the PV source circuit. However, the circuit connecting the optimizers together and then continuing to the DC disconnect is now referred to as the **DC-to-DC converter circuit**, as shown in Figure 8-6. This is a controlled circuit and does not require the solar variability correction when calculating wire size.

(DC-to-DC Converter Circuit)

Systems that incorporate micro inverters do not have any DC circuits. The NEC considers the wiring connecting the panel to the micro inverter as part of the internal circuit of the panel. All wires connecting the micro inverters together and leaving to attach to other equipment (the AC combiner box or AC disconnect) are considered part of the inverter output circuit, as illustrated in Figure 8-7.

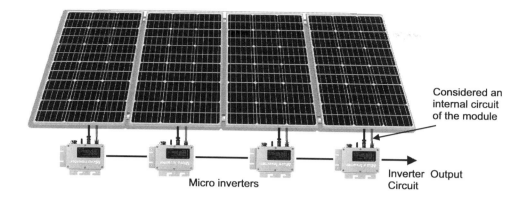

Figure 8-7: Circuits in a system that incorporate micro inverters

DC Wiring Circuits

Normally the panels will arrive pre-wired with pigtails sized with large enough wire to accommodate the maximum possible current from each string and terminated with MC 4 connectors (normally #12 AWG).

But it is very common that jumpers must be created on site to extend the strings to the combiner or junction box. Since connecting in series only affects the voltage of the string, the ampacity of a single panel will be the ampacity of the entire string.

Since most panels on the market today have an Isc rating of below 12.8 A, the the 20 amp rating of #12 AWG conductors will be more than adequate (12.8 A x 1.25 x 1.25 = 20 A). If the panel came with #12 AWG wire, then #12 AWG wire should do for the jumpers.

In a typical array, the PV string conductors will be supported on the rails of the array located behind the solar panels (in the shade). Since they are more than 7/8 inch (23 mm) off the deck of the roof and not exposed to sunlight, no temperature adjustment should be required.

PV string circuit cables can be installed without the use of conduit or raceways. These conductors should be supported and secured by staples, cable ties, straps, hangers, or similar fittings designed and installed so as not to damage the cable, at intervals not exceeding 4.6 ft (1.4 m) and within 12 inches (305 mm) of every outlet box, junction box, cabinet, or fitting.

All other PV circuits should be installed in conduit.

Sizing Overcurrent Protection

The sizing for overcurrent protection devices, such as fuses or circuit breakers located in the PV source circuit, is accomplished in a manner similar to determining the amp rating for the wires. It is important to ensure that the overcurrent device is not rated for ampacity greater than that of the wire to which it is connected.

For example, do not connect a 60 A breaker to a wire with a rated capacity of 55 A. The effect of this would be that the wire would fail before the breaker tripped - effectively providing no overcurrent protection at all. It is permitted to incorporate a 50 A breaker on a wire with a rating of 55 A, provided both meet the maximum circuit ampacity (including the safety factors).

For example, if the Isc of a string is 8.71 amps, then the wire must be sized to handle a current of:

8.71 A (Isc) x 1.25 (solar variability) x 1.25 (safety margin) = 13.6 amps

Always round up to the next available size when selecting a fuse or circuit breaker, but make sure that in rounding it up, the breaker does not exceed the ampacity limits of the wire to which it is connected.

In this example, a 15 amp DC breaker or fuse would be selected (rounding up from 13.6 amps). The wire selected (#12 AWG PV wire) has an ampacity rating of 25 amps (if termination ratings are unknown). So the breaker will work since it will trip long before the wire has reached its ampacity limit.

But what if a #8 AWG USE-2 wire had been selected for a circuit carrying 52.06 amps (after adjustments)? This size wire has an ampacity rating of 55 amps. But the next available breaker on the market (rounding up) is rated at 60 amps. If a 60 amp breaker is attached to a 55 amp rated wire, the effect is the same as if there were no breaker at all (for the wire would exceed its ability to carry current before the breaker trips). This is clearly not acceptable.

Two options would be to install a 50 amp breaker (which might then nuisance trip under normal operating conditions), or a better solution would be to install a larger #6 AWG wire (with an ampacity rating of 75 amps).

Like wire, overcurrent protection device ratings are determined under standard test conditions. Prolonged exposure to ambient temperatures greater than 25°C may affect its ability to handle current, causing the element within the fuse to melt under loads lower than the system was designed to handle.

So care should be taken to locate overcurrent protection in shaded areas, away from direct sunlight. Check the manufacturer's specifications on how to derate the system when placing these devices in a high-temperature environment.

Special NEC Note: When using smaller wire sizes, the NEC sets limits on the size of fuses or breakers that can be used with these wires. For #14 AWG wire, the fuse must be 15 A or less. For #12 AWG wire, the fuse must be 20 A or less. For #10 AWG wire, the fuse or breaker must be 30 A or less.

Since the overcurrent device in the combiner box selected is only rated to a maximum of 20 amps, it is probably safer to use wire, at least #12 AWG in size (larger wire can nearly always be substituted, but not smaller).

Separating AC & DC Conductors

Many large ground mounted PV arrays have tracking systems that use AC power to drive the tracker systems. It may be tempting to install the AC conductors in the same support system (conduit or raceway) as the DC source circuit conductors.

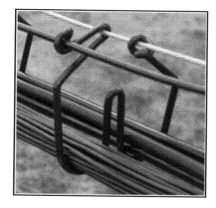

FIGURE 8-8: CABLE RING WITH THREE CABLE SECTIONS

(FROM CAB PRODUCTS)

Even though both conductors (the AC and DC) are part of the same PV source circuit, they must be separated by a partition, an example illustrated in Figure 8-8.

Wiring the Combiner to the DC Disconnect

The PV source circuit in a grid-tied system runs from the combiner box or junction box to the DC disconnect. In a stand-alone system, it runs from the combiner box to the charge controller.

Within the combiner box, the wires leading from the multiple strings are connected together in parallel (negative to negative, positive to positive). Within the junction box they simply transition from one conductor to another.

The voltage carried upon the wire leaving the combiner box will remain the same as that generated by each individual string, but the current will increase to the sum of the current from all the strings entering the combiner box.

PV System Wiring Checklist

✓ Make sure the proper wire type has been selected for its location (USE-2, TWN, etc)
✓ Determine the Isc of each string, multiply by 1.25 (variability factor) and by 1.25 (safety factor)
✓ Derate the ampacity of the circuit due to ambient air temperature and for conduit exposed to sunlight.
✓ Derate the ampacity of the circuit due to conduit fill.
✓ Select the proper cable size based on the derated ampacity of the circuit.
✓ Circuits in systems operating at 50 volts or less may not use smaller than 12 AWG copper conductors
✓ Factor in voltage drop calculations

If the ampacity rating of each circuit running from each string has been determined correctly (the Isc x 1.25 x 1.25), then the ampacity rating of the wire used in the circuit leaving the combiner is the sum of all the cables leading into the combiner.

For example, if there are two strings entering a combiner box, each with an Isc of 9.43, then each string conductor would be sized to handle 9.43 A x 1.25 x 1.25 =14.73 amps.

The two strings combined in parallel results in:

$$2 \times 14.73 \text{ A} = 29.46 \text{ amps}$$

This is the maximum amps that may run over the conductors leaving the combiner box.

Circuits that incorporate PV optimizers are sized based on the maximum output current of the optimizer. Since these optimizers are connected together in a string, the maximum current of one optimizer is the maximum current of the string.

SolarEdge power optimizers, for example, limit the output current of their units to 15 A. So the ampacity rating of the DC-to-DC converted circuit on this string must be:

$$15 \text{ A (max optimizer output)} \times 1.25 \text{ (NEC safety factor)} = 18.75 \text{ A}$$

Note that no solar variability adjustment is required since the optimizers limit the current output of the string.

DC Circuits in Metal Conduit

All DC circuits that exceed 30 volts or 8 amperes that are run within buildings must be contained within metal conduit (usually EMT). It is thought that the metal conduit adds an extra level of protection over and above PVC, made necessary because of the added danger DC arcing presents.

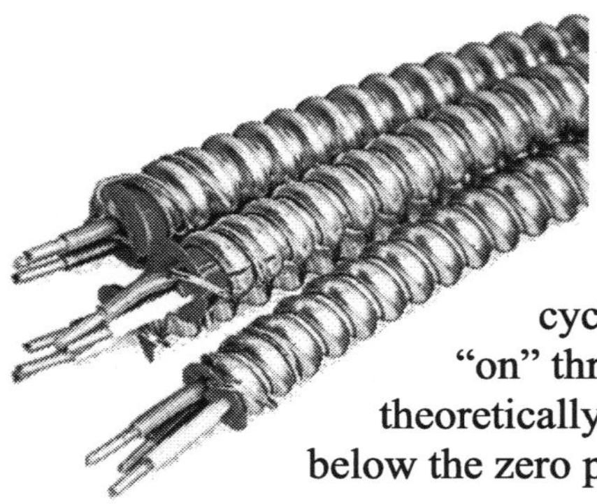

FIGURE 8-9: METAL-CLAD (MC) TYPE CABLE

When an AC circuit arcs, the current alternates "on" and "off" (where the waveform crosses the zero point) once per cycle. The DC is different in that it is always "on" throughout the arc. The DC arcs are theoretically harder to stop because they don't drop below the zero point.

The 2017 NEC allowed for the use of **metal-clad (MC) cable** as well. MC cables, as depicted in Figure 8-9, may make DC cable runs within buildings easier to install where tight bends make the installation of EMT conduit difficult.

> MC Cable

Conduits and Raceways

Except for the PV string circuit (where pre-connectored pigtails link the panels together), the wire used in a PV system should be contained within **raceways** to protect it from exposure to the elements, contact from other conductors or grounds, or contact with humans and/or animals. A raceway may include conduits, tubing and square wireways.

> Raceway

Installing Conduit

When selecting the conduit for a circuit, it is important to ensure that it is:
- the proper type (metal, PVC, liquid tight, etc) for the location,
- the proper size for the number and size of wire that will be pulled through,
- supported properly.

There are essentially five types of conduit permitted by the NEC for PV systems. These include:

- **Electrical Metallic Tubing (EMT):** This is a thin-walled metal tubing that is easy to bend, inexpensive and widely available. For this reason it is used in the vast majority of PV installations, particularly for the PV output circuit. EMT fittings are typically either compression or set screw attached.

> Electrical Metallic Tubing (EMT)

- **Rigid Metal Conduit (RMC):** RMC is a thick-walled, rigid galvanized steel raceway, almost always made of galvanized steel. The end of each section is threaded, and require threaded couplings during assembly. It is much heavier than EMT and more difficult to bend. RMC also requires field threading when sections are cut, as illustrated in Figure 8-10. RMC does, however, provide a higher level of physical protection to the wires than EMT. The cost of RMC is about twice that of EMT.

> Rigid Metal Conduit (RMC)

Figure 8-10: Field threading RMC conduit

- **Flexible Metallic Tubing (FMT):** FMT is a ribbed tubing which, as the name indicates, is quite flexible. This is more expensive than EMT but is often more practical when working in confined spaces or dealing with multiple tight bends. FMT is permitted only within buildings, as it is not water resistant.

> Flexible Metallic Tubing (FMT)

- **Rigid Polyvinyl Chloride (PVC):** PVC non-metallic conduit is permitted, as long as it meets all the proper rating requirements (such as

> Rigid Polyvinyl Chloride (PVC)

Chapter 8: Wiring the System   Page 231

TABLE 8-1: NUMBER OF CONDUCTORS PERMITTED IN CONDUIT

EMT Conduit Trade Size	Wire Size (THHN & THWN)								
	10 AWG	8 AWG	6 AWG	4 AWG	3 AWG	2 AWG	1 AWG	1/0	0/4
1/2"	5	3	2	1	1	1	1	1	
3/4"	10	6	4	2	1	1	1	1	1
1"	16	9	7	4	3	3	1	1	1
1 1/2"	38	22	16	10	8	7	5	4	2
2"	63	36	26	16	13	11	8	7	4

SCHEDULE 40

SCHEDULE 80

LIQUID-TIGHT FLEXIBLE NONMETALLIC

UV resistance). PVC is also inexpensive, widely available and moisture resistant. It is often used for PV output circuits where the cable is buried within the ground. But PVC expands and contracts when the temperature changes, so expansion fittings are required for long cable runs. There are two main types of PVC, **schedule 40** PVC and **schedule 80**. Schedule 40 is thinner and usually white in color. Schedule 80 is made with thicker walls (although the outside dimensions are the same) and is usually a dark gray color. PVC fittings attach with glue-like solvents.

- **Liquid-Tight Flexible Nonmetallic (LFNC):** LFNC is the flexible partner for PVC. This watertight conduit is appropriate for outdoor applications within confined spaces where flexibility is required.

Conduit Fill

When sizing the conduit, it is important to make sure it is large enough to handle the anticipated number of wires that will be housed inside the conduit. The space within a conduit is never completely filled with cables for a number of reasons.

These include:
- to allow for heat from the conductors to dissipate,
- for ease of pulling (both during the initial pull and later, should more conductors need to be added).

The NEC allows that a conduit be filled to a maximum of 26% of its capacity during its initial installation, and up to 40% fill when adding conductors to existing conduit. Calculating conduit fill can be a bit tricky. Table 8-1 can give a bit of guidance as to how many conductors can be placed in various trade sizes of EMT conduit.

Temperature adjustments to the ampacity characteristics of wires based on conduit fill should also be accounted for (calculations outlined in Chapter 5).

Conduit Support

When supporting conduit (on a structure or rack), EMT conduit must be supported within 3 ft (.91 m) of each outlet box, junction box, cabinet, or fitting, and every 10 ft (3 m) thereafter. Non-metallic conduit (such as PVC) requires more frequent support, as outlined in Table 8-2.

 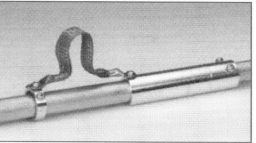

FIGURE 8-11: CONDUIT EXPANSION FITTINGS, PVC (LEFT) AND EMT (RIGHT)

Flexible metal conduit and liquidtight flex must be supported at intervals not to exceed 4 1/2 feet (1.37 m) and within 12 inches (305 mm) on each side of every outlet box or fitting.

Conduit Expansion

All conduit will expand and contract when exposed to different temperatures. PVC tends to expand much more dramatically than does metal conduit. For that reason the NEC requires expansion fittings such as those illustrated in Figure 8-11 if the expected length will change more than ¼ inch due to thermal expansion and contraction.

Conduit Size	Distance Between Supports
½ inch - 1 inch	3 feet
1 1/4 inch - 2 inch	5 feet
2 ½ inch - 3 inch	6 feet
3 ½ inch - 5 inch	7 feet
6 inch	6 feet

TABLE 8-2: SUPPORT DISTANCES FOR NON-METALLIC CONDUIT

PVC tends to expand about five times as much as steel conduit. This is one reason why EMT is generally used on rooftop installations rather than PVC. The NEC is fairly vague in its requirements for expansion fittings on EMT, stating only that it should be provided "where necessary." Often designers simply multiply the expansion determined for PVC by a factor of 0.2 (or one-fifth).

Assume, for example, that there is an installation where there are 200 feet (61 m) of PVC conduit running along the edge of a flat roof. The temperature extremes for this site range from a low of - 10° F (-23° C) to 120° F (49° C). This is a temperature change of 130° F (54° C).

Temperature Change (°F)	Length Change of PVC Conduit (inches / 100 ft)
5	0.2
10	0.41
25	1.01
50	2.03
80	3.24
100	4.06
120	4.87
130	5.27
140	5.68
150	6.08
200	8.11

TABLE 8-3: EXPANSION CHARACTERISTICS OF PVC RIGID NONMETALLIC CONDUIT

(FROM NEC TABLE 352.44 (A))

The NEC, in Table 352.44 (A) (reproduced in part in Table 8-3) indicates that the PVC conduit will expand at a rate of 5.27 inches/100 ft (134 mm / 30.5 m). Since this conduit run is 200 ft (61 m) in length, expansion joints must be added that will allow at least 10.54 inches (270 mm) of expansion and/or contraction within the run.

If the conduit used was EMT, the expansion adjustment would be one-fifth as much, or 10.54 inches x 0.2 = 2.1 inches.

These expansion calculations are independent of the size (diameter) of the conduit.

The temperature change underground is assumed to be relatively small. So no expansion fittings are required for buried PVC conduit. However, care must be taken if the PVC raceway is assembled and solvent-welded outside of the trench and in the sun. It is then dropped into the trench and immediately covered with soil. As it cools the conduit may contract, causing the weakest joint to pull apart underground. Allow the pipe to cool within the trench before covering it.

Conduit Location

The type of conduit selected for a given location is largely a factor of the environment into which it will be placed.

High humidity, underground, or other areas subject to corrosion often use PVC conduit. For systems in areas where the conduit is exposed and there are wide swings in temperature and/or UV exposure, an installer may find that EMT or RMC conduit is the best choice.

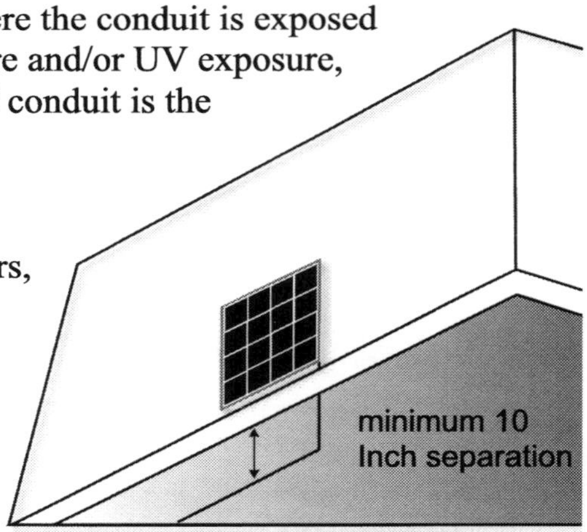

FIGURE 8-12: MAINTAIN 10 INCH MINIMUM CLEAR DISTANCE BETWEEN ROOF DECK AND CONDUIT

Inside a building, the conduit should be routed along building structural members, such as beams, columns, frames and trusses.

When PV conductors are run beneath roofs, they should not be installed within 10 inches (254 mm) of the roof decking or sheathing, as illustrated by Figure 8-12. This is to avoid a situation where a firefighter may cut through the roof of a building and hit energized wires. They should be installed on the underside of support members when possible.

On flat roofs, conduit runs between sub arrays and DC combiner boxes should be designed and installed to follow the shortest path between the two connection points. With this goal in mind, DC combiner boxes should be located as close as possible to the array.

Wires placed in conduit exposed to sunlight and mounted directly on roof surfaces should take into account the heat ampacity adjustments outlined in Chapter 5.

Bending Conduit

With non-metallic conduit, it is generally necessary to purchase pre-manufactured bend pieces to turn corners or go around obstacles in a conduit run. EMT conduit, however, can be bent using a tool such as a **one-shot bender** similar to the one illustrated in Figure 8-13.

FIGURE 8-13: MANUAL ONE-SHOT EMT CONDUIT BENDER

(FROM GARDNER BENDER)

ONE-SHOT BENDER

Conduit benders are designed to bend the metal without crimping it. When bending the conduit, it is important to take into account the sweep of the bend when calculating distances.

For instance, when making a 90° bend in a ½ inch (12.7 mm) conduit, subtract the **take-up** length from the desired **stub height**.

TAKE-UP

STUB HEIGHT

When bending a ½ inch (12.7 mm) conduit, the take-up length (the length of conduit required to make a 90° bend without crimping the metal) is 5 inches (127 mm). For ¾ inch (19 mm) conduit, the take-up length is 6 inches (152 mm). For 1 inch (25.4 mm) conduit, the take-up length is 8 inches (203 mm).

If the end of the conduit must be 12 inches (305 mm) off the ground, subtract 5 inches (127 mm) from 12 inches (305 mm) and begin making the bend at the 7 inch (178 mm) mark, as illustrated in Figure 8-14.

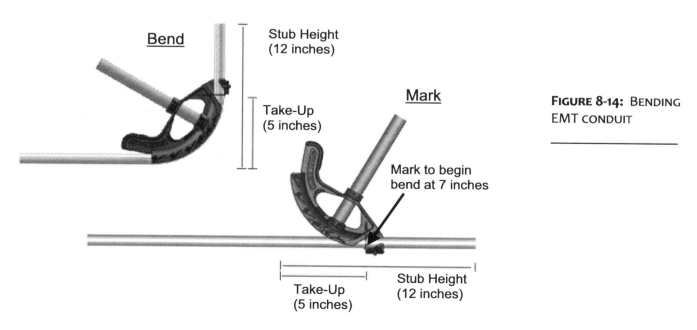

FIGURE 8-14: BENDING EMT CONDUIT

CONDUIT RUN

CONDUIT BODY

FIGURE 8-15: TYPICAL CONDUIT BODY

In any one **conduit run** (the section of conduit between two pull points), the NEC restricts the amount of bending to a total of 360° before it is necessary to insert a pull box, or a **conduit body** (such as the unit shown in Figure 8-15). A conduit body is typically a box with a removable cover that integrates into the conduit run.

Conductors can be pulled through one portion of the box and out through the open cover. The conductors are then fed back through the open cover and out the other side of the conduit body. Once the cable pull is complete, the cover is re-attached to the conduit body box. Conductors cannot be spliced within a conduit body - unless the unit is specifically rated for that purpose.

The NEC limits the number of bends because each bend makes it more difficult to pull the conductors through the conduit. With 360° of bends or more in a single run, it becomes extremely difficult if not impossible to pull the cables around the corners. This limit can be made up of any combination of bends (such as four 90° bends, or three 90° bends and two 45° bends, etc.).

Pulling Wires through Conduit
For short conduit runs with only a couple of conductors, the wires can usually be pushed through the conduit from one end to the other. However, when the conduit has several bends and more than two conductors must be installed, a **fish tape** may be required to pull the wires through the conduit.

FISH TAPE

The fish tape normally has a hook on one end (as shown in Figure 8-16), which is pushed through the conduit.

Once the fish tape has been pushed through the conduit, the conductor(s) are attached to the hook, secured with electrical tape, and then pulled back through the conduit. A **trailing pullstring** is often pulled along with the conductors. This string can later be used to pull more conductors through the conduit without the need to re-fish the run.

TRAILING PULLSTRING

FIGURE 8-16: USING A FISH TAPE TO PULL WIRE THROUGH CONDUIT

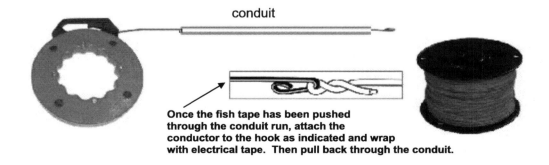

Once the fish tape has been pushed through the conduit run, attach the conductor to the hook as indicated and wrap with electrical tape. Then pull back through the conduit.

For larger installations, a **cable tugger** can be attached to the conduit at the end of the pull, and then the rope is fed into the tugger's pulley, drawing the rope and cable through the conduit. The cable is carefully fed into the opposite end of the conduit pull as the tugger is used to draw it through the pipe.

Tugger units such as the one depicted in Figure 8-17 are able to exert up to 10,000 pounds (45,500 Newtons) of pulling force.

It may be quite difficult to pull cables through conduit, especially for long runs or runs with several bends. Installers often apply **wire-pulling compound**, or **cable lube**, to the conductors to allow them to slide through more easily.

There are a number of types of cable lube available on the market. Regardless of the type selected, apply the lubricant on the head of the wires entering the conduit, and spray or squirt a bit down the conduit. As the wires are pulled, continue to apply a small amount of lube to the wires as they enter the conduit.

CABLE TUGGER

FIGURE 8-17: CABLE TUGGER

(FROM GRAINGER INDUSTRIAL SUPPLY)

WIRE-PULLING COMPOUND

CABLE LUBE

Underground Installation

The temperature changes underground are generally assumed to be relatively small. So conduit expansion and contraction is assumed not to take place in buried runs. However, if a PVC conduit is assembled in the sun, then placed in a trench, it may contract as it cools. Allow the conduit to cool to ground temperature before filling and compacting soil into the trench.

Backfill around all conduit must be smooth, granular soil with no rocks. The complete depth requirements for wire buried under ground are outlined in the NEC Table 300.5.

For photovoltaic installers, typical depth requirement for buried cable include:

- 24 inches (610 mm) deep for **direct buried cable** (although if at all possible, all buried cable should be in conduit)
- 18 inches (460 mm) deep when enclosed in PVC (in trench or under a driveway)
- 6 inches (150 mm) deep when enclosed in metallic conduit, 18 inches deep if running under a driveway

DIRECT BURIED CABLE

Many AHJs require schedule 80 PVC where the conduit emerges from underground, whether or not there is exposure to vehicles. They may also require that RMC, rather than EMT conduit, be used in locations when there are vehicles present.

FIGURE 8-18: METALLIC WARNING TAPE FOR BURIED CONDUCTORS

The NEC requires that the location of all service conductors that are not encased in concrete must be marked with a warning tape. These tapes, as shown in Figure 8-18, should be placed in the trench at least 12 inches (305 mm) above the buried conduit.

As PV systems are considered a separate service, assume this provision applies to all buried conductors from the PV system.

Conduit Durability

As most PV systems are designed to last 25 years or longer, care should be used in selecting and installing conduit to ensure it lasts at least as long.

Issues include:
- UV resistance. Metal raceways do not degrade in sunlight, however PVC and liquid tight will be affected. For this reason it is recommended that metal conduit be used on rooftops or other areas that may be exposed to sunlight.
- Moisture resistance. Metal conduit may rust or corrode when exposed to moisture. Most EMT is made of galvanized metal that should resist corrosion in all but the harshest of environments (such as salt water spray at the beach).
- Fire resistance. Metal conduits will survive exposure to fire better than PVC.

DC Disconnect

The PV source circuit typically terminates at a DC disconnect designed to interrupt the flow of DC power before it reaches the inverter.

DC disconnects are often incorporated as part of the inverter unit it is designed to isolate (as shown in Figure 8-19), or can be a separate piece of equipment.

FIGURE 8-19: DC DISCONNECT BUILT INTO A SOLAREDGE INVERTER.

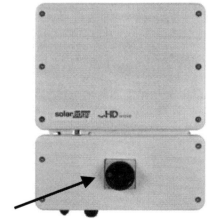

DC disconnect

Location

The NEC (690.15) requires that any PV circuit with current greater than 30 A incorporate a DC disconnect to isolate the equipment attached to that circuit. The disconnection device must be either incorporated into the device, or readily accessible and within 10 ft (3 m) of the equipment it is protecting.

Wiring the Inverter Input Circuit

The inverter input circuit on a grid-tied system runs from the DC disconnect to the inverter. The only change from the PV output circuit is that a disconnecting device has been introduced into the system.

No ampacity adjustments are required in the transition from the PV output circuit to the inverter input circuit (other than perhaps temperature adjustments, if the system transitions from a high heat environment to a lower temperature environment at this point).

On inverters where the DC disconnect is incorporated into the unit, the inverter input circuit is pre-wired and nothing is required from the installer.

Wiring the Inverter Output Circuit

Everything changes within the inverter. Not only is the waveform changed from DC to AC, but the voltages and corresponding ampacities change as well.

While voltages anywhere up to 600 Vdc (for residential) and even higher for commercial systems may enter the inverter in a grid-tied system, and voltages typically at 48 Vdc in a stand-alone system, the voltage leaving will typically be at 240 Vac (for a single-phase system). As power remains constant, amps clearly must be different on both sides of the inverter.

Grid-Tied Systems

The wire leaving the inverter must be sized to 1.25 the ampacity rating (the NEC safety margin) of the maximum current output of the inverter. For example, if the output power of the inverter is rated at 6,600 W, producing 240 Vac, then the continuous current leaving the inverter would be 27.5 amps. The wire must then be rated at 1.25 x 27.5 A = 34.375 amps.

If, in this example, the conductor specified for this circuit is THWN, a quick look at NEC Table 310.16 reveals that #10 AWG (rated to 35 amps) can be used if the inverter terminations and the breaker in the service panel both are temperature rated to 75° C. If either is only rated to 60° C, then #8 AWG THWN (rated at 40 amps at 60° C) wire must be used. Best practice would dictate selecting #8 AWG conductors for this circuit.

Micro Inverter Systems

In a micro inverter system, the inverter is connected directly to the solar panel, so there is no PV source circuit (in fact, no DC circuits at all). The conductors leaving the micro inverters are considered the inverter output circuit.

Normally the inverter output circuit would be based on the ampacity of the branch circuit created by connecting together the micro inverters, then adding the NEC safety factor of 1.25 for wire sizing.

An installer might calculate how many micro inverters are required, connect them together in a branch circuit, then calculate the wire size based on the output current of each micro inverter times the number of micro inverters in the circuit, then multiply it by the NEC 1.25 safety factor.

However, micro inverters are not connected in series, but in parallel. Making such connections manually would be difficult and time consuming. So micro inverter manufacturers provide a proprietary cable with parallel connectors (Figure 8-20) that make field connections simple and easy (simply click the proprietary cable into the micro inverter).

FIGURE 8-20: ENPHASE MICRO INVERTER Q CABLE CONNECTION

Since the connection cable has already been selected, it has predetermined ampacity limits. The Enphase Q cable, for example, is #12 AWG with a current limit of 20 amp.

If connected to an Enphase IQ8+ 72-2-US micro inverter, each with a continuous power output of 290 W, then each unit has the potential to generate 1.21 A or current (290 W / 240 Vac = 1.21 A).

Given the 20 A limit of the cable, the maximum number of micro inverters that can be connected to this proprietary wire with this selected micro inverter is 13:

$$13 \times 1.21 \text{ A} = 15.73 \text{ A} \times 1.25 = 19.66 \text{ A}$$

Any wire connected to this proprietary cable (at a junction box, for example) should be at least large enough to accommodate the 20 A capacity of the proprietary cable - even if the current installation does not require its full capacity (for example only 5 micro inverters are currently connected). One of the advantages of micro inverter systems is that they can be easily expanded. Allowing for the maximum capacity of each branch circuit is good wire design practice.

If there was only one circuit in the system, it would be perfectly acceptable to pass this circuit through an AC disconnect and then connect it to a 20 A double pole breaker in the service panel.

AC COMBINER BOX

However, larger micro inverter systems with more than one branch circuit require an **AC combiner box** such as the one illustrated in Figure 8-21. Enphase, the most popular brand of micro inverters installed worldwide, incorporates their array monitoring system in the combiner box. Even smaller systems with only one branch circuit generally require a combiner box as part of the design.

Once combined, a single inverter output circuit leaves the combiner and is connected to the AC disconnect.

AC Disconnect

The AC disconnect is designed to isolate the inverter from the electrical grid.

FIGURE 8-21: ENPHASE COMBINER BOX WITH COMMUNICATION (IQ GATEWAY) INTEGRATED

In a solar PV system, the AC disconnect is usually mounted to the exterior wall between the inverter and utility meter. The AC disconnect is sized based on the output current and voltage of the inverter.

The AC disconnect should:
- isolate the PV system from the electric utility grid,
- be visible and accessible,
- be externally operable, meaning the operator does not have to open it up and be exposed to live parts in order to turn the system on or off,
- be located within 3 meters (10 feet) of the equipment it is designed to isolate (normally the inverter - although a disconnect switch built into the inverter will often serve this purpose),
- be located within 3 meters (10 feet) of the point where the building connects with the grid (normally the utility meter),
- be clearly marked "Distributed Generation Disconnect Switch" (or similar) with letters 3/8 inch or larger,
- be lockable.

Check with the utility to find out what requirements they have regarding the AC disconnect. They may have very specific rules regarding its location and perhaps even the specific unit which must be used.

This disconnect allows the utility to disconnect the PV system from the grid, so their workers may safely work on their portion of the system.

Line/Load Connections

All disconnects generally connect a power source (the line side) to equipment that use that power (the load side). It is important to ensure the proper connections are made. The unit will often come labeled, indicating which connection is for the line, and which for the load.

In a PV array system, the AC disconnect is normally placed between the connection to the grid and the inverter. In this situation, the inverter will be considered the load, and the grid will be considered the line. It is a bit

FIGURE 8-22: SMALL FUSED AC DISCONNECT INDICATING LINE AND LOAD SIDE CONNECTIONS

Line Side:
Only connection lugs are energized when disconnect open

Load Side:
No power to fuses when disconnect open

confusing because it would be normal to think of the inverter as a power source as well. However when the grid is disconnected, the inverter automatically shuts down - however, the grid side is still energized.

In Figure 8-22, when the disconnect is switched off (open), most of the unit is no longer energized. Only the connection lugs from the line side (power source) are still "hot."

Connecting to the Utility

The inverter output circuit terminates at the point where the PV system connects to the electrical service from the utility. These can either be **load side connections**, or **supply side connections** (often referred to as line side connections).

Load Side Connections

For most residential and smaller commercial systems, the connection to the utility service will be made on the customer side of the main disconnect. This is referred to as a load side connection.

In most cases, this takes place by simply adding a breaker to the main service panel and connecting the conductors from the inverter to this breaker. Since more power is now entering the service panel box than when it was only being supplied from the utility, care must be taken to ensure that the current rating of the box (the busbar rating) is sufficient enough to handle the increased load.

If the sum of the breakers from the utility and the PV system do not exceed the busbar rating of the service panel, then the connection may be made anywhere on the panel where there is free space for a double pole breaker.

Scenario 1: Busbar Rating not Exceeded

A customer's existing service panel is rated at 200 amps. The service from the utility (the rating at the main disconnecting breaker) is 100 amps. The breaker from the PV system has been sized and determined to be 20 amps.

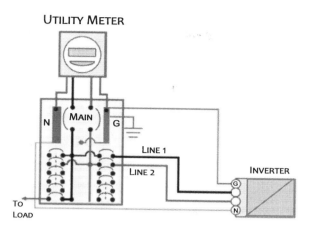

FIGURE 8-23: SINGLE-PHASE INVERTER LOAD SIDE CONNECTED TO STANDARD RESIDENTIAL SERVICE PANEL

Solution: Not a problem. The combination of the two systems feeding the service panel are 120 amps (100 A from the utility, 20 A from the PV system), which is less than the rating of the service panel busbar (200 A). The PV system can be connected to any space on the busbar where there is space for the double pole breaker feeding power from the inverter, as illustrated in Figure 8-23.

The NEC allows load side connections up to 120% of the busbar rating of the service panel. If this **120% Rule** is used in the connection, the point of connection of the PV system must be located at the opposite end of the busbar from the main disconnect, as indicated in Figure 8-24.

120% RULE

Scenario 2: Busbar Rating is Exceeded

A customer's existing service panel is rated at only 100 amps. The service from the utility (the rating at the main disconnecting breaker) is 100 amps. The two-pole breaker from the PV system has been sized (inclusive of the 1.25 safety factor) and determined to be 20 amps.

Solution: The 120% Rule will allow this connection. The combination of the two systems feeding the service panel are 120 amps (100 A from the utility, 20 A from the PV system), which is 120% of the rating of the service panel box. The PV system can be connected to the busbar, but only at a location at the opposite end of the main breaker.

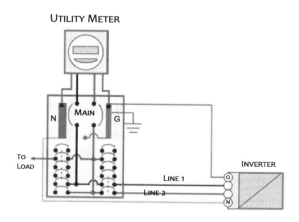

FIGURE 8-24: SINGLE-PHASE INVERTER LOAD SIDE CONNECTED TO STANDARD RESIDENTIAL SERVICE PANEL USING 120% RULE

When installed in this manner, a label must be placed next to the breaker that reads:

Warning: Power source output connection
Do not relocate this overcurrent device.

Locating the breakers at opposite ends ensures that no point on the busbar will be subject to the full sum of the current from the supplying breakers. If the main breaker is located in the center of the panel, then the PV system can be attached at either the top or the bottom (but not both from multiple inverters).

When connecting three-phase PV systems to three-phase service panels, the 120% rule is applied in exactly the same manner as with a single-phase system. The sum of the breakers feeding the panel (from the inverter and the main) cannot exceed 120% of the busbar rating of the service panel.

The NEC also allows for a load-side connection within the service panel if the sum of the breakers feeding the loads, plus the input breaker from the inverter, do not exceed the rating of the main breaker.

For example, if a 100-amp service panel was fed by 100-amp service from the utility and a 40-amp service from the inverter, as long as the sum of the breakers serving the load were less than 60 amps, a load-side connection could be made within this panel

Scenario 3: Adding a Main Lug Load Center

A customer's existing service panel is rated at only 100 amps. The service from the utility (the rating at the main disconnecting breaker) is 100 amps. The two-pole breaker from the PV system has been sized and determined to be 30 amps.

Solution: The 120% Rule has been exceeded and this connection cannot be made in the main distribution panel as described. The combination of the two systems feeding the service panel are 130 amps (100 A from the utility, 30 A from the PV system), which is 130% of the rating of the service panel box.

One option would be to replace the main breaker with a 90 amp breaker, reducing the total load to the service panel to within the 120% limit.

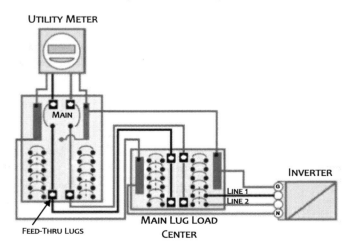

FIGURE 8-25: PV SYSTEM LOAD SIDE CONNECTED TO A MAIN LUG LOAD CENTER CONNECTED TO THE MAIN SERVICE PANEL THROUGH FEED-THRU LUGS

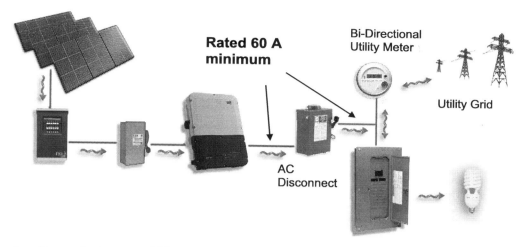

FIGURE 8-26: SUPPLY SIDE GRID CONNECTION

Another solution would be to install a **main lug load center**, connecting it to the main service panel through the **feed-thru lugs** at the bottom of the unit. A main lug load center, often called a **secondary load center**, is similar to the main panel, but does not have a main disconnect. Connect the PV system within an appropriately rated **secondary load center**, rather than the main service panel, as illustrated in Figure 8-25.

Supply Side Connections

Connecting the PV system somewhere between the main disconnect and the utility meter (or the utility's demarcation point) is referred to as a supply side, or **line side connection**.

Local utility requirements will govern where supply side PV connections can be made. The utility must be notified to turn off power to the facility before the connection is made and it generally requires a licensed electrician.

Some reasons to opt for supply side connection may include:
- undersized service panel (does not qualify for the 120% Rule),
- service panel may incorporate a main breaker with ground fault protection (GFPE) that does not allow back-fed circuits. Connecting load side will damage the main breaker.
- the utility may require it,
- larger systems can be accommodated,
- penetrating the wall to make a connection in the service panel may be difficult and/or costly.

When connected on the supply side, the PV system is considered an additional service at the facility, and is therefore subject to the NEC provision that all service connections be rated at a minimum of 60 amps.

The conductors, as well as the PV system disconnect, must be rated for at least 60 amps, even if the anticipated load is below that threshold, as shown in Figure 8-26.

MAIN LUG LOAD CENTER

FEED-THRU LUGS

SECONDARY LOAD CENTER

LINE SIDE CONNECTION

Chapter 8: Wiring the System

TAP

FIGURE 8-27: INSULATED PIERCING CONNECTOR

(FROM RITELITE SYSTEMS LTD)

FIGURE 8-28: LINE TAP SUPPLY SIDE CONNECTION ABOVE THE MAIN BREAKER IN A RESIDENTIAL SERVICE PANEL

(FROM CIVIC SOLAR)

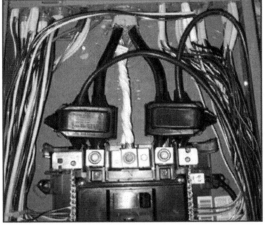

Insulated piercing connectors, or **taps**, as shown in Figure 8-27, allow the connection from the PV array to be "spliced" to an existing conductor coming from the utility without having to cut or disconnect the existing conductor. However, making these connections to live wires is not a recommended (or safe) practice.

Such connections are routine in residential systems when the system will exceed the busbar rating. These can be made within the service panel enclosure, but on the supply side of the main disconnect. An example is pictured in Figure 8-28.

It is more common, however, in supply side connected PV systems to cut into the conduit between the main disconnect and the meter and install a junction box where the PV system can be connected to the conductors from the utility, as shown in Figure 8-29.

When installed in this manner, the AC disconnect for the PV system is located between the connection point and the inverter(s).

FIGURE 8-29: ELECTRIC JUNCTION BOX, ALLOWING PV SYSTEM TO TAP INTO THE SERVICE FROM THE UTILITY

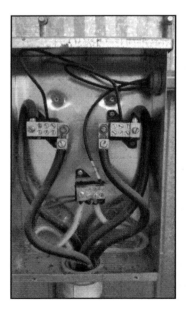

The wiring and disconnect should be sized to handle at least 125% of the maximum potential current leaving the inverter(s). Note that this current is based on the capacity of the inverter, not the array to which it is attached.

At minimum, in this example, this new connection must use 60 A rated conductors and a 60 A fused disconnect.

The conductors between the AC disconnect and the tap are unprotected service-entrance conductors. They are subject to all the same fault currents that might affect the existing service-entrance conductors.

For this reason, the conductor must be sized according to the **10-foot tap rule** or the **25-foot tap rule**, whichever applies.

> 10-FOOT TAP RULE
>
> 25-FOOT TAP RULE
>
> FEEDER TAP CONDUCTOR

10-Foot Tap Rule

If the length of the **feeder tap conductor** (the wire that connects the tap to the AC disconnect), is less than 10 feet (3 m), then the minimum current rating must be calculated in the following manner:

$$10\% \times [\text{Feeder Breaker} + (\text{Inverter Output} \times 1.25)]$$

For example, if the feeder breaker was rated at 200 A and the inverter output current was 15 A, then:

$$10\% \times [200\text{ A} + (15\text{ A} \times 1.25)] = 10\% \times 218.75\text{ A} = 21.875\text{ A}$$

Had the 10% rule not applied, the ampacity calculation for this circuit would have been 15 A (inverter output current) x 1.25 = 18.75 A. With the 10% rule, the new minimum ampacity for the conductor must be at least 21.875 A.

Had the 10% calculation resulted in a number smaller than 18.75 A (in this example), then the larger of the two amp ratings would apply.

But the NEC requires a minimum 60 A rating for a supply side connection, so in this example the conductor between the AC disconnect and the tap must be rated for at least 60 A.

25-Foot Tap Rule

If the length of the feeder tap conductor is between 10-25 feet, then the minimum current rating must be calculated in the following manner:

So again, if the feeder breaker was rated at 200 A and the inverter output current was 15 A, then:

$$33.33\% \times [200\text{ A} + (15\text{ A} \times 1.25)] = 33.33\% \times 218.75\text{ A} = 72.9\text{ A}$$

The longer the conductor, the more resistance. This increased resistance will impact the ability of the feeder breaker to trip in the event of a fault, so the conductor must be larger to avoid potential damage.

Many string inverters and micro inverters contain arc fault protection devices. There are also combiner box integrated solutions available to system designers. These devices must, however, be rated for PV systems.

Except in the most rare of situations, no tap conductor located inside a building and incorporated within a PV system should be longer than 25 feet.

Grounding and Bonding System

In most cases, the racking system selected will be made of metal (a conducting material). The aluminum frames on the solar panels, as well as the racking and conduit can act as a lightning rod (especially when placed on a rooftop).

Other stray and unwanted random power can also find its way onto these conducting materials (short circuits, down electric lines, etc). For this reason, all metallic parts of the system must be bonded together and connected to the grounding system to provide any stray current a safe pathway to ground.

BONDING

Bonding the conductive components of the system together and connecting them to ground is an extremely important part of a PV system design and installation.

IT IS A MATTER OF LIFE AND SAFETY!

This is an area that really requires the expertise of someone who is knowledgeable in all things electrical. While the novice (after studying the requirements of the NEC) may be able to design and install the grounding system – it is advisable at a minimum to have it inspected by an expert. Any flaw in the system could result in fire, serious injury, or even death.

While a properly installed bonding and grounding system will provide some limited lightning protection to the system, its main purpose is to provide protection from shocks to any person who may come in contact with an exposed metal component that may have inadvertently become energized.

Bonding

All metal components, such as the aluminum frame of PV panels, the racks, metal conduit, metal disconnect boxes, etc, must be bonded together. Once electrical continuity has been established between all the metal in the system, it is then connected to the grounding system.

FIGURE 8-30: A BONDING JUMPER MAKING AN ELECTRICAL CONNECTION BETWEEN TWO RAIL SECTIONS

In a typical PV system, the aluminum frames of the solar modules are attached to the metal rails supporting them with bonded bolts and/or washers designed to cut through the anodized coating of the metal to make a solid electrical connection.

If multiple rails are required to support the panels, then they must be bonded together, normally with a **bonding jumper**, such as the example illustrated in Figure 8-30.

BONDING JUMPER

Equipment grounding conductors (EGC) are then used to connect the various metal components to the grounding system. It is common to use #6 AWG bare copper wires for equipment grounding conductors, as shown in Figure 8-31. The NEC requires that equipment grounding conductors smaller than #6 AWG must be protected from physical harm by routing them inside raceways or conduits.

> EQUIPMENT GROUNDING CONDUCTOR (EGC)

> **FIGURE 8-31:** BONDING RAILS WITH A BARE COPPER #6 AWG WIRE USING STAINLESS STEEL LUG

The equipment grounding conductor continues from the array to the combiner box, along with the current carrying conductors within the conduit.

At each junction box or disconnect, the EGC bonds to the metal box, then continues on until it is terminated at the inverter. This portion of the bonding and grounding system is referred to as **equipment grounding.**

> EQUIPMENT GROUNDING

The sizing for equipment ground conductors can be found in NEC Table 250.122, a portion of which is recreated in Table 8-4.

Challenges in Equipment Grounding:

Properly installing an equipment grounding system can be challenging. Issues often include:
- Most panels and rails have an anodized surface that is non-conductive. Care must be taken to break through this layer to the metal surface below.
- A bond between two different metals (typically aluminum and copper) will suffer **galvanic corrosion**, weakening the bond over time. Stainless steel can be used to connect these dis-similar materials.
- It is time consuming (adding to the installed cost of the system).
- It is difficult to do correctly, making an improperly grounded system the most common of installation mistakes.
- Bare copper equipment grounding conductors can be unsightly.
- Installers must use connectors that are listed for grounding connections.

> GALVANIC CORROSION

The equipment grounding system continues from the DC side of the system to the AC side of the inverter, bonding to any metal boxes or disconnects it may

Rating or Overcurrent in circuit ahead of Equipment (not exceeding)	Size (AWG) Copper	Size (AWG) Aluminum
15 Amps	14 AWG	12 AWG
20 Amps	12 AWG	10 AWG
60 Amps	10 AWG	8 AWG
100 Amps	8 AWG	6 AWG
200 Amps	6 AWG	4 AWG

TABLE 8-4: MINIMUM SIZE OF EQUIPMENT GROUNDING CONDUCTORS

(FROM NEC TABLE 250.122)

Chapter 8: Wiring the System

FIGURE 8-32:
THE EQUIPMENT
GROUNDING SYSTEM

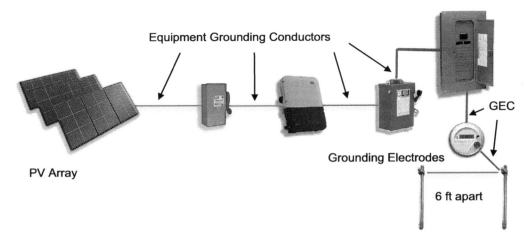

GROUNDING ELECTRODE
CONDUCTOR

GROUNDING
ELECTRODE

pass through, and terminates at the grounding bus bar in the electrical service panel. The service panel grounding bus bar should already be connected with a **ground electrode conductor (GEC)** to an existing AC **grounding electrode** (or ground rod).

A grounding electrode is generally a stainless steel and copper or zinc-coated rod at least 5/8 inch (15.87 mm) in diameter. At least 8 ft (2.44 m) of the rod must be in contact with the earth (buried).

The NEC 250.53 (2) states that if the resistance to earth is less than 25 ohms, then a single grounding electrode for the system will be acceptable. If two ground rods are used, then no resistance measurement is required.

Few PV installers (or code inspectors for that matter) have the equipment or patience to test the earth resistance. So often a second ground rod is installed if it is not already in place.

The paralleling efficiency of ground rods increases by separating them by twice the length of the longest rod. At a minimum the two ground rods must be 6 feet (1.8 m) apart, but bonded together, generally with #6 AWG bare copper wire.

A simple equipment grounding system is illustrated in Figure 8-32.

In older photovoltaic systems, the DC circuits were solidly grounded within the inverter. There was no direct connection between the DC grounded conductor and AC grounded conductor (they are physically separated within the inverter by the transformer).

When this is the case (which in today's world is quite rare), the NEC requires that a separate DC grounding system be installed, as shown in Figure 8-33, and then bonded to the AC grounding system.

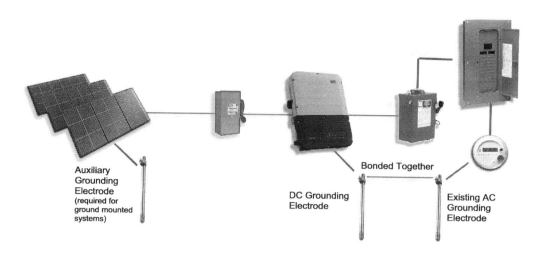

FIGURE 8-33:
GROUNDING SYSTEM FOR SOLIDLY GROUNDED INVERTER

The DC side of the grounding system is connected to the earth from the inverter using a grounding electrode conductor (GEC). The GEC should be no smaller in size than the largest conductor used in the PV system.

So, for example, if a #2 AWG copper conductor was used to carry current (the positive and negative wires) from the array to the inverter, then the grounding electrode conductor must be a #2 AWG copper conductor (or larger). The GEC must be a minimum #8 AWG copper or #6 AWG aluminum, regardless of the size wire used in the remainder of the PV system.

This portion of the grounding and bonding system (connecting the equipment grounding to the ground rod) is referred to as **system grounding**.

SYSTEM GROUNDING

Most modern inverters are functionally grounded, rather than solidly grounded - so the DC equipment grounding system is physically connected to the AC equipment grounding system. As a result, no separate DC system ground is required.

Ground Faults

When current is flowing through the grounding and bonding system, there is a problem. Somehow an electric current is energizing part of the system not intended to be energized. This is known as a **ground fault**.

GROUND FAULT

A **ground fault protection device (GFPD)** is required for PV systems that are mounted on the roof of a dwelling unit. The purpose of a GFPD is to avoid fires and damage should large amounts of stray current find its way onto the system (through a lightning strike, for example).

GROUND FAULT PROTECTION DEVICE (GFPD)

All inverters listed under UL 1741 incorporate ground fault protection. This GFPD device is designed to open the circuit (shut everything off) if excessive current flows through the grounded conductor.

Chapter 8: Wiring the System

According to the NEC, a GFPD must:
- detect ground faults in PV arrays
- interrupt the fault current
- indicate that a ground fault has occurred
- disconnect the faulted part of the array

ARC FAULT DETECTION DEVICE

In addition to ground fault protection, the NEC also requires that PV systems operating at 80 Vdc or more must also incorporate an **arc fault detection device (AFDD).** This will be contained within the UL1741 listed inverter as well.

When connections become corroded or are poorly installed, an arc fault may occur, resulting in a high power discharge of electricity (an arc) between the conductors. This can result in fire or damage to the system. Arc fault detection devices monitor the system for signs of arcing and automatically shut down the inverter if a fault is detected.

Chapter 8 Review Questions

1. In order to avoid fault and possible fires, connectors on all the solar panels in an array must be:
   A) MC4 connectors.
   B) the identical type and brand or listed as intermatable.
   C) factory terminated.
   D) UL 1703 listed.

2. When extending the source circuit from the solar panel to the junction box, an installer must normally create a:
   A) PV jumper.
   B) bonded junction connection.
   C) pigtail.
   D) field splice.

3. The wires that connect the solar panel to the junction/combiner box is known as the:
   A) PV combiner circuit.
   B) PV source circuit.
   C) PV input circuit.
   D) PV output circuit.

4. The wires that connect optimizers together in a system that uses optimizers and a fixed voltage inverter (SolarEdge, for example) is known as the:
   A) PV combiner circuit.
   B) DC-to-DC converter circuit.
   C) Inverter input circuit.
   D) Inverter output circuit.

5. Which of the following is considered the maximum CURRENT a panel can produce under standard test conditions?
   A) Open Circuit Voltage (Voc)
   B) Maximum Power Voltage (Vmp)
   C) Short Circuit Current (Isc)
   D) Maximum Power Currents (Imp)

6. An array of 24 panels, each with an Isc of 8.35 amps, has been configured into 4 strings of 6 panels each. The wires from these strings are combined in a combiner box. What is the minimum ampacity rating required for the wire used in the PV Output Circuit?
   A) 13.05 amps
   B) 33.4 amps
   C) 41.75 amps
   D) 52.19 amps

7. For the same array as described in the previous question (#6), assume the PV source circuit wiring has been placed in conduit that is exposed to sunlight and is mounted directly to the deck of the roof. These conduits run to a combiner box mounted so that it is never exposed to sunlight. What is the minimum ampacity rating of each circuit breaker that is connected to each PV source circuit within the combiner box?
   A) 10 amps
   B) 15 amps
   C) 35 amps
   D) 60 amps

8. If the wire from a PV array is placed in a conduit and exposed to sunlight, the ampacity rating required for that wire must:
   A) be adjusted by a 1.25 correction factor.
   B) increase.
   C) be adjusted if there are less than 4 conductors in the conduit.
   D) not exceed 23 amps.

9. All DC circuits that exceed 30 volts or 8 amperes and are run within buildings must:
   A) be rated for 90° C (high temperature wire).
   B) not exceed 50 Vdc.
   C) be contained within metal conduit.
   D) be labeled as "DC Circuit" every 10 feet and within 1 foot of each wall penetration.

10. EMT conduit stands for:
    A) electrical metallic tubing.
    B) exterior metallic tubing.
    C) environmentally maintained tubing.
    D) electro-mechanical tubing.

11. According to the NEC, when wire that is enclosed in PVC pipe is buried in a trench, it must be located a minimum of:
    A) 24 inches (610 mm) deep.
    B) 18 inches (460 mm) deep.
    C) 12 inches (305 mm) deep.
    D) 6 inches (150 mm) deep.

12. When bending EMT conduit:
    A) bends cannot extend beyond 90 degrees.
    B) subtract the take-up from the stub height.
    C) no more than 360 degrees of bends are allowed in a single circuit.
    D) ensure the bend is properly crimped.

13. EMT conduit must be supported:
    A) every 3 feet, except for buried conduit runs.
    B) every 10 feet and within 3 feet of each box, fitting or cabinet.
    C) every 4 ½ feet and within 12 inches on each side of each box, fitting or cabinet.
    D) at a distance not to exceed the take-up subtracted from the stub height.

14. A _____ is used to pull conductors through a conduit.
    A) fish tape
    B) conduit body
    C) cleat
    D) PFAS

15. The NEC requires that any DC circuit with current greater than 30 A incorporate a DC disconnect to isolate the equipment attached to that circuit. The disconnection device must be:
    A) located within 10 ft (3 m) of the equipment it is protecting.
    B) readily accessible.
    C) clearly labeled.
    D) all of the above.

16. An array of 24 panels, each with an Isc of 8.35 amps, has been configured into 4 strings of 6 panels each. The wires from these strings are combined in a combiner box. The PV output circuit connects the combiner box to the DC disconnect which is attached to the 9 kW single-phase 240V grid-tied inverter.

    What is the minimum ampacity rating of the wire contained within the Inverter Output circuit?
    A) 37.5 amps
    B) 41.75 amps
    C) 46.875 amps
    D) 52.19 amps

17. When working with micro inverters, the wires connecting the inverters together are referred to as:
    A) the PV source circuit.
    B) the inverter input circuit.
    C) the combiner circuit.
    D) a branch circuit.

18. An AC Disconnect should comply with all of the following **EXCEPT**:
    A) isolate the PV system from the electric utility grid.
    B) be located at least 10 feet away from the point where the PV system connects with the grid.
    C) be externally operable.
    D) be clearly marked "Distributed Generation Disconnect Switch."

19. When the connection from the PV array is spliced to an existing conductor coming from the utility, this connection is known as a:
    A) tap.
    B) busbar connection.
    C) pigtail.
    D) feed-thru.

20. Which of the following is **NOT** a common and legitimate reason for opting to make a supply side connection rather than a load side connection?
    A) The array is too small for a load side connection.
    B) There is no room in the service panel for a load side connection.
    C) The utility requires it.
    D) The service panel is equipped with a main GFPE breaker that does not allow backfeed.

21. When determining the ampacity rating of the connection between the service entrance cable and the AC disconnect (located within 10 feet) in a **SUPPLY SIDE** connection, the designer must use the:
    A) ampacity of the service entrance circuit.
    B) maximum ampacity output of the inverter x a 1.25 safety margin.
    C) the ampacity rating of the AC disconnect.
    D) the 10-foot tap rule.

22. When an unintended pathway is created between a PV array and a grounded surface, this is known as:
    A) grounding and bonding.
    B) an arc fault.
    C) a ground fault.
    D) short circuit current.

Lab Exercises : Chapter 8

8-1. Terminating MC4 Connectors

The purpose of this exercise is to become familiar with field terminations of MC4 connectors.

a. Take a short length of PV wire (either AWG #10 or AWG #12).
b. Strip about ¼ inch of insulation off both ends of the wire.
c. Crimp the male barrel contact to one end of the wire, and crimp the female barrel contact to the other end of the wire.
d. Slide the female MC4 connector over the male barrel contact until it clicks into place. Repeat on the other side with the male MC4 connector over the female barrel contact.
e. Using the two MC4 assembly wrenches, tighten the MC4 connectors into place.
f. Test the jumper with a multimeter, performing a continuity test to ensure the connections are good.

> **Lab Exercise 8-1:**
>
> *Supplies Required:*
> - Length of PV cable (either AWG #10 or AWG #12)
> - Male and female MC4 connectors with barrel contacts.
>
> *Tools Required:*
> - multimeter
> - MC4 field connector kit

8-2. Bending EMT Conduit

The purpose of this exercise is to practice manually bending EMT conduit using a one-shot conduit bender.

Create a 90 degree bend to a stub height of 15 inches.

a. Mark the conduit, subtracting the take-up length from the stub height.
b. Place the one-shot bender in the proper location on the mark.
c. Bend the conduit to 90 degrees.
d. Check the height to ensure the bend places the end of the conduit 15 inches from the floor.
e. Insert a 45 degree bend into the conduit.

> **Lab Exercise 8-2:**
>
> *Supplies Required:*
> - 1/2" EMT conduit
>
> *Tools Required:*
> - Measuring tape
> - 1/2" one-shot conduit bender

Chapter 9

Energy Storage Systems

Chapter Objectives:

- Understand lead-acid and lithium-ion batteries and the advantages and disadvantages of each.
- Explore how batteries cycle.
- Comprehend the effect of temperature on battery systems.
- Understand how batteries are rated.
- Examine how batteries are charged and discharged.
- Describe safety issues while working with battery systems.
- Determine storage requirements and locations of battery banks.
- Note several common PV battery systems on the market today.
- Determine how V2G will affect future PV installations.

Energy storage systems (ESS) integrated into a solar PV system have rapidly become the norm. From 2010 to 2020 the US PV industry saw the growth of **PV + storage** grow by an annual rate of 50% and prices decline by 88%.

In the third quarter of 2023, 11% of all residential storage systems in the U.S. were paired with storage.

Major Types of Storage Systems

There are many forms of energy storage that can be deployed in systems both large and small. But at the individual customer level, battery storage systems remain the sole practical option.

Batteries come in many shapes and sizes for a variety of applications (from watch batteries to car batteries and more).

Traditionally, PV systems that incorporated storage relied on **lead-acid deep cycle** batteries. But in recent years, **lithium-ion** batteries have come to dominate, capturing more than 90% of the global grid battery storage market.

Lead Acid Deep Cycle

Unlike a traditional car battery, deep-cycle lead-acid batteries are designed to provide a relatively constant stream of power for a long period of time. Car batteries, on the other hand, are designed to provide a short burst of power to start the vehicle's engine. After that, the alternator takes over and recharges the battery.

ENERGY STORAGE SYSTEMS (ESS)

PV + STORAGE

LEAD-ACID DEEP CYCLE

LITHIUM-ION

A car's battery can go its entire life without being drained more than 20% of its overall power. Traditional vehicle batteries are rated by their **cold cranking amps (CCA)** and their **reserve capacity (RC).**

Cold Cranking Amps (CCA)

Reserve Capacity

The CCA refers to how many amps a battery can produce for 30 seconds at 0°C. RC refers to the number of minutes a battery is able to produce 25 amps while maintaining a voltage rating of 10.5 volts or greater.

Deep cycle batteries are designed to discharge deeply (as much as 80% or more) often, without dramatically affecting the life of the battery or its performance. The reserve capacity of a deep cycle battery is typically two to three times that of a typical automotive battery. Conversely, a deep-cycle battery will only have about half as many cold cranking amps available.

The most common forms of lead-acid batteries include:
- **flooded (wet),** liquid electrolyte, either with removable caps or sealed,
- **gelled,** the electrolyte is suspended in a gelatin-like medium. Protects from spilling,
- **absorbed glass mat (AGM),** sometimes called "starved electrolyte" or dry batteries, because the fiberglass mat is saturated but there is no excess liquid present in the cell.

Flooded

Gelled

Absorbed Glass Mat (AGM)

SAFETY NOTE: It should be assumed that all lead-acid batteries (sealed or not) will vent off toxic fumes at some point during their life cycle. For this reason they should be stored in a well-ventilated location.

Lithium-Ion Batteries

Lithium-ion battery technology is not new. It was first developed in 1912, but not commercially available until the 1970s. Lithium is the lightest of all metals, having the greatest electrochemical potential and the largest energy density by weight.

Advantages of lithium-ion batteries over lead-acid batteries include:
- higher energy density by weight, so smaller/lighter units can provide more power.
- more power available, as it is practical to drain lithium-ion batteries to 85% depth of discharge or more without a significant impact on the life of the battery.
- extended life cycle. Tests show that after 2,000 cycles, a lithium-ion battery will deliver 75% of its original capacity, as compared to 500-1000 cycles for lead-acid batteries.
- more resistant to cold temperatures, operating well in temperatures as low as -20° C.
- essentially maintenance-free. Also spill proof, as they contain no liquid.

- low self-discharge. Shelf life (when unused) can be as long as 5-10 years. They discharge at a rate of 1-3% per month, compared to 4-6% per month for lead-acid batteries.
- faster charging. Lithium-ion batteries can charge at a rate about four-times that of lead-acid. This allows for quicker recovery when the battery bank is used as an emergency backup.
- better **roundtrip efficiency**. Roundtrip efficiency is the percentage of power that can be retrieved from a battery versus the amount of power that went into the battery during the charging cycle. With lead-acid batteries, the roundtrip efficiency is typically around 80% (meaning that for every 100 Wh that go into the battery, only about 80 Wh can be retrieved). A typical lithium-ion battery has a roundtrip efficiency of around 95%.
- the capacity is independent of discharge rate.

Disadvantages of lithium-ion batteries include:

- a higher initial cost.
- almost 70% of the global lithium deposits are concentrated in South America's ABC (Argentina, Bolivia and Chile) region. Most of the world's cobalt, another critical component in some lithium-ion batteries, is mined in the Congo. This may cause supply chain disruption as this technology becomes more popular.
- a lithium-ion battery can be permanently damaged if it is charged at low temperatures (below freezing). A lead-acid battery, however, can still accept a low current charge in cold weather.
- recycling lithium costs five-times as much as producing it initially. As a result, only a small number are currently recycled (as compared to 98% of all lead from lead-acid batteries that is currently recycled).
- when damaged, lithium-ion batteries can spontaneously combust.
- lithium-ion cells and batteries are typically not as robust as lead-acid batteries. They require protection to ensure the current is maintained within safe limits. As a result, they require protection circuitry incorporated to ensure they are kept within their safe operating limits.

Battery Cycles

When discussing batteries, a **cycle** refers to a single instance when the battery goes from fully charged, to maximum discharge (which will vary from cycle to cycle) and then is fully recharged again.

Car batteries, for instance, are designed simply to start the engine. They only discharge about 2-5% of their energy during each cycle. If they discharge more than this, the life of the battery will be shortened dramatically.

DEPTH OF DISCHARGE (DoD)

The amount of energy that is drained during the cycle is referred to as its **depth of discharge (DoD)**. If a battery bank is drained of 80% of its power before being recharged, that is considered an 80% DoD. If only half of the power is used before recharging, then the system is operating at a 50% DoD (and so on).

Deep Cycle Lead-Acid

The lower the average depth of discharge a lead acid battery bank experiences, the longer it will last. As indicated in Table 9-1, a specific battery that cycles at 20% DoD may last 3,200 cycles, but the same battery cycling at 70% DoD may only be expected to last for 1,450 cycles.

TABLE 9-1: AN EXAMPLE OF A DEEP-CYCLE BATTERY'S LIFE EXPECTANCY AT VARIOUS DOD SETTINGS

DOD (depth of discharge)	Number of cycles	Life Span
80%	1250	4.2 years
70%	1450	4.8 years
60%	1700	5.7 years
50%	2050	6.8 years
35%	2600	8.7 years
20%	3200	10.7 years

Most lead-acid batteries are designed to operate at a 50% DoD. However battery banks incorporated into grid interactive systems (DC-coupled and AC-coupled) are typically set to an 80% DoD, as they do not cycle as often as a stand alone system.

Most inverters designed to work with batteries allow the operator to determine the preferred depth of discharge. External low-voltage disconnects can also be incorporated into the system to control the depth of discharge of the battery bank.

SULFATION

Lead acid batteries should not be left discharged for more than 2 days as **sulfation** (a buildup of sulfur deposits within the battery's cells) will take place in a discharged battery. Also, the greater the DoD, the more risk of increased sulfation.

STATE OF CHARGE (SoC)

The **state of charge (SoC)** of a battery is another method used in indicating how much energy remains in a battery. It is the inverse of a battery's depth of discharge. For example, a battery drained to a 40% DoD is said to be at a 60% state of charge (0% SoC = empty; 100% SoC = full).

Lithium-Ion

Most modern lithium-ion batteries have DoDs ranging anywhere from 80% to 95%. Some even claim to discharge 100% with extremely long life cycles (2,000 to 10,000 cycles). As they do not contain sulfuric acid, sulfation is not a problem in lithium-ion batteries.

Effect of Temperature on Batteries

Deep Cycle Lead-Acid
Batteries are affected by temperature. Lead acid batteries will generally last longer when kept cool. For every 5.5°C (10°F) over 25°C (77°F) under which a lead-acid battery operates, the life of the battery can be cut in half. For this reason, lead-acid batteries should be stored in a cool or climate-controlled location. Ideal battery storage temperature conditions should be below 25°C (77°F) and above 0°C (32°F).

As temperatures fall, the energy capacity of the battery also falls. This is why a car may have trouble starting on a cold winter's morning. The rated charge capacity of a lead-acid battery is measured at 25°C (77°F). The ability of a battery to fully charge falls about 1% per degree the battery temperature is below 20°C (68°F). The capacity of the battery will fall by 50% when the temperature reaches -27°C (-22°F).

Deep cycle lead-acid batteries tend to undercharge at cold temperatures and overcharge at high temperatures at a rate of about +/- 1% per degree measured from 20°C (68°F). Overcharging or undercharging will decrease the life of the battery.

Lithium-Ion
Lithium battery manufacturers often state that the operational temperature of their batteries range from -20°C (-4°F) to 55°C (133°F) (varies depending on brand and model). This, however, refers only to the discharging of the battery - not the charging.

Charging a Lithium battery in temperatures below 0°C (32°F) should be avoided unless the battery is equipped with a battery management system designed to manage cold temperatures. Otherwise the battery may be damaged or it may reduce its lifespan.

While the lithium-ion battery may be discharged in lower temperatures, the chemical process is affected and the result will be a lower capacity.

Lithium-ion batteries should be ideally stored in cool, dry conditions at a temperature of 15°C (60°F).

How Batteries are Rated

Deep Cycle Lead-Acid
The capacity of deep cycle batteries is rated by nominal voltage (6 Vdc, or 12 Vdc, or 24 Vdc, etc) and in **amp-hours (Ah)**. An amp-hour suggests that the battery can deliver one amp of power (at its nominal voltage) for one hour (or 10 amps for six minutes, or ½ amp for two hours, and so on).

(AMP-HOUR (AH))

The capacity of the battery can be determined by multiplying the nominal voltage by the amp-hour capacity. The result will be a battery capacity rated in watt-hours (Wh). For example:

$$12 \text{ Vdc (nominal)} \times 35 \text{ Ah} = 420 \text{ Wh}$$

The watt-hour rating of a lead-acid battery may not translate into that amount of useable energy. Typically, the depth of discharge is not factored into the rating, nor is lower capacity the result of rapid discharge rates.

Because of the internal resistance of batteries, the faster the energy is discharged, the less capacity that is available within the battery.

TABLE 9-2: HOW C-RATING AFFECTS THE CAPACITY OF A SURRETTE DEEP CYCLE SOLAR SERIES 5000 BATTERY

Capacity	Cap/AH	Current (I)
Capacity at the 100 hour rate	2490	24.9
Capacity at the 72 hour rate	2349	32.62
Capacity at the 50 hour rate	2172	43.44
Capacity at the 24 hour rate	1837	76.5
Capacity at the 20 hour rate	1766	88.3
Capacity at the 15 hour rate	1642	109.5
Capacity at the 12 hour rate	1536	128
Capacity at the 10 hour rate	1466	146.6
Capacity at the 8 hour rate	1377	172.2
Capacity at the 6 hour rate	1254	209
Capacity at the 5 hour rate	1183	237
Capacity at the 4 hour rate	1095	274
Capacity at the 3 hour rate	989	330
Capacity at the 2 hour rate	848	424
Capacity at the 1 hour rate	600	500

For example, a battery, such as the one referenced in Table 9-2, that has a 2,490 Ah capacity when fully discharged over a 100 hour period of time, may only have 1,766 Ah available if the energy is used in only 20 hours – and only about 989 Ah of storage capacity if the energy is drained in just three hours.

A battery's **rated capacity** is generally set at the 20-hour level, so the storage capacity of one system can easily be compared to another.

RATED CAPACITY

RATE OF DISCHARGE

C-RATE

The **rate of discharge** of a battery is referred to as its **C-rate**.

A 1C discharge rate of current indicates that all the amp-hours within the battery are discharged in one hour. A 2C discharge rate would indicate the battery will be drained in 30 minutes. A 0.5C rate (also designated as C/2) would drain the battery in two hours. Generally manufacturers recommend that batteries not be discharged at a rate greater than 1C.

Lithium-Ion
The capacity of lithium-ion batteries designed for PV installations is typically listed based on how many kilowatt-hours (kWh) of electricity they can store, rather than amp-hours. In fact it is often difficult to determine the amp-hours and the voltage at which they operate.

There are two measurements to be aware of when selecting a lithium-ion battery:

- **nameplate capacity**. The maximum amount of energy the battery can hold.
- **usable capacity**. The maximum amount of energy the battery can discharge without exceeding the manufacturer's recommendation for depth of discharge.

NAMEPLATE CAPACITY

USABLE CHARGE

When sizing a system that incorporates lithium-ion batteries, the installer should focus on the usable capacity of the energy storage system.

Lithium-ion battery capacity can also be affected by discharge rate, but this is typically controlled within the battery management system.

Charging Batteries

Lead-Acid Batteries
Manufacturers recommend a **rate of charge** of 0.33C for lead acid batteries (from empty to full in about 3 hours), but they can be charged at a higher rate for a portion of the charging cycle without creating oxygen and water depletion.

RATE OF CHARGE

Tests indicate that a healthy lead-acid battery can be charged at a rate up to 1.5C as long as the current is reduced as it reaches about an 80% state-of-charge (SoC).

The charging process of a deep cycle lead-acid battery is generally controlled by a pulse width modulating (PWM) charge controller. These controllers charge in phases. When the batteries are low, the charge controller will divert all the energy from the solar array into the battery bank. If a load is using energy at the same time – that energy is powered from the battery bank. This phase of the battery charging cycle is referred to as **bulk charging** (or constant current).

BULK CHARGE

Once the battery bank reaches a certain state of charge (usually 80-90%), the controller moves into the stage known as **absorption** (or tapering) **charging**. During this phase, the voltage remains the same (constant voltage) but the current is gradually reduced until the battery bank reaches full charge.

ABSORPTION CHARGE

Once the battery has reached a full charge, the voltage is reduced and the system is in maintenance mode – or providing a **floating charge** to the battery bank. This is also sometimes referred to as a trickle or maintenance charge. The float voltage is typically from 13.2 Vdc to 13.8 Vdc at 25° C.

FLOATING CHARGE

The PWM charge controller regulates the charge of the battery bank by rapidly turning on and off (pulsing many times a second) the energy flowing from the array to the battery bank. During bulk charging, the full power is directed to the battery bank. But during the tapering and floating phase, the pulsing of the current effectively slows down the rate at which energy is flowing through the system.

Manufacturers like to promote that PWM charge controllers have additional benefits, including the ability to:

- recover lost battery capacity and **desulfate** a battery (where the battery is overcharged with an **equalization charge** to clean the battery of sulfur and to drive all the cells in a battery to their optimum voltage).
- dramatically increase the charge acceptance of the battery.
- maintain high average battery capacities (90% to 95%) compared to on-off regulated state-of-charge levels that are typically 55% to 60%.
- reduce battery heating and gassing.
- automatically adjust for battery aging.
- compensate for changes in battery temperature to avoid over (when too hot) or under charging (when too cold).
- self-regulate for voltage drops and temperature effects in solar systems.

Lithium-Ion

Lithium-ion batteries normally charge using a **constant current, constant voltage** process, as illustrated in Figure 9-1. In this process, the battery accepts full current until a voltage set point is reached. At this voltage the battery may only be charged to 80% of its capacity.

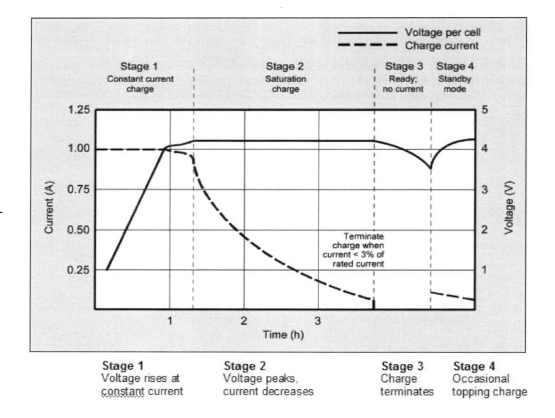

FIGURE 9-1: CONSTANT CURRENT, CONSTANT VOLTAGE PROCESS OF CHARGING LITHIUM-ION BATTERIES

For the remaining 20%, the voltage is maintained as the current decreases over time until the battery reaches 100% capacity. These two phases are similar to the bulk and absorption charging phases in a lead acid battery. However, lithium-ion batteries can be damaged if left charging over a long period of time, as is the case in float mode. It is also unnecessary to "trickle charge" lithium-ion batteries as they have very low rates of self-discharge.

Charging a lithium-ion battery too quickly may result in it aging prematurely. The accepted charging rate for lithium-ion batteries is between 0.5C (C/2) and C1, taking between one to two hours to fully charge the battery.

While waiting two hours to recharge a battery bank in your home may not be an issue, most people would consider it too long to recharge an electric vehicle during a long trip.

The constant current portion of the charging cycle can happen relatively quickly, getting the battery to an 80% state of charge. It is the remaining 20% (constant voltage) that takes the majority of the charging time.

It does not affect a lithium-ion battery to only partially charge the battery. For this reason, many medium-sized EVs can obtain an 80% state of charge using a Level 3 charging station (DC fast charging) in 17 to 52 min.

Self Discharge

When a battery is not in use, it can still lose some of its charge over time. This is known as **self discharge**, or sometimes referred to as standby loss or shelf loss. The rate at which power is lost differs depending on the battery type, and increases as the battery ages.

> SELF DISCHARGE

The rate of self-discharge for lead acid batteries depends on the temperature where the battery operates or is stored - the higher the temperature, the more loss. At a temperature of 27° C (80° F), a lead acid battery will self-discharge at a rate of approximately 4% per month.

So, a battery with a 250 amp-hour rating would self-discharge at a rate of approximately ten amp-hours per month. A lead acid battery being stored at 18° C (65° F) will only discharge at a rate of about 3% per month.

Lithium ion batteries self discharge at a rate about half that of lead acid batteries (only about 1-3% per month depending on temperature).

Battery Safety

Battery systems create a number of unique safety concerns.

Lead-acid batteries often release hydrogen gas, which can be explosive when confined and exposed to an ignition source. Flooded batteries release hydrogen continuously during charging, while VRLA batteries release hydrogen only when overheated and/or overcharged. A flooded battery will emit approximately 60 times more hydrogen than a similar sealed battery.

There are a number of safety and health hazards associated with batteries.

These include:
- electrical hazards,
- fire and explosion hazards,
- chemical hazards,
- other related hazards.

Care must be taken to protect those working on the battery system as well as those who may be in, or occupying, the facility where it is located.

Typical safety concerns imposed by batteries include:
- electrical shock from touching the exposed terminals of the batteries or conductors attached to them.
- short circuits that can cause a sudden release of the stored energy of the battery, resulting in an arc or explosion.

Battery Safety

Battery systems create a number of unique safety concerns. First, remember that these banks store DC electricity, and can be dangerous if touched by unprotected flesh or a metal object.

Warning: Subjecting a battery to abuse or conditions for which it was never designed can result in uncontrolled and dangerous failure of the battery. This may include explosion, fire and the emission of toxic fumes.

Some good safety practices when working with batteries include:
- ensure that the batteries are not damaged or leaking before working with the battery bank.
- do not expose a battery to open flame, smoking or electrical sparks. A battery can produce flammable hydrogen gas, and a small spark can cause an explosion.
- remove all metal jewelry and metallic clothing.
- use only insulated tools.
- wear safety goggles/face shield and gloves.
- battery acid in the eye should be flushed with clean water, then get professional help immediately. Acid on the skin should be rinsed with cool, clean water for at least 20 minutes.
- spilled liquid from damaged or dropped batteries should be treated with baking soda to neutralize the acid.
- batteries can be quite heavy, so use proper lifting techniques (lift from the legs) and wear steel-toed boots (in case a battery is accidentally dropped).

- overcharging, which can result in damage to the battery, leaking of acid, or in severe cases, explosion due to the excessive release of hydrogen gas.
- injury from lifting or moving the heavy batteries in an improper way.
- burns from exposure to spilled sulfuric acid.

Lithium-ion batteries also present a limited number of safety concerns. These include:
- overcharging. Charging voltages over 65 volts can cause damage to a 48-volt lithium-ion system
- temperature. Avoid charging batteries under extreme cold (below 0°C) or extreme heat.
- mechanical damage. Lithium-ion batteries can experience **thermal runaway** (burst into flames) if physically damaged. Protect batteries in proper enclosures and/or rooms.

> THERMAL RUNAWAY

Battery Standards

As batteries are integrated into more and more systems, battery safety becomes a growing concern. To address these various technologies and stationary battery safety, UL developed the UL 1973 standard in 2010. The equivalent European standard is *IEC 61427: Secondary cells and batteries for photovoltaic energy systems (PVES) – General requirements and methods of test.*

In 2015, UL released its standard UL 9540 for energy storage systems. An energy storage system (ESS) certified to UL 9540 is comprised of a UL 1973-certified stationary battery pack used in conjunction with a UL 1741-certified inverter. Achieving certification to both individual UL product standards is required to obtain the UL 9540 system certification. This certification applies to PV AC-coupled and DC-coupled systems.

The NFPA 855, *Standard for the Installation of Stationary Energy Storage Systems* was first released in 2020. This installation code addresses safety issues inherent in ESS systems. It includes specific criteria for facility ventilation, signage, fire protection systems, and emergency operations protocols.

The NFPA 1 is the US code that addresses fire and life safety issues for the public and for first responders. The 2021 revision of NFPA 1 includes requirements in Chapter 52 extracted from *NFPA 855, Standard for the Installation of Stationary Energy Storage Systems*.

Battery Bank Location

Care should be taken in determining the location requirements for energy storage systems. This is not something that can be treated as an afterthought, but should be considered during the initial site inspection if batteries are to be incorporated into the system.

Batteries pose a considerable safety concern, and their safe storage and maintenance is addressed in rules developed by OSHA (Occupational Safety and Health Administration).

Two primary fire codes (International Fire Code (IFC) and NFPA 1: Fire Code) define the appropriate construction and supporting infrastructure that must be provided for storage battery rooms, such as shown in Figure 9-2.

These provisions include requirements for:

FIGURE 9-2: SECURE WELL LIT, WELL VENTILATED BATTERY STORAGE ROOM.

- ventilation,
- access,
- signage,
- smoke detection,
- spill control, drains and washing stations,
- wall fire rating,
- and more.

NFPA 855 further outlines that storage systems in a residential setting can only be installed in:

- attached garages separated from the dwelling unit (maximum size allowed in this location is 80 kWh)
- detached garages or other non-occupied buildings (maximum 80 kWh)
- outdoors, either attached to the exterior wall or mounted on the ground, located at least 3 feet (1 m) from any doors or windows (maximum 80 kWh)
- or in an enclosed utility closet or storage room (maximum size allowed within the building is 40 kWh).

Each ESS unit can be up to 20 kWh in size, and adjacent units must maintain at least 3 feet (1 m) of separation to mitigate the risk of fire propagation between units.

Many vendors offer secure and vented cabinets designed to store and protect the system, as illustrated in Figure 9-3.

FIGURE 9-3: VENTED BATTERY STORAGE UNIT FOR SMALLER PV SYSTEM

(FROM THE GREEN POWER COMPANY)

These enclosures typically offer features such as:

- venting (as hydrogen gas is lighter than air, venting should be at or near the top of the enclosure),
- cable management,
- access restriction (locks),
- spill containment,

- protection from environment (moisture, heat, dropped tools, etc).

AC Coupled and High Voltage Battery Systems

The announcement of the **Tesla Powerwall** battery system "re-energized" discussions regarding the use of lithium-ion batteries within PV systems and how they might be incorporated into AC-coupled systems.

> TESLA POWERWALL

Historically, the NEC has limited the voltage of battery systems within buildings to less than 50 Vdc (basically 48 V nominal systems). In the 2017 NEC a new section, Article 706, was added that allows for and addresses permanently installed energy storage systems (ESS) with higher voltage battery systems.

Manufacturers quickly responded with **high voltage battery systems (HV)** that operate at voltages as high as 1,000 Vdc.

> HIGH VOLTAGE BATTERY SYSTEMS

The advantage of the high-volt storage systems is that they do not impose the same level of **conversion losses** incurred by reducing higher voltages from the array to lower battery voltages, and then converting them to higher grid voltage levels. Additionally, these high voltage storage systems can be used with less complex (and therefore less expensive) inverters.

> CONVERSION LOSSES

The NEC (706.20 (B)) restricts the voltage of an ESS within dwelling units to 100 volts. But systems can exceed this limit (up to the 600 V limit imposed on residential units) if the live parts of the system are not accessible during routine maintenance.

These higher voltage systems are generally **self-contained energy storage systems (ESS)**, which are assembled, installed, and packaged as a single unit. Typically, this type of ESS is manufactured and sold as a single product and listed to the UL 9540 standard.

> SELF-CONTAINED ENERGY STORAGE SYSTEM (ESS)

All the components (such as the batteries, inverter, charge controller, and overcurrent protection) of a self-contained system are located within a single box and only accessible by an authorized technician (not the homeowner or untrained installer).

Another acceptable option for exceeding the 100 Vdc limit for an ESS is to use pre-engineered matching and listed components that are connected in a "plug-and-play" fashion (designed to be compatible with each other by the manufacturers). These components can be from a single manufacturer, or from multiple manufacturers but designed to work as a complete system.

FIGURE 9-4: TESLA POWERWALL 2

(*FROM TESLA*)

Examples of popular AC-coupled battery storage systems on the market today include:

The Tesla Powerwall 2

The Tesla Powerwall 2 (Figure 9-4) is an example of a self-contained system, with all the components (battery, controller, and an inverter) built in. The battery has 13.5 kWh of usable capacity and has a maximum continuous power output of 5 kW. The lithium ion battery also features a 90% roundtrip efficiency.

It offers a 10-year warranty and can be expanded (scaled) up to nine (9) units. It also incorporates its own battery-based inverter into the unit. It weighs 114 kg (251 lbs).

The Tesla Powerwall 2 is one of the most popular residential AC-coupled solutions on the market. As the unit is self-contained with an inverter, it generates only AC power (no DC version is available).

The retail price of the unit ranges from $8,000 - $9,000 (2023 U.S. prices). The Powerwall 2 is only available through Tesla's network of certified installers.

LG Chem RESU Battery

Many installers match the LG Chem RESU 10H battery with a battery-based inverter from another manufacturer to create an AC-coupled system. This solution (Figure 9-5) is an example of the option where installers combine matching and listed components from multiple manufacturers that are designed to work together.

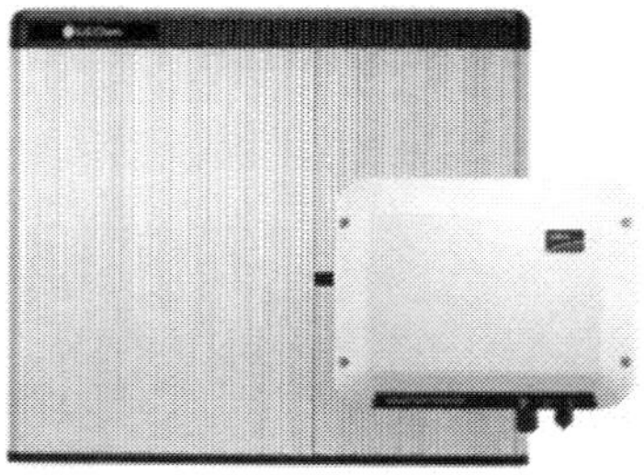

FIGURE 9-5: LG CHEM BATTERY STORAGE UNIT MATCHED WITH A SMA INVERTER

(FROM SMA AMERICA)

The RESU 10H has a maximum power rating of 5.0 kW to go along with 9.6 kWh of usable capacity. The lithium ion battery also features a 94.5% roundtrip efficiency and can be utilized to 95% depth of discharge (DoD). It weighs about 111 kg (245 lbs).

The battery alone (no inverter, charger or transfer switch) retails for between $6,000 - $8,000 (2023 U.S. prices).

A number of inverter manufacturers have created battery-based inverter/charges designed to integrate with the LG Chem RESU battery bank.

Enphase AC Battery

In 2020, Enphase began selling an AC-coupled solution that integrates with their micro inverters, called Encharge (Figure 9-6). It comes in two sizes, the Encharge 10 provides a total usable energy capacity of 10.08 kWh with twelve embedded micro inverters that combine for a 3.84 kW power rating. The smaller Encharge 3 has a total usable energy capacity of 3.36 kWh and includes four embedded micro inverters with a 1.28 kW power rating.

These self-contained units also include a battery-based inverter (which converts the AC power from the existing micro inverters to DC power for the battery and back to AC) as well as a lithium ion battery pack. Enphase claims that their battery functions at a 100% depth of discharge (DoD) and has an 89% roundtrip efficiency.

Figure 9-6: Enphase AC Battery system

(From Enphase)

The complete system integrates with their AC combiner box (for wire management and communications) as well as their "smart switch" which functions as a transfer switch to disconnect from the grid to avoid islanding.

In 2023 the Encharge 10 cost between $6,000 - $7,000 with an additional $2,500 for the transfer switch (plus installation).

Electric Vehicle Bi-Directional Charging

If a residential homeowner is going to spend tens of thousands of dollars for a energy storage system - then why not drive it?

Electric vehicles contain very large batteries. The 2022 Ford-150 Lightning standard range battery comes with 98 kWh of usable capacity. Ford offers a (rather pricey) proprietary bi-directional charging station that allows owners to use the F-150 Lightning as a battery backup for the home.

Bi-directional charging is technically possible and as Ford demonstrates, currently available. However, uniform standards amongst all the various EV manufacturers as well as warranty issues for the vehicles may delay the dream of using the EV as an energy storage system for the home or business.

In 2023, Tesla announced deals that Ford and General Motors that will result in two of the nation's largest automakers relying on Tesla's charging technology for their electric vehicles, pushing the U.S. auto industry toward a de-facto standard.

Vehicle to Grid (V2G)

Vehicle-to-grid (V2G) is a system in which plug-in electric vehicles can communicate with the power grid. The vehicle can either return electricity to the grid (selling power during periods of peak load demand) or throttle their charging rate during periods of high energy cost.

(VEHICLE-TO-GRID (V2G))

V2G storage capabilities can also enable EVs to store and discharge power from solar arrays, serving as the battery backup for the home or business. Known as **Vehicle-to-home (V2H)**, this system uses the battery bank in the electric (EV) or plug-in hybrid (PHV) vehicle as the battery bank for a grid interactive system.

(VEHICLE-TO-HOME (V2H))

It is important that these systems do not drain the battery of the vehicle to a point where it cannot be driven when needed. So, these system typically limit discharge so the battery will always be charged to 70-90% of its capacity.

Not all EVs and PHVs are compatible with these systems. Nissan has been an early adopter of this technology. As of 2018, all Nissan Leafs on the market can be discharged with vehicle-to-grid stations.

Electric Vehicle Charging Stations

Recharging an electric vehicle (EV) is accomplished by connecting to **electric vehicle supply equipment (EVSE)**. These charging stations can take a number of forms.

Level 1, 120-Volt Charging

This option is the simplest form of charging and requires nothing but an ordinary 120 Vac outlet on a 20 amp circuit. These units, similar to that pictured in Figure 9-7, are supplied by the vehicle's manufacturer. **Level 1 charging** stations normally draw about 1.4 kW when charging.

These systems are typically located in homes or employee parking lots and take 6-10 hours to fully charge a vehicle.

Advantages
- Low installation cost
- Low impact on utility peak demand charges

Disadvantages
- Charging is slow - around 3-5 miles (5-8 km) of range added per hour of charging

Level 2, 208/240-Volt Charging

Level 2 charging is much faster than Level 1 charging, but requires a 208 Vac three-phase or 240 Vac single-phase power connection.

Level 2 charging stations, such as Figure 9-8 can theoretically provide up to 80 amps of current to EVs, although most operate only at about 30 amps. The rating of the unit will determine the ampacity rating of the circuit to which it is connected.

These systems are typically located in homes, shopping centers, or commuter parking lots and take 1-3 hours to fully charge a vehicle.

Advantages:
- Charge time is significantly faster than Level 1. EVs will get between 10-20 miles (16-32 km) of range per hour of charge.
- More energy efficient than Level 1.

Disadvantages:
- Installation costs are much higher than Level 1. They may potentially impact utility peak demand charges by creating a significant spike in demand.

DC Fast-Charging

DC fast-charging (sometimes called Level 3) equipment delivers high power directly into an EV's battery system, bypassing the need for a rectifier (that converts AC power from the grid to DC power that can be used in batteries). Charging speeds can be very fast, normally about an 80% charge within 30 minutes.

These systems, as shown in Figure 9-9, are typically located in designated charging facilities and take about 30 minutes to fully charge a vehicle.

FIGURE 9-9: DC FAST-CHARGING EV CHARGING STATION

Chapter 9 Review Questions

1. Which battery technology has come to dominate world markets, capturing about 90% of the market share in recent years?
   A) deep cycle lead acid
   B) AC coupled
   C) DC coupled
   D) lithium-ion

2. Traditional car batteries are rated in:
   A) amp-hours.
   B) cold cranking amps.
   C) roundtrip efficiency.
   D) maximum rate of discharge.

3. Deep cycle lead acid batteries are rated in:
   A) amp-hours.
   B) cold cranking amps.
   C) roundtrip efficiency.
   D) maximum rate of discharge.

4. Which of the following is **NOT** a common type of deep cycle lead acid battery?
   A) sulfated
   B) gelled
   C) flooded
   D) absorbed glass mat

5. Which of the following is **NOT** an advantage in selecting a lithium-ion battery over a deep cycle lead acid battery?
   A) higher energy density by weight
   B) easier to recycle
   C) more resistant to cold temperatures
   D) extended life cycle

6. Which of the following battery types have been known to spontaneously combust when a cell is damaged?
   A) deep cycle lead acid
   B) lithium-ion
   C) absorbed glass mat
   D) nickel cadmium

7. If a deep cycle battery is at a 60% DoD:
   A) it has 40% of its charge remaining.
   B) it has 60% of its charge remaining.
   C) it has 40% of its lifetime cycles remaining.
   D) it has 60% of its lifetime cycles remaining.

8. Typically, lithium-ion batteries can operate at a DoD of:
   A) between 80% - 95%.
   B) 10 years.
   C) 15°C (60°F).
   D) 2,000 to 10,000 cycles.

9. In a lead-acid battery, the battery's charge capacity:
   A) will decrease by about 1% per degree C below 20°C.
   B) is unaffected by temperature.
   C) will decrease by about 2% per degree C above 25°C.
   D) will undercharge in extremely high temperatures.

10. Charging a lithium-ion battery at temperatures below 0°C:
    A) will result in overcharging the battery without a temperature sensor installed between the battery and the inverter.
    B) will have no effect, as lithium-ion batteries are unaffected by temperature.
    C) can damage the battery.
    D) will result in undercharging the battery without a temperature sensor installed between the battery and the inverter.

11. The capacity of a battery is generally measured in:
    A) Ah at C/20.
    B) number of cycles at DoD.
    C) nominal voltage.
    D) peak charge at MPPT.

12. A battery C-rating of C/20 means:
    A) at that rate of discharge a battery will be fully drained in 20 minutes.
    B) at that rate of charging, it will take 20 hours to fully charge a battery bank.
    C) at that rate of discharge, it will take 20 hours to fully drain the battery.
    D) at the current rate of discharge, optimum capacity will be reached at a 20% DoD.

13. A battery bank with a capacity of 600 Ah will:
    A) provide 20 amps of power for 600 hours.
    B) provide 20 amps of power for 15 hours to a 50% DoD.
    C) provide 10 amps of power for 15 hours to a 20% DoD.
    D) provide 600 continuous amps over a 24 hour period.

14. The maximum amount of energy a lithium-ion battery can discharge without exceeding the manufacturer's recommendation for depth of discharge is referred to as the battery's:
    A) C-rate capacity.
    B) float capacity.
    C) nameplate capacity.
    D) usable capacity.

15. If a battery is charged at a constant rate from empty to full over a four-hour period, it has a:
    A) rate of charge of 0.4C.
    B) rate of charge of 4C.
    C) rate of charge of 0.04C.
    D) rate of charge of 0.25C.

16. Lithium-ion batteries generally charge using:
    A) a constant current, constant voltage process.
    B) pulse width modulation.
    C) an MPPT charge controller.
    D) an equalization charge.

17. When a battery is not in use it will generally lose some of its charge over time. This is known as:
    A) equalization.
    B) sulfation.
    C) self discharge.
    D) absorption.

18. Which of the following is generally NOT a safety concern when working with lead acid batteries?
    A) lifting injuries
    B) leaking acid
    C) thermal runaway
    D) excessive electrical discharge due to short circuit

19. Which of the following is generally NOT a safety concern when working with lithium-ion batteries?
    A) lifting injuries
    B) leaking acid
    C) thermal runaway
    D) excessive electrical discharge due to short circuit

20. In the US, AC-coupled and DC-coupled systems must comply with which manufacturing energy storage system standard?
    A) ANSI 1703
    B) UL 9540
    C) NFPA 70
    D) NEMA 3R

21. In the US, the standard that outlines where energy storage systems may be placed in a building is:
    A) UL 9540
    B) UL 1703
    C) UL 1741
    D) NFPA 855

22. Fire codes only allow lithium-ion storage systems greater than 40 kWh but less than 80 kWh to be placed:
    A) in attached garages separated from the dwelling unit.
    B) in detached garages or other non-occupied buildings.
    C) outdoors, either attached to the exterior wall or mounted on the ground, located at least 3 feet (1 m) from any doors or windows.
    D) all of the above.

23. Rules regarding the safe storage and maintenance of batteries are developed by:
    A) OSHA
    B) NEMA
    C) ANSI
    D) IEEE

24. In 2017 the NEC added Article 706 that addresses rules for permanently installed Energy Storage Systems. This applies to ESSs operating at more than 50 volts AC or 60 volts DC. What is the MAXIMUM voltage limit for a system installed in a single-family home where live parts ARE accessible during routine maintenance?
    A) 50 volts AC or 60 volts DC
    B) 100 volts
    C) 600 volts
    D) voltage is unlimited

25. What is the MAXIMUM voltage limit for an ESS installed in a single-family home where live parts ARE NOT accessible during routine maintenance?
    A) 50 volts AC or 60 volts DC
    B) 100 volts
    C) 600 volts
    D) voltage is unlimited

26. Which of the following will NOT decrease the life expectancy of a battery bank in a PV system?
    A) increase the depth of discharge.
    B) increase the heat where they are stored.
    C) increase the number of cycles.
    D) increase the system voltage.

27. When a plug-in electric vehicle can communicate with the power grid and provide power from the EV to the utility is known as:
    A) level 1 charging.
    B) level 2 charging.
    C) level 3 charging.
    D) V2G.

28. DC fast-charging is also known as:
    A) level 1 charging.
    B) level 2 charging.
    C) level 3 charging.
    D) V2G.

Lab Exercises : Chapter 9

9-1. Compare Tesla Batteries to Available Lead-Acid Batteries

Lab Exercise 9-1:

Supplies Required:
- none

Tools Required:
- Calculator
- Access to the Internet

The purpose of this exercise is to determine the lifetime cost of incorporating a Tesla battery system into a home and compare that with currently available lead-acid battery technologies.

a. Convert the kWh capacity of Tesla's Powerwall into amp-hour capacity (you will need to determine the voltage at which the system operates).

b. Research available deep-cycle lead-acid battery systems with a similar amp-hour rating.

c. Determine how many of these batteries are required to create a 48-volt battery bank. How do the costs compare?

d. Assume the lead-acid batteries only have a 5-year life expectancy at a 60% depth of discharge. Is the Tesla system price competitive?

Chapter 10

Mounting Systems

Chapter Objectives:

- Examine the various PV mounting options and how to determine the best option for a given installation.
- Design a rooftop mounting system.
- Explore loading issues, wind exposure and occupancy categories, roof zones and setbacks as factors that affect the design of a rooftop racking system.
- Note the proper way to attach a PV panel to a racking system.
- Review various rooftop system racking configurations.
- Design a ground mounted system.
- Explore the various pier options within ground mounted systems.
- Understand how soil conditions, site access, setbacks and conductor accessibility affects the design of a ground mounted system.

At its most basic, a mounting system is a metal assembly that secures the solar panels together and then secures the array to a structure (or the ground). There are dozens of commercial racking systems available on the market.

The brand and style selected will be determined by the site, budget, and preferences of the designer and/or owner. Typically the system will come with very detailed installation instructions and each system will present its own unique installation challenges.

Racking System

A great advantage of using a complete packaged racking system (rather than designing one) is that the systems have already been completely engineered to account for live and dead load weight issues, structural integrity, environmental conditions, aesthetics and maintenance issues. Using a "pre-engineered" system also may be critical in obtaining a permit in some jurisdictions.

Broadly speaking, racking systems will fit into one of four broad categories:
- roof mounted,
- ground mounted,
- pole mounted,
- or building integrated.

In most cases, the racking system selected will be made of metal (a conductive material). The aluminum frames on the solar panels, as well as the racking, can act as a lightning rod (especially when placed on a rooftop). Other stray and unwanted random power can also find its way onto these conducting materials (short circuits, down electric lines, etc). For this reason, all metallic parts of the system must be bonded together and connected to the grounding system to provide any stray current a safe pathway to ground.

Roof Mounted Systems

The most popular roof mounted system for residential units is the **flush-to-roof** racking system (also called **standoff mounting**). Most of these systems are designed to work on sloped roofs with asphalt shingle (the most widely used system in the U.S.), but specialized systems are available to mount on metal, slate and even tile roofs.

At many installation sites a rooftop may be the only location on the site with the available space and sunlight for a PV system. However, there are concerns that must be addressed before deciding to place the array on the roof.

- Check the condition of the roof to ensure it is structurally able to support the added weight imposed by adding an array. If unsure, consult with a structural engineer.

- Many systems cannot be installed in areas where high winds may result in wind loading that exceeds a predetermined psf. Check with local officials to determine if this is the case.

- Regions with extremely heavy snow may not be suitable for these types of installations. Not only do they increase the load on the roof, but heavy snows may be difficult to clear from the panels on a roof and adversely impact the operation of the system.

- In instances where the roof structure is suspect, local authorities may require a stamped **engineered drawing** to attest to the condition of the structure.

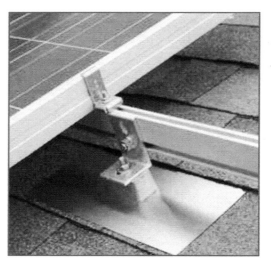

FIGURE 10-1: FLUSH-TO-ROOF MOUNTED FOOTING WITH FLASHING INCORPORATED INTO THE DESIGN

(FROM QUICK MOUNT PV)

Roof mounted systems incorporate **footings** that attach to the roof and hold the array parallel to the surface at a distance of between 2 to 10 inches. Great care must be taken when attaching these footings to ensure that the penetrations holding them in place are waterproofed properly to avoid future roof leaks.

Many systems incorporate flashing (such as the system illustrated in Figure 10-1) to help avoid leaks.

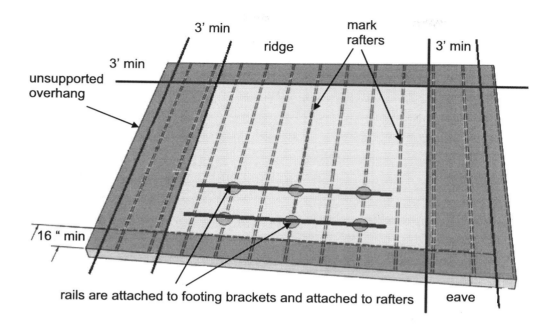

FIGURE 10-2: FLUSH-TO-ROOF MOUNTING LAYOUT ON ROOF

The footing mounts must be attached to the support rafters (shown in Figure 10-2), not simply to the plywood decking of the roof. It is therefore critical to mark the location of the rafters. It may help to snap a chalk line on top of the shingles to indicate where the rafters are located during the installation process. If it is impossible to hit a rafter with each footing mount, it is common to attach a 2 x 4 or 2 x 6 brace between the two rafters, then attach the mount to the brace. Metal rails are then attached to the footing units.

The lag bolts that secure the footing mounts to the roof rafters must be able to hold up to the forces exerted by the dead load of the array (usually 3-5 lbs per sq ft) as well as the live loads (which might be much greater, as much as 25-50 lbs per sq ft).

Calculating Space Required for the Array

Just how many panels can fit on any given roof will, in large measure, depend on how unobstructed (with dormers, vent pipes, chimneys, access hatches, and skylights) the surface is.

Assuming a flat or pitched roof with no obstructions, a good rough rule of thumb for systems that use crystalline panels is about 10 watts per square foot. So if the south-facing slope of a roof measures 30 x 20 ft = 600 sq ft (10 x 6 m = 60 sq m), a rough estimate would be that a 6,000 watt system could be installed within this space. Systems that use thin film panels require about 30% more space.

To get a closer estimate of the space required for a system, start with the fact that under standard test conditions, a panel's rating assumes that the irradiance of the sun will be 1,000 W/m^2.

So the area required for any given array can be calculated as:

Total Area Required = Array Power Output / (1000 W/m² x Panel Efficiency)

For example, if the output required from the array has been determined to be 6.8 kW, and the panel selected has a conversion efficiency of 18%, then:

Total Area Required = 6,800 W / (1000 W/m² x 0.18) = 37.78 m²

The array would require 37.78 m² (or about 407 sq ft). Space would also have to be allocated on the roof for required minimum setbacks as well as access paths on flat roofs.

But knowing the area required for the array may not indicate just how many panels will fit on a specific site. To calculate this, it will be necessary to know:
- the size of the portion of the roof where the array will be sited.
- setback requirements for the location.
- the size of the panel selected. Do not forget to include the space required for mounting hardware (usually about ½ inch (13 mm) between panels). This can have a serious accumulated impact on larger arrays.
- the number of panels in the array.
- the angle of the array for ground mounted or flat-roof systems (although the angle may already have been factored in calculating the array size while determining distances required to avoid inter-row shading).
- orientation of the panels, either portrait or landscape. This will largely be determined by the racking system selected.

For example: assume a 5.8 kW system is required to meet a household's needs. A 290 W panel was selected (20 panels total, 5,800 W / 290 W = 20), and the array is configured into two strings of 10 panels. A review of the panel's specifications show that it measures 65.43 x 38.98 inches (1,662 x 990 mm).

When doing the site visit, the south-facing portion of the roof was measured and found to be 21 x 35 ft (6.4 x 10.7 m), with no obstructions (pipes, vents, chimneys, etc).

To calculate the area available to mount an array, subtract the required clear perimeter aisles from the overall dimensions of the roof. The International Fire Code requires three feet (.91 m) of clear access area on the sides and ridge of the roof. Good wind loading design requires a minimum of 10 inches (254 mm) at the eave.

So in this example, a roof space measuring 21 feet (252 inches) by 35 feet (420 inches) can only accommodate an array measuring:

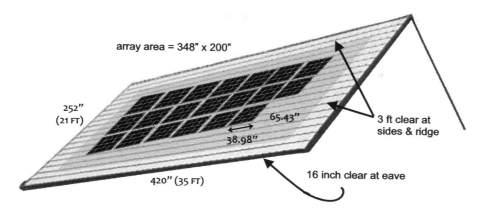

FIGURE 10-3: PLACING 20 PANELS ON A 35 FT X 21 FT ROOF AREA

420 inches width - 72 inches (36 inches x 2 sides) = 348 inches width
252 inches height - 46 inches (36 inches + 10 inches) = 206 inches height

The selected panel is 65.43 inches in height and 38.98 inches in width. The orientation will be portrait, with ½ inch mounting connectors attached between each panel (adding to the width of the array).

So the maximum number of panels in each direction equals:

348" (array area width) / 38.98" (panel width) + .50" (connector) = 8.81 panels
206" (array area height) / 65.43" (panel height) = 3.14 panels

The logical choice might be to try to arrange the panels in two rows of 10 panels (as there are to be 20 in the array). However only eight (8) panels can fit horizontally. Fortunately three rows will fit (barely), so the array could be arranged as illustrated in Figure 10-3, in two rows of seven panels, and one row of six (20 in total).

Note that the physical layout does not have to match (nor rarely does it) the logical string calculations, in this case two strings of 10 panels.

Loading Issues

When placing an array on top of a roof, it is important that the structure can support both the dead weight (simply the physical weight of the panels and the racking system), as well as the live weight (which includes snow loads, wind loads, and in some cases, seismic loads).

Dead Loads

Most roofs built or renovated since 1970 have had to comply with building codes that are more than adequate to support the **dead load** weight of a solar array.

DEAD LOAD

Local building codes will provide guidance as to the amount of weight allowed on roofs within that jurisdiction. In most locations the local

[INTERNATIONAL BUILDING CODE (IBC)]

authorities will follow the recommendations found within the **International Building Code (IBC)**.

Most PV systems only add between 2.5 - 4 psf (pounds per square foot) of weight to a roof, about the same weight as a second layer of shingles. A structurally sound roof should have no problem supporting this weight. Many jurisdictions also impose a 45 psf weight limit at each attachment point. Most commercially available systems are designed to fall well below this limit.

Live Loads

[LIVE LOAD]

Any load on the roof that is not constant, is considered a **live load**. These live loads may be the result of seismic activity, but more commonly they are exerted by the force of wind moving against and over the array. Local building codes typically have sections that deal with wind loading, but the most comprehensive guide for estimating wind loads on structures is **Standard No. 7** of the American Society of Civil Engineers (ASCE).

[ASCE STANDARD #7]

Live loads include:

[WIND LOAD]

- **wind loads**. The amount of wind present will vary from place to place. Check with the local authorities (AHJ) to determine the wind load calculations they recommend. Remember that wind not only presses down on the structure, but also pulls up (like the wing of an airplane). So mounting hardware, such as lag screws must be sized in such a way as to keep the array in place under all conditions that may occur on the site.

[SNOW LOAD]

- **snow loads**. In most cases the amount of snow that will accumulate on the rooftop should have been calculated in the original building design. Depending on how the array is mounted, the array may affect where the snow accumulates (in drifts). Some additional calculations may be necessary. A structural engineer may be required if snow loading appears to be an issue.

[SEISMIC LOAD]

- **seismic loads**. Racking systems that are physically attached to the roof (with mounting brackets or lag screws) generally do not require additional seismic load consideration. Ballasted systems may require additional seismic load support. Again, check with the local AHJ for seismic requirements for specific jurisdictions.

Pull-Out and Shear Loads

The mounting hardware of a PV system on a roof will be subject to a number of stresses. These can be lateral, or **shear loads**. Or the stresses can be vertical, or **pull-out loads**.

[SHEAR LOADS]
[PULL-OUT LOADS]

Wind Loading

Wind loading on solar panels depends on three (3) basic factors: wind speed, the height of the panel above the roof, and the relative location of the panel on the roof.

When calculating pull-out loads, there are a number of free online calculators that will do the calculations, but they do require a bit of site-specific information, some of which may not be intuitive.

These include:
- wind exposure category,
- occupancy category, roof type (hip, gable, sloped or flat),
- three-second wind gust speed (many sources are available online to obtain these speeds for specific locations),
- roof zone,
- tributary area of the footer,
- dimensions of the roof,
- roof slope,
- height of array off the ground.

These variables for a specific installation will be required to calculate the **wind uplift force** (or **pull-up force**) for a given installation.

Wind Exposure Category
The amount of wind hitting a building is affected by the characteristics of the area immediately surrounding the structure, or its **surface roughness**. An open field or lake will be subject to much more wind than will a building in the middle of a forest.

The **wind exposure categories** that impact the placement of solar arrays include:
- Exposure B. Urban and suburban areas, wooded areas or other terrain with many closely spaced obstructions.
- Exposure C. Open terrain with scattered obstructions with heights less than 30 ft (9.15 m) located more than 1,500 ft (450 m) from the building site.
- Exposure D. Flat unobstructed areas exposed to wind flowing over open water, smooth grasslands for a distance of not less than 5,000 ft (1,500 m) from the building site.

Occupancy Category
Building codes often factor in the use or activity which takes place within a building when determining just how stringent the code must be for that structure. For example, a carport is not as critical as an emergency shelter, so need not be built to the same specifications.

Occupancy categories includes:
- Risk Category 1 - buildings that are a low hazard to human life in event of failure, such as agricultural facilities.
- Risk Category 2 - residential structures, as well as all other buildings that don't fit into categories 1, 3, or 4.

Chapter 10: Mounting Systems

- Risk Category 3 - buildings that are a significant hazard to human life, such as public assembly buildings, power-generating stations and water treatment facilities.
- Risk Category 4 - essential buildings, such as hospitals, and fire, rescue ambulance and police stations.

Roof Zone

Not all parts of the roof area are affected in the same way by wind flowing over it. The edges will experience greater pull-out loads than will the center of the roof, and the corners will experience even greater loads.

A typical roof is divided into zones, as indicated in Figure 10-4. If the PV system is designed with proper perimeter setbacks, then the array should fall within zone 1.

The wind exposure category, as well as the **roof zone,** will affect how much wind passes over a specific panel. Angle and orientation of the roof will also play a role.

Assume that all factors have been taken into account and it is found that at a particular site experiencing 50 mph winds, the array is subjected to 6.4 psf wind load. If the wind speeds double, then the force applied to the array will increase by a factor of four (the effect is squared). So, at 100 mph winds, the force on the array from the wind would be 25.4 psf. If increased to 150 mph, the wind load would increase 9-fold to 57.6 psf.

Reducing the amount of area of the array that each footer supports will reduce the amount of stress placed on the underlying rafter.

Tributary Area of Footing

The **tributary area** on a roof is the area supported by a particular rafter, or in this case, footer. It is calculated by finding the area halfway between the

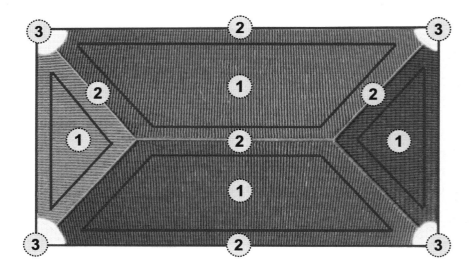

FIGURE 10-4: WIND LOADING ZONES ON A ROOF

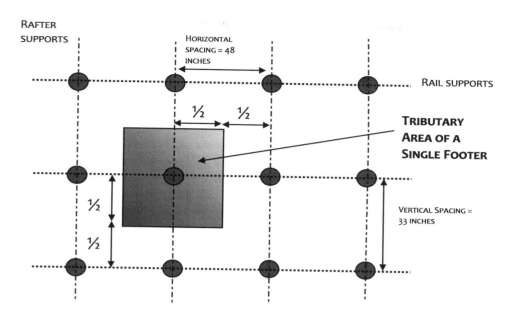

FIGURE 10-5: CALCULATING TRIBUTARY AREA OF A SINGLE FOOTER

supports, then calculating the square footage within that area, as shown in Figure 10-5.

For example, assume an array is to be placed on a roof with rafters placed on two-foot centers (spaced every two feet). Footers for the array are placed every 4 ft (1.2 m) or every other rafter horizontally, and the vertical rails are spaced every 33 inches (838 mm). Most solar panels are about 65 inches (1,650 mm) long, but this will vary depending on power output and manufacturer. Then:

Tributary Area = 4 feet (48 inches) x 2.75 feet (33 inches) = 11 sq ft or
Tributary Area = 1.2 m x .838 m = 1 sq meter

If the size of the array and the uplift force is known, then the pull-up force per footer can be calculated in the following manner:

Example:
- 9.6 kW system
- 2 x 4 pine truss roof on 24" (609 mm) centers with composite shingle roof
- 32 modules (Solar World 300W) @ 18 sq ft per module = 577.28 sq ft
- 26 mounting brackets (every other rafter)
- 25 psf wind force

577.28 sq ft x 25 lbs/sq ft uplift force = 1,443.23 lbs of force / 26 mounting footers = 555 lbs per footer

The pull-out (withdrawal) capacity of the lag bolt is a factor of the size of the bolt, the length (how many inches it is secured into the wooden rafter) and the type of wood from which the rafter is made. Common examples of pullout capacity are shown in Table 10-1.

TABLE 10-1: PULL OUT CAPACITY OF COMMON LAG BOLTS

Lag Bolt/Screw Size	Allowable Withdrawal (pounds/inch thread)		
	Pine	Spruce	Douglas Fir
#10 screw	90 (lbs/inch)	146 (lbs/inch)	173 (lbs/inch)
#18 (5/16 inch)	205 (lbs/inch)	235 (lbs/inch)	266 (lbs/inch)
#20 (3/8 inch)	235 (lbs/inch)	269 (lbs/inch)	304 (lbs/inch)

So, if 5/16 inch (7.9 mm) lag bolts were selected for this example, then the minimum length required would be 3 inches (76 mm):

555 lbs pull-out load per footer / 205 lbs/inch (force rating pine roof truss with 5/16 lag bolts) = 2.7 inches (round up to 3 inches)

For a greater margin of safety, larger and/or longer lag bolts can be used. The lag bolt length, however, should be at least ½ inch (13 mm) less than the combined thickness of the sheeting and rafter. This will avoid splitting the wood should the bolt exit the bottom of the rafter.

A typical roof construction in the U.S. is illustrated in Figure 10-6.

Pull-out loads could also be reduced by placing the supports closer together. In this example, the spacing of the footers could be every rafter, rather than every other rafter. The load per rafter can also be reduced by staggering the footers, as illustrated in Figure 10-7. Spacing requirements will vary depending on the code requirements.

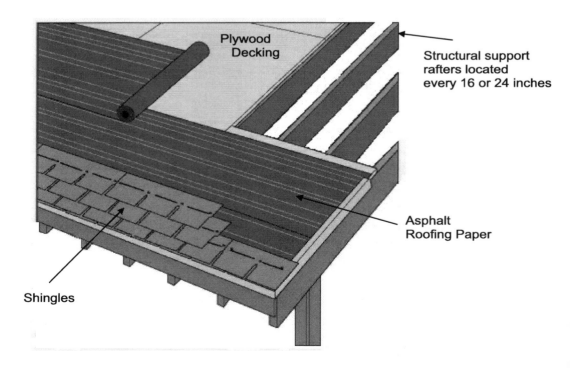

FIGURE 10-6: STRUCTURAL CROSS SECTION OF A TYPICAL SHINGLE ROOF

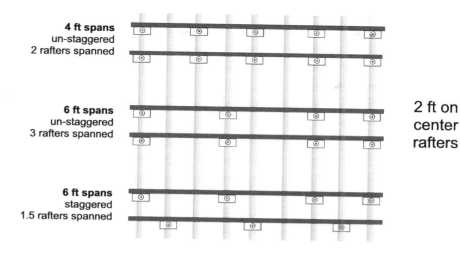

FIGURE 10-7:
REDUCING LOAD BY STAGGERING FOOTERS ON RAFTERS

Often a steeper sloped roof will require a tighter spacing of support brackets (as illustrated in Table 10-2).

Be sure to treat all penetrations with an appropriate sealant, even when using flashed mounting assemblies. Codes regarding flashing for solar array footers can be found in IBC 1503.2.

Setbacks

The International Fire Code (IFC) sets some very specific requirements when placing solar panels on the roof of residential and/or commercial buildings. These rules do not apply to non-habitable detached buildings such as carports, garages, or barns.

In order to allow firefighters access to the roof and space to cut ventilation holes, all buildings must be provided with at least two pathways from the lowest roof edge to the ridge on each roof plane where an array is mounted. At least one roof access pathway must be provided on the street or driveway side of the roof.

The pathways must be:
- at least 36 inches (914 mm) wide,
- over areas capable of supporting firefighters accessing the roof,

	Maximum Horizontal Anchor Spacing (on center)		
Roof Slope	16 inch	24 inch	32 inch
Flat to 6:12 (0-26°)	5 feet 4 inches	6 feet 0 inches	5 feet 4 inches
7:12 to 12:12 (27°-45°)	16 inches	24 inches	32 inches
13:12 to 24:12 (46°-63°)	16 inches	24 inches	32 inches

TABLE 10-2: MAXIMUM ANCHOR SPACING OF SUPPORTS ON ROOF MOUNTED PV SYSTEM

- located in areas with minimal obstructions, such as vent pipes, conduit, or mechanical equipment.

If the PV array occupies less than 33% of the total roof area, or the building has an automatic sprinkler system installed and the array occupies less than 66% of the total roof area, then an 18-inch (457 mm) setback on both sides of the ridge is acceptable. Otherwise there must be a 36 inch (914 mm) setback at the ridge of the roof as well.

Modules installed on dwellings must not be placed on the portion of a roof that is below an emergency escape and rescue opening. There must be a pathway not less than 36 inches (914 mm) wide to the emergency escape and rescue opening.

The IFC does not specify the setback at the eave, but good practices dictate at least 10 inches (254 mm) to avoid excessive wind loading.

The IFC also restricts the size of rooftop arrays. No single section may be larger than 150 x 150 ft (45 x 45 m), with minimum 4-ft (1.2 m) access pathways between the array sections. There must also be clear 4-ft (1.2 m) access pathways to all skylights, roof access hatches, or smoke vents.

For larger commercial and multi-family buildings, where each side of the building is 250 ft (76 m) or longer, there must be a 6-ft (1.8 m) wide clear perimeter around the entire array.

Flat roof systems on commercial buildings can be quite large. They must be designed in a way that allows access to the entire roof by firefighters in an emergency, as indicated in Figure 10-8.

FIGURE 10-8: COMMERCIAL ROOFTOP SYSTEM DESIGNED TO MEET 2012 INTERNATIONAL FIRE CODE

The International Fire Code (IFC) requires:
- a minimum 6-foot-wide clear perimeter around the edges of commercial roofs larger than 250 x 250 ft (76 x 76 m). However, if either axis of the roof is 250 feet or less, the perimeter pathway can be reduced to a minimum of 4 feet.
- a 4-ft (1.2 m) clear area around the access hatch.
- that there must be at least one 4-ft (1.2 m) wide clear pathway between the access hatch to the roof and the perimeter pathway.
- that no section of the array can be larger than 150 x 150 ft (45 x 45 m), with 4-ft (1.2 m) pathways between the sections.
- that designers allow for smoke ventilation areas between the array sections. This can be accomplished in any of three ways. These include: pathways that are at least 8 ft (2.4 m) wide, pathways that are 4 ft (1.2 m) wide and border skylights or smoke and heat vents, or pathways that are 4 ft (1.2 m) wide with 4 x 8 ft (1.2 x 2.4 m) "venting cutouts" located every 20 ft (6 m) "on alternating sides of the pathway."

For odd shaped roofs, exceptions, alternatives, etc, refer to the IFC for additional guidance.

Attaching Panels to Rails

Once the footing brackets are secure, and the rails attached, it is time to attach the panels.

Because the panels are mounted so closely to the roof decking, it is generally necessary to mount them using a **top-down mounting clamp,** such as the one illustrated in Figure 10-9.

FIGURE 10-9: TOP DOWN MOUNTING BRACKET

These clamps, as well as the other fasteners required to assemble the racking system, will be included in the system provided by the manufacturer. They will have been sized and engineered to meet all of the "pull out" and other structural requirements of the unit (another good reason to stick with pre-designed racking systems).

To best support the panel, the rails should always be oriented to cut across the short side of the panel, rather than the long side (as illustrated in Figure 10-10). On a roof-mounted system, the layout of the rafters will likely determine whether the array is mounted in a landscape or portrait orientation, as a single rail should be supported by multiple rafters.

Torque | TORQUE

The tightening of a mechanical fastener, such as a bolt on a solar panel racking system or a lug in an electrical panel, is referred to as **torque**. Most equipment will come with torque specifications, instructing the installer as to the level of tension required for a particular connection.

Torque is usually measured (in the US) in foot-pounds (ft.lb.) and elsewhere in newton metres (N·m).

It is important to get the tension, or torque of the connection right. If the bolt is tightened too much, or over torqued, it can cause:
- galled hardware: where the bolt/nut weld together. The bolt may have to be broken in order to remove it.
- stressed hardware: where the material is damaged and may lead to premature failure.
- deformation: where the bolted connection becomes crushed and damaged.

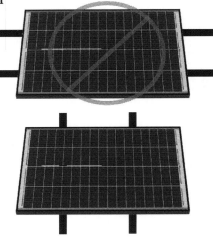

FIGURE 10-10: PROPER ORIENTATION OF PANEL ON RAIL

Connections that are too loose, or under torqued can lead to:
- hardware that come apart under environmental stresses (like wind and snow)
- electrical connections that become hot, and can lead to arcing or fires.

A special torque wrench like the one shown in Figure 10-11 can be used during installation to insure that the connections are made to the proper torque specification.

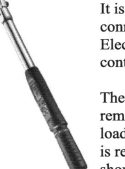

FIGURE 10-11: DIGITAL TORQUE WRENCH

It is also recommended that properly torqued mechanical connections be marked with with a UV rated paint marker. Electrical connections should be marked with a QC (quality control) gel, as indicated in Figure 10-12.

The marks provide a quick visual indication that the bolts remain in place. Changes in temperature, wind and other live loads will often cause connections to loosen over time. This is referred to as **vibrational loosening**. All connections should be inspected periodically and retorqued as needed.

VIBRATIONAL LOOSENING

Galvanic Corrosion

When two dissimilar metals (such as the aluminum frame of a solar panel and the copper ground wire) come in contact with each other, **galvanic corrosion** may occur.

GALVANIC CORROSION

Galvanic corrosion is an electrochemical process. An electrical current will pass between the two dissimilar materials, causing one to break down. Metals that are prone to this corrosion are known as anodes; metals resistant to it are known as cathodes. The risk of galvanic corrosion depends on the chemical properties of the metals as well as environmental conditions such as heat or humidity.

FIGURE 10-12: TORQUED LUGS MARKED WITH QC GEL

The most effective way to minimize galvanic corrosion is to use connectors that are made of the same material to which they are attached. If this is not possible, then fasteners such as bolts and screws should be made of the metal less likely to corrode (like stainless steel). Many fasteners are also treated (galvanized) to minimize corrosion.

Rubber or fiber washers may also help to keep metal surfaces separated, however these may break down over time due to UV or intense heat and moisture.

Rail-less Systems:
Every second that can be saved in the installation process reduces the cost of the PV system. With this in mind, a number of manufacturers have developed mounting systems that use the structure of the panel rather than a support rail.

Rail-less systems, such as the one illustrated in Figure 10-13, attach the mounting bracket directly to the side of the panel. Systems of this type occasionally require a special panel be used that is designed to receive the mounting bracket.

(RAIL-LESS SYSTEMS)

While these systems save money on parts, shipping and even installation, rail-less systems do take a bit more planning up front to properly lay out and level the mounts. Wire management can also be an issue, as the rails cannot be used to support the cables.

FIGURE 10-13: RAIL-LESS SOLAR PANEL MOUNTING SYSTEM

Shared Rail Systems
Another innovation in mounting design is the **shared rail system**. In these systems, the rails are mounted on the perimeter of the panels and interior rails are shared between two panels, as illustrated in Figure 10-14.

(SHARED RAIL SYSTEM)

As with rail-less systems, advocates point to the lower part count that can result in faster installations. Detractors argue these systems are difficult to

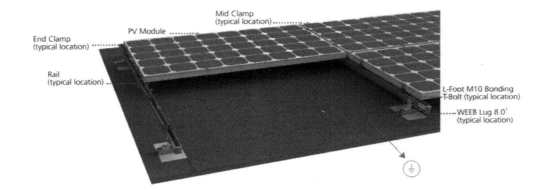

FIGURE 10-14: SHARED RAIL SOLAR PANEL MOUNTING SYSTEM

(FROM EVEREST SOLAR SYSTEMS)

mount on uneven surfaces and often do not incorporate bonding, so bonding jumpers must be installed.

Flat Roof Systems (Ballasted)

Flat roofs, typical on commercial structures, are notorious for leaking even under the best of circumstances. Adding hundreds of mounting penetrations necessary for a standard railed PV system would almost certainly result in leaks. As a result, most flat roof installations utilize a ballasted (weighted) mounting system such as the one illustrated in Figure 10-15.

FIGURE 10-15: BALLASTED MOUNTING SYSTEM FOR FLAT ROOF

These systems typically utilize concrete blocks placed in integrated pans to hold the system in place. They are engineered to withstand wind loads as high as 150 mph (240 kph).

Most systems of this type are designed to avoid the issue of inter-row shading. If the system does not integrate row offset distances, make sure this is planned for in the layout design.

> EAST/WEST MOUNTING SYSTEMS

East/west flat roof mounting systems (shown in Figure 10-16) are designed to address this problem, but are also gaining in popularity for a number of reasons, even when shading is not an issue.

FIGURE 10-16: EAST/WEST FLAT ROOF MOUNTING SYSTEM

Advantages of these systems include:
- can fit more modules on a roof (avoids inter-row shading).
- more consistent energy production throughout the day.

These systems do tend to have a fairly low profile, so wire management and standing water can be an issue.

Squirrel Guards

Many animals find the space between the solar array and the roof a delightful location to set up house. Birds, raccoons, and most particularly squirrels often nest in this warm, dry location. They can cause significant damage to the installation and a potential safety hazard to the home.

Squirrels and other rodents are especially troublesome as they tend to chew on electrical wires. In short order they can chew through the insulation, causing arcs or short circuits. Accumulated debris from nests can also pose a moisture issue or even a fire hazard.

Figure 10-17: Squirrel guard installed on rooftop solar array

(From Solatrim.com)

To avoid these problems, it is often a good idea to install a squirrel guard (or critter guard) around the perimeter of a rooftop installation, as shown in figure 10-17. There are a number of products on the market designed specifically for this purpose.

Avoid any product that might block the flow of air under the array, as this air flow is necessary to cool the solar panels.

Grounding and Bonding the Solar Array

The solar array itself (panels, rails, footings, etc) is part of the electrical system and therefore must be bonded together and connected to the system ground.

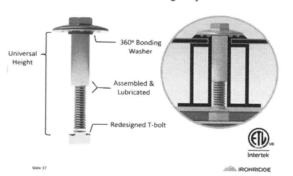

Figure 10-18: A top-down mounting clamp with bonding

Most commonly the panels are bonded to the rails with the use of bonding bolts and washers, as shown in Figure 10-18.

The "teeth" of the unit bite into the panel as well as the rail, making a solid electrical connection.

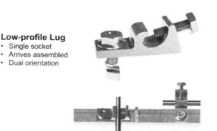

Figure 10-19: Bonding lug

The rails are then typically bonded together using a bonding lug similar to the unit shown in Figure 10-19. A #6 AWG bare copper wire connects from the bonding lug to the main grounding system (often in a junction or combiner box.

Ground Mounted Systems

Where there is space available, placing the array on the ground avoids a number of the problems inherent in roof-mounted systems.

Advantages of a ground mounted system include that it:
- can be oriented as needed, rather than based on the orientation of the building.
- avoids putting penetrations into the roof, which all too often result in leaks.
- is easier to access for maintenance (such as snow removal).
- operates in a cooler environment, rather than on a hot roof.

Various systems will employ different methods to achieve basically the same goal (to support the array in a safe and stable way). But all methods will need to address similar issues.

Foundation Types

PIER

Ground mounted arrays are installed on galvanized steel and/or aluminum support structures. There are a number of options when selecting the **piers**, or support poles (piles) used in ground mounted PV systems.

These include:
- driven pile pier,
- earth screw,
- helical anchor,
- ballasted,
- drilled and grouted piles.

DRIVEN PILE

Driven pile piers normally offer good lateral and vertical support in soils that are firm and well compacted (silt and clay). Where they are practical, they typically offer the most economical option. Driven piles, however, are not usually suitable if the site has soil that consists of coarse gravel or rock. Also, equipment access limitations typically limit driven pile foundations to slopes less than 15°.

GROUND SCREW

Ground screw piers are normally suitable within a wider range of soil and site conditions. When pilot holes are pre-drilled, they can be installed in rocky soils and even bedrock. Earth screws are typically a bit more expensive than driven pile, but offer good pullout resistance and can be installed on slopes up to 30°.

Ground screws can be easily removed and moved if the need arises. Also they are relatively easy to adjust the pier height to adjust for variations in terrain.

A **helical anchor** consists of a helical bearing plate attached near the bottom of a narrow shaft, as illustrated in Figure 10-20. They are normally a more expensive option than either driven piles or earth screws.

Helical anchors are often used in sites with soft soil, such as clean sand or clay. The bearing plate provides good pullout resistance, however the narrow shaft does not provide much lateral bearing capacity.

If the site is environmentally sensitive, helical piles may prove to be the preferred solution. They do not require excavation or soil removal and installation creates a minimal vibration disturbance. They also avoid drill cuttings (soil raised to the surface as the result of drilling a hole). This prevents soil mixing and maintains the natural landscape.

HELICAL ANCHOR

FIGURE 10-20: HELICAL ANCHOR SUPPORT FOR GROUND MOUNTED PV SYSTEM

Ballasted systems, either precast or pour-in-place concrete foundations, may be the best option where drilling into the soil is either undesirable or impractical.

Ballasted systems, for example, are often used in **brownfields** or landfills where disturbing the soil may cause damage to the environment. Even abandoned parking lots can be suitable locations for ballasted systems and do not require damaging the surface of the lot.

BROWNFIELD

A ballasted system can be more easily moved if necessary. However, ballasted ground mounts can not be placed on a site with more than a 5° slope, and are often quite expensive when compared with alternative mounting options.

Drilled and grouted concrete piers have traditionally been the "go to" foundation of choice for small to medium sized projects. They involve drilling a hole to below frost level, placing a metal post into that hole, and then filling it with concrete.

DRILLED AND GROUTED

FIGURE 10-21: SINGLE POST FOUNDATION SUPPORT

Advantages of concrete piers are that minimal equipment is required for installation, and they can be relatively shallow compared to driven steel piles. The disadvantages are that they use concrete (not an environmentally preferred option), are labor intensive and take days to cure.

Chapter 10: Mounting Systems

FIGURE 10-22: DOUBLE POST FOUNDATION SUPPORT

Ground mounted systems typically employ either single post or double post support structure.

Single post foundations support the racking with a single row of posts located in the center or rear center of the array, as seen in Figure 10-21. Most systems of this type employ cantilevered struts and braces.

Double pole foundations (Figure 10-22) utilize two rows of foundations that support piers, typically referred to (in the northern hemisphere) as the southern pier (front) and the northern pier (back).

Concrete Footings

Many ground mounted arrays are supported on metal poles that are imbedded in concrete. Holes for these supports must be dug to below the **frost line** (the point where soil no longer freezes). This avoids the shifting and heaving that may occur as the soil freezes (and expands) and melts (and contracts).

FROST LINE

The frost depth will vary from location to location (again, check with the local building department), from as much as 100 inches (250 cm) in northern Minnesota, to as little as five inches (125 mm) in central Georgia. In any case, most jurisdictions will require that footings be placed at least 12 inches (305 mm) below **unsettled soil** (soil not previously disturbed by construction).

UNSETTLED SOIL

Concrete is placed below and around the metal pole within the hole. Often installers will drill a hole near the base of the pole and insert a metal pin (as illustrated in Figure 10-23) that prevents the pole from sliding out of or twisting in the concrete, should it shrink slightly around the pole as it sets up.

Set the pipe into the hole and pour concrete around the pipe until it completely fills the hole. Also pour concrete into the pipe to secure the re-bar inserted in the bottom portion of pipe. Make certain the pipe is vertically plumb and allow the concrete to set for at least 24 hours.

FIGURE 10-23: PLACING A VERTICAL SUPPORT FOR A GROUND MOUNTED PV ARRAY

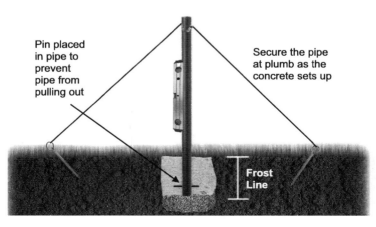

If the pipe mast is filled with concrete, be sure to leave at least one foot of hollow pipe at top for attaching connectors and/or assemblies.

Site Preparation

A great deal of work must be done to prepare the site for a ground mounted system.

Assess the Site

Identify the property lines. There may be required setbacks from the edge of the property. The area where an array can legally be placed should have been defined early in the planning process, but confirm this before beginning the installation. Existing drawings should be available from the property owner, the surveyor, or the local building department.

Check also if any **easements** exist on the site that may impact the placement of the system.

EASEMENT

Call and have all underground utilities identified and marked.

Another issue that may arise is storm water runoff. Many sites have been engineered to manage the flow and accumulation of surface water, so ensure the system will not disrupt the civil engineering already incorporated into the site.

Site Access

For smaller installations, access to the site might not pose a large obstacle, although getting cement to the site might present a challenge. But larger installations usually involve large pieces of construction equipment.

Pathways, room to operate, loading and unloading areas all should be available. Plan that during construction the equipment will disrupt the site. Mud can be a serious problem, bogging down equipment and generally making a mess of things. It may be that temporary (or permanent) gravel access pathways will need to be created.

During construction and after, security should be a concern. Ground mounted systems by their very nature are more accessible than rooftop systems. This makes them a target for thieves, vandals, or even children at play.

A fence around the system is not only a good idea, but often required by code in many jurisdictions. Codes require that there must be a clear 10-ft (3 m) pathway around the perimeter of the array, so any fence should be outside this area.

Grounding of the perimeter fence should be done at least once, within 50 ft (15.25 m) of the point where power from the array leaves the site to connect with the grid. These conductors may be overhead or buried.

Bonding and grounding of the fence should be accomplished in the following manner:
- Drive an 8-ft (2.4 m) grounding electrode at the point of grounding inside the perimeter of the fence.
- Bond the rod to a fence post using a conductor sized appropriately for the system, but not smaller than a #8 AWG copper conductor.
- Any barbed wire or wire mesh that is part of the fence itself should be bonded to the fence post near the ground rod, using the same sized copper bonding jumper as determined above.
- The gates in the fence shall be bonded to the fence post using the same size copper bonding jumper as described above, and that fence post bonded and grounded to another grounding electrode. If the gate is located within 50 ft (15.25 m) of the conductors leaving the array, then the grounding electrode at the gate can serve as the sole ground. If not, then multiple ground rods will be required. If a gate in the perimeter fence is positioned underneath the power lines, then fence posts on both sides of the gate must be bonded and grounded as described.

Conductor Accessibility

The NEC requires that all conductors and connections be secured away from the reach of those not trained or qualified to work on the system (such as the neighbor's children, for example).

Article 690.31(A) states that circuits operating at 30 volts or more located in a "readily accessible location" must be protected. While most conductors in a PV installation are within conduit, the PV source circuit typically is composed of loose (accessible) conductors running from the panel to the combiner box.

Rooftop installations are generally interpreted as not readily accessible, so this provision typically applies to ground mounted systems.

The conductors can be protected by:
- containing the PV source circuit in a raceway.
- building a fence around the array (no longer accessible).
- locating the conductors at least 8 ft (2.4 m) above the ground surface.

The 2017 NEC added a fourth option, stating that conductors connected with MC connectors (such as the standard MC4 connector) are also considered inaccessible, as separating them requires a tool.

(INACCESSIBLE)

However, a number of inspectors are reluctant to accept standard terminations as being **inaccessible**. Specific ground mounted installations may have to protect PV source circuit wiring.

Online Design Tools for Racking Systems

Since 2007, companies such as Mounting Systems have offered online tools engineered to streamline the process of designing racking systems for solar arrays.

Since that time, most racking system manufacturers offer online design tools (such as advertised in Figure 10-24) that factor in a location's wind loading, snow loading, rafter span and other factors - then generate an engineer-reviewed and code-compliant design, as well as a complete bill of materials.

Once site details have been entered (zip code, slope of roof, size of array, etc), the system will ask for specific product options, such as whether the system is ground mounted, roof mounted, flush-to-roof, or ballasted. It will also ask for the specific panel that has been selected for the project, pulling the structural details of that panel from a database to size the system.

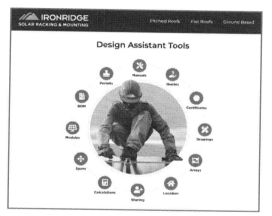

FIGURE 10-24: EXAMPLE OF ONLINE RACKING DESIGN TOOL

(FROM IRONRIDGE)

Based on size and load, the online system will generate a system drawing indicating span lengths, support locations, as well as engineered diagrams that can be submitted to the permitting authorities. Normally the systems will also generate a complete bill of materials required to complete the racking system.

Chapter 10 Review Questions

1. A racking system that follows the pitch of the roof and is off-set from the roof deck by four to 10 inches is referred to as a(n):
   A) flashed -to-deck racking system.
   B) building integrated racking system.
   C) flush-to-roof racking system.
   D) flat-roof racking system.

2. Which of the following is **NOT** part of a typical roof-top racking system?
   A) helical piers
   B) top-down mounting clamp
   C) footings
   D) rails

3. When mounting a racking system on a rooftop, in most cases ensure that:
   A) the array is located at least 36 inches from the sides and ridge of the roof.
   B) the mounting brackets are secured to the rafters.
   C) the rails are mounted perpendicular to the run of the rafters.
   D) all of the above.

4. Roof-top mounting systems generally use top-down mounting clamps because:
   A) they are generally less expensive than bottom-up mounting clamps.
   B) there is usually very little room to work behind the panels on a roof-top installation.
   C) they are required by the NEC for residential roof-top installations.
   D) all of the above.

5. Rails on mounting systems should be oriented to:
   A) attach to the perimeter of the panels.
   B) run perpendicular to the short side of the panel.
   C) run perpendicular to the long side of the panel.
   D) run perpendicular to the long side when oriented in a landscape configuration, but along the short side when oriented in a portrait configuration.

6. Live loads refer to all the following **EXCEPT**:
   A) the weight of the racking system.
   B) the lift from wind passing over a panel.
   C) the weight of snow and/or rain on the rooftop.
   D) additional stress placed on the roof during an earthquake.

7. When two dissimilar metals (such as the aluminum frame of a solar panel and the copper ground wire) come in contact with each other, _____ may occur.
   A) galled hardware
   B) vibrational loosening
   C) dielectric malformation
   D) galvanic corrosion

8. Which of the following is NOT a common problem if too much torque is applied to a connection?
   A) galled hardware
   B) stressed or damaged connections
   C) deformed connections
   D) connections that become hot, and can lead to arcing or fires

9. Holes dug for footings for a ground-mounted racking system must:
   A) extend below the frost line.
   B) extend at least 20 inches below undisturbed soil.
   C) extend a minimum of 5 inches beyond the circumference of the support pole.
   D) all of the above.

10. Which of the following is NOT a common pier used in ground-mounted PV arrays?
    A) driven pile
    B) ground lug
    C) earth screw
    D) helical anchor

11. For ground mounted arrays, fire codes generally require:
    A) there must be a clear 10-ft (3 m) pathway around the perimeter of the array.
    B) that the array be located within 10-ft (3 m) of the property line.
    C) that utility easements be present on the site.
    D) holes for supports must be dug to below the frost line.

12. A conductor is considered inaccessible if:
    A) it is contained within a conduit or raceway.
    B) a fence has been built around the array.
    C) it is located at least 8 ft (2.4 m) above the ground surface.
    D) all of the above.

Lab Exercises : Chapter 10

Lab Exercise 10-1:

Supplies Required:
- Commercial racking system designed to mount at least 3 solar panels

Tools Required:
- multimeter
- screwdriver
- Other tools will vary based on the racking system used.

10-1. Assembling a Commercial Racking Unit

The purpose of this exercise is to become familiar with a commercial racking system and install a minimum of three panels.

a. Follow the assembly instructions provided by the manufacturer to assemble a racking system.
b. Using a multimeter, test the voltage output of each panel.
c. Mount a minimum of three panels on the racking system.
d. Connect the panels together in series. Using a multimeter, test the voltage output of each panel.
e. Place the array in direct sunlight and re-test the voltage output of each panel, as well as the voltage of the string.
e. With the multimeter set to resistance, place the probes on two parts of the metal components of the racking to test for continuity (bonding).

Chapter 11
Designing a Stand-Alone System

Chapter Objectives:

- Identify the components of a stand-alone PV system and factors that go into their selection.
- Determine the current and voltage of the system.
- Calculate how to size an array for a stand-alone system.
- Calculate how to size a battery bank for a stand-alone system.
- Understand how to install and test a lead acid battery bank.
- Calculate wire and overcurrent protection sizes based on the ampacity of the various circuits.
- Determine the minimum ampacity requirements for each circuit within the system.
- Understand the concept of ripple and its effect on installed equipment.

In recent years, the vast majority of residential PV systems designed and installed within the developed world have been grid-tied. However, stand-alone systems, such as the example in Figure 11-1, are still installed in places where the grid is simply not available at an affordable cost, or where independent power is desired or mandatory.

Some examples of places where stand-alone systems may be required include:
- isolated cabins or homes,
- out buildings on a large ranch or farm,
- temporary traffic signals used during highway construction,
- mobile phone towers,
- railway switches,
- remote water pumps or electric fencing,
- and more.

Designing a Stand-Alone System

Stand-alone systems rely entirely on the power contained in the battery bank to run all the loads of the system. The photovoltaic panels serve mainly to recharge the batteries. This is an important concept to understand when sizing the battery bank to meet the load demands.

While the independence of a stand-alone system may be appealing, there are very real downsides to this design.

FIGURE 11-1: BLOCK DIAGRAM OF A PV STAND-ALONE SYSTEM

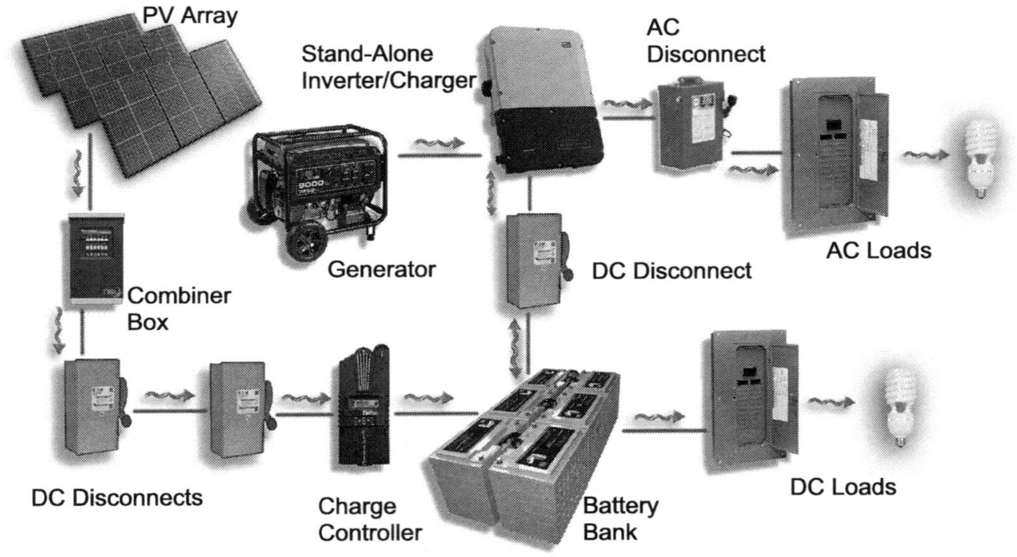

These include:
- lower voltages that make voltage drop of much greater concern,
- higher costs due to the addition of a battery bank and charge controller,
- increased maintenance due to the addition of a battery bank,
- more system inefficiency. Batteries are not 100% efficient. Only about 85% of the energy put into a lead-acid battery can be extracted as useful power due to internal losses. A larger array will be required to get the same amount of usable power, as compared with a grid-tied system.

Stand Alone System Components

By definition, a stand-alone system does not interact with the utility grid. So the system will need a source of energy when the sun is not shining.

For this reason, stand-alone systems require the incorporation of a bank of batteries (and often a generator as a backup for the batteries). In order for these batteries to charge properly, they will also require a charge controller that controls how power from the array is fed into the battery bank.

Since most household loads use AC electricity to operate, the system will need an inverter to convert the DC electricity from the battery to AC power to be used in the loads.

However, there may be DC loads present as well. Many appliances and lighting loads, such as televisions, computers, and LED light fixtures, are native DC loads.

These and other appliances are fed from multistage power-conversion equipment that first rectifies the incoming AC into DC (Figure 11-2). Usually, this also incorporates a second DC-to-DC converter stage that converts the rectified DC voltage into a lower regulated voltage as required by the end load. Each of these conversions wastes electricity.

Feeding these DC loads from a DC power source avoids the waste involved in these energy transformations.

Since there is no electrical grid with which to interact – there is no need for an AC disconnect to isolate the system from the grid.

FIGURE 11-2: AC-TO-DC ADAPTER FOR COMMON DC NATIVE LOADS

Determining the Current and Voltage of the System
Within residential PV systems, the NEC limits the DC voltage to 600 Vdc.

However, there has traditionally been an additional limitation imposed by the NEC for systems that incorporate a battery bank. Within all dwellings, the maximum voltage permitted for a battery bank has long been 60 Vdc / 50 Vac. Codes have since changed to allow for higher voltage energy storage systems (ESS), but these have largely affected the grid interactive market.

As a result, most products available for stand alone PV systems have been developed to fall below these limits. So typical stand-alone PV systems will have nominal voltages of either 12 Vdc, 24 Vdc, or 48 Vdc (all remaining under the 60 Vdc limit).

In calculating the nominal voltage of the various components within a stand-alone PV array, it is the battery bank that controls the process. The output voltage from the array must be compatible with the nominal voltage of the battery bank. The input voltage of the inverter must also be compatible with the nominal voltage of the battery bank.

As with a grid-tied system, the size of the array (and thus the current) will be determined by the required load of the home. Since there is no outside source of power (other than, perhaps, an emergency generator), the array must produce all the power required to run the home each day.

Since the voltage is low (typically 48 nominal volts), the ampacity flowing through the wires is high and voltage drop can be a significant issue within the DC circuit.

Charge Controller
Charge controllers (typical specifications shown in Figure 11-3) can perform a number of functions, but their primary role is to protect the battery bank as it is charged.

FIGURE 11-3:
SPECIFICATIONS FOR A
TRISTAR MPPT SOLAR
CHARGE CONTROLLER

(From Morningstar, Inc)

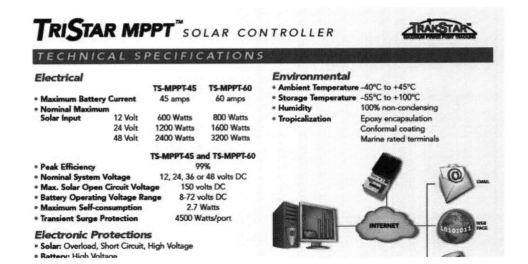

Without a charge controller, a fully charged battery bank might overcharge as the sun continues to shine and no load is drawn from the system. Most charge controllers also monitor the battery bank to ensure it does not drain down to below a certain level (50% DoD, for example).

Note: No charge controller is required in systems where the voltage of the source circuit and voltage of the battery bank are matched (both nominal 48 Vdc, for example) and the maximum charging current when multiplied by one hour is less than three percent of the battery bank's rated capacity.

In other words, if a 48 Vdc array were charging a 48 Vdc battery bank with a 200 Ah rated capacity, and if the maximum current from the array were less than six amps (200 Ah x 3%), then no charge controller is required. This avoids the need for a charge controller in systems that are essentially "trickle charging" a battery bank.

When selecting a charge controller, there are a number of factors to consider. Aside from all the features discussed earlier (such as MPPT, efficiencies, reliability, etc), the charge controller must be compatible with:
- the array's nominal voltage,
- the array's system size (watts),
- the maximum solar open circuit voltage,
- the maximum battery bank current,
- the battery bank's nominal voltage.

Another factor to consider when selecting a charge controller is the maximum input voltage the unit can handle.

For example, some charge controllers designed to work with either 12 Vdc, 24 Vdc or 48 Vdc battery banks can receive input voltages of as much as 600

Vdc. These higher voltages allow for the array to be configured at a much higher voltage than the 48 V limit of the battery bank.

The higher the voltage, the lower the amps for the same amount of power generated. Lower amps allows for smaller wire. The cost savings on wire may be significant when the array is located a significant distance from the battery bank.

Charge controllers of this type are often referred to as **MPPT charge controllers**, and simply perform a DC-to-DC conversion. In this case, they step down the voltage from the output of the array to the input limitation of the battery bank.

A quick look at the specification sheet for the Tri-Star MPPT-45 depicted in Figure 11-3 shows that it can accept up to 150 Vdc from the array, then can step down to be compatible with a 12 Vdc, 24 Vdc or 48 Vdc battery bank. When it is configured with a 48 Vdc battery bank, it can be connected to (up to) a 2.4 kW solar array.

Diverted Loads

Normally, when the battery bank is fully charged, the controller simply opens the circuit between the array and the bank (shutting down the flow of electricity). While this protects the batteries from overcharging, it also wastes potential power.

Some stand-alone systems incorporate a load-dump (or a **diverted load**). A typical system might divert unneeded power to a hot water heater (or a similar high-capacity load), so the energy not required at the moment to charge the battery bank is not wasted.

The diverted load must be large enough to handle the maximum power output of the array when the battery bank is full and no loads are being serviced.

A few considerations when sizing a diverted load include that the:
- diverted load current must not exceed the maximum current rating for the charge controller,
- diverted load must have a voltage rating greater than that of the battery bank,
- diverted load power rating must be at least 150% of the maximum power output rating of the array,
- wiring and overcurrent protection to the diverted load must be rated at least 150% of the maximum current rating of the charge controller.

Chapter 11: Designing a Stand-Alone System   Page 311

FIGURE 11-4:
SPECIFICATIONS OF A STAND-ALONE INVERTER

(FROM MAGNUM ENERGY)

INVERTER SPECIFICATIONS	ME2012	ME2512	ME3112
Input battery voltage range	9 - 16 VDC	9 - 16 VDC	9 - 16 VDC
Nominal AC output voltage	120 VAC ± 5%	120 VAC ± 5%	120 VAC ± 5%
Output frequency and accuracy	60 Hz ± 0.1 Hz	60 Hz ± 0.1 Hz	60 Hz ± 0.1 Hz
1 msec surge current (amps AC)	60	100	120
100 msec surge current (amps AC)	37	45	50
5 sec surge power (real watts)	3700	5000	6000
30 sec surge power (real watts)	3450	4500	4800
5 min surge power (real watts)	3100	3500	3950
30 min surge power (real watts)	2400	2900	3500
Continuous power output at 25° C	2000 VA	2500 VA	3100 VA
Maximum continuous input current	266 ADC	333 ADC	413 ADC
Inverter efficiency (peak)	95%	91%	90%
Transfer time	16 msecs	16 msecs	16 msecs
Search mode (typical)	5 watts	5 watts	5 watts
No load (120 VAC output, typical)	20 watts	23 watts	25 watts
Waveform	Modified Sine Wave	Modified Sine Wave	Modified Sine Wave

Selecting a Stand-Alone Inverter

Selecting a stand-alone inverter differs from the selection of a grid-tied inverter in a number of key ways. Specifications for a typical stand-alone inverter can be found in Figure 11-4.

Factors to consider include:
- input voltage rating,
- generator compatibility,
- wave form,
- continuous power output rating
- peak power output rating,
- ground fault protection,
- output AC voltage.

Input Voltage

The input voltage rating of the inverter must match the nominal voltage of the system's battery bank. This DC input rating will typically be 12-volt, 24-volt or 48-volt.

Inverter/Charger

Many stand-alone inverters are designed to accept AC power from an outside source (such as a generator or even the electrical grid) and convert this power into a DC current that is then used to charge the battery bank (when the solar array proves to be inadequate for the job). Such a unit is referred to as an **inverter/charger**.

(INVERTER/CHARGER)

Depending on the system configuration, it may be necessary to specify a unit with this feature. Generally this is done so the battery bank can be sized to handle "normal" conditions rather than "worst case".

Wave Form

Modern stand-alone inverters incorporate an output wave form of either modified sine wave or pure sine wave (square wave inverters are nearly extinct). Pure sine wave inverters more closely mimic the analog wave form of grid electricity, but are typically more expensive than modified sine wave inverters. Modified sine wave inverters may not work well with some sensitive electronics.

Continuous Power Output Rating

The power rating of a grid-tied inverter is based on the output of the array. All the power from the array flows through the inverter to the loads. But it does not have to service all the load demand, since the grid is also available to supply power.

In a stand-alone system, the inverter must supply 100% of all the AC load demand. There is no grid to pick up the slack. So the size of the inverter is determined by the load demand, not by the size of the array.

If, during the load analysis of the home, it is determined that 2,800 watts of power might be required at any one time (the maximum power draw) within the home, then a stand-alone inverter with a continuous power rating of at least that amount will be required to service the system load demands.

Just as it is harmful to a car's engine to drive a vehicle at its maximum speed for a long period of time, it is not good to run an inverter at its maximum output rate for long periods of time. So, the continuous power output rating of the selected inverter should be greater than 125% of the maximum power draw of the home.

Peak Power Output Rating

Any load with a motor uses more energy during start-up than it does when operating normally. All stand-alone inverters are capable of handling larger loads for brief periods of time than at their continuous load rating.

The surge rating of the inverter should be adequate to handle those brief periods of increased load demand. If not, then the load may not start - or overcurrent protection within the inverter may trip, shutting down the unit.

Most stand-alone inverters have a 5 second surge capacity of nearly twice its continuous power rating. For normal household loads, this should be adequate. If the stand-alone system is designed to exclusively run large inductive loads (with a high surge at start-up), then a larger inverter might need to be selected.

TABLE 11-1: SAMPLE DESIGN RATIO CALCULATIONS FOR AN ARRAY AT 28° ALTITUDE

	Solar Insolation (hr/day)	Load Demand (kWh/month)	Design Ratio
January	2.88	851	295 kW
February	3.84	1152	300 kW
March	4.24	1103	260 kW
April	5.37	505	94 kW
May	5.8	481	83 kW
June	6.11	733	120 kW
July	6.14	652	106 kW
August	6.02	641	106 kW
September	5.53	499	90 kW
October	4.09	416	102 kW
November	3.31	816	247 kW
December	2.5	409	163 kW
Average	**4.65**	**688**	**148 kW**

Ground Fault Protection

In grid-tied inverters, ground fault protection (GFP) is always incorporated into the inverter. This is not always the case with a stand-alone inverter. So it may be necessary to incorporate a GFP device elsewhere in the system.

AC Output Voltages

Finally, as with a grid-tied inverter, it is necessary that the unit provide AC current that is compatible with the load demands of the home. In the United States, this will generally be 120 Vac or 240 Vac. Smaller inverters often are limited to 120 Vac. It is assumed that there will be no large loads requiring 240 Vac incorporated into the system.

Sizing a Stand-Alone System Array

Since the amount of sunlight available varies throughout the year, as do load requirements, the design of a stand-alone array must take into account the "worst case" situation. For example, load demands might be highest during the winter, when solar insolation is at its minimum for this location.

When calculating the size of a grid-tied array, the designer incorporates monthly averages into the calculation. For example:

> 688 kWh (average monthly load) / 30.5 days per month =
> 22.56 kWh per day/ 4.65 (average solar insolation) =
> 4.85 kW / .90 (system derate) = 5.39 kW

For a stand-alone system, the design must utilize the month with the highest **design ratio** (month's load/month's insolation), not the annual averages. In Table 11-1, this month would be February. This is referred to as the **critical design month**.

(DESIGN RATIO)

(CRITICAL DESIGN MONTH)

	Solar Insolation (hr/day)	Load Demand (kWh/month)	Design Ratio
January	3.38	851	251 kW
February	4.25	1152	271 kW
March	4.11	1103	268 kW
April	4.74	505	106 kW
May	4.66	481	103 kW
June	4.74	733	120 kW
July	4.83	652	155 kW
August	5.13	641	125 kW
September	5.27	499	95 kW
October	4.36	416	95 kW
November	3.84	816	212 kW
December	2.97	409	138 kW
Average	**4.36**	**688**	**158 kW**

TABLE 11-2: SAMPLE DESIGN RATIO CALCULATIONS FOR AN ARRAY WITH A WINTER BIAS, AT 58° ALTITUDE

1152 kWh (critical monthly load) / 30.5 days per month =
37.77 kWh per day/ 3.84 (critical month solar insolation) =
9.83 kW / .75 (system derate) = 13.11 kW

Note that the array for the stand-alone system is significantly larger than a grid-tied system would be for the same property. This is due to the fact that the design ratio for the critical design month (in this case, February) is twice that of the annual average.

Also, an additional 15% inefficiency has been added to the system to account for the relatively low roundtrip efficiency of power flowing through a lead-acid battery.

The **roundtrip efficiency** of a battery is the difference between the amount of power that goes in (in charging) and the amount that leaves (available to power the load). Lead-acid batteries normally experience internal losses of between 10%-30%. For example, 100 Wh of energy used to charge the battery bank may only result in 85 Wh of power leaving the battery bank.

ROUNDTRIP EFFICIENCY

Lithium-ion batteries generally perform better, with about a 90%-95% roundtrip efficiency.

As a result of roundtrip efficiency, it is a good practice to increase the derate factor of a system that incorporates lead-acid batteries by at least 15%.

Since stand-alone system sizes are based on the critical design month, rather than the annual average, it may be advantageous to set the altitude of the array to maximize power output during the critical design month, even if it lowers overall annual production, as demonstrated in Table 11-2.

If the angle of the array is shifted towards a winter bias, from 28° altitude to 58° altitude, the resulting array size needed to meet the critical design month's load requirements is smaller, even though the total annual production is less.

The new calculation would be:

> 1152 kWh (critical monthly load) / 30.5 days per month =
> 37.77 kWh per day/ 4.25 (critical month solar insolation) =
> 8.89 kW / .75 (system derate) = 11.85 kW

String Calculations for a Stand-Alone System

Voltage increases as units are connected in series, so if the panel's nominal voltage is 24 Vdc, connecting two of these panels in series will result in a nominal 48 Vdc system. Three in series will result in a nominal 72 Vdc system, and so on.

Traditionally, charge controllers and inverters used in stand-alone systems operate with input at 48 nominal volts. So in such a system, two 24-volt panels would be connected in series (strings), then combined in a combiner box.

Stand-alone systems that incorporate an MPPT charge controller (buck-boost) will require string calculations similar to those conducted for grid-tied inverters, using the maximum input voltage of the charge controller as the upper limit of the voltage for each string. No minimum voltage will need to be calculated, as the MPPT charge controller does not need a minimum voltage to function.

Sizing a Stand-Alone Battery Bank

When sizing a battery bank, the following system parameters must be known (as indicated in Table 11-3):

- size of load to be serviced (AC daily load),
- system voltage,
- how long it will operate independent of additional power from the solar array,
- depth of discharge of the batteries,
- efficiency of the inverter,
- and the temperature at which the batteries will operate.

Sample Design

Design a stand-alone PV system for a small weekend cabin. A load analysis shows that on a typical visit, about 6 kWh per day of power is used. It has been determined that there are no special DC load requirements (everything will run off the AC system).

[(AC Daily Load	/ Inverter Efficiency)	+ DC Daily Load]	/ DC System Voltage	= Amp hrs per Day
[(6 kWh	/.90)	+ 0 Wh]	/ 48 Vdc =	138.89 Ah
Amp hrs per Day	x Days of Autonomy	/ Depth of Discharge	/ Battery Ah Capacity	= Battery Strings
138.89 Ah	x 2	/.40	/ 360 Ah =	1.93 (2)
DC System Voltage	/ Battery Voltage =	# of Batteries in Series	X Battery Strings	= Total Batteries
48 Vdc	/ 6 Vdc =	8	x 2 =	16

TABLE 11-3: BATTERY BANK SIZING WORKSHEET

Determine the Load

The first step in sizing a stand-alone battery bank is to determine the daily load requirements of the system. This should include both AC and DC loads.

Determine the Voltage of the System

A number of factors will go into determining the voltage of the system, but the size of the overall system may be the most compelling. Designing a system where the continuous current is greater than 100 amps can be costly, (in wire) as well as present a safety hazard.

A 1,000 watt (1 kW) array configured at 12 volts will result in a circuit with a current of 83 amps (1,000 W / 12 V = 83 A).
Similarly, a 4 kW array configured at 12 volts would result in a circuit with a current of 333 amps. Very large wire would be required to safely transport that current from the array to the charge controller.

But if the 4 kW array was reconfigured to 48 volts, then the circuit designed to carry this power would only need to deal with 83 amps of current (4,000 W / 48 V = 83 A).

By increasing the voltage of the system, it is possible to reduce the current, saving money on wire, overcurrent protection, disconnects, etc. The lower amps will also make the system somewhat safer.

Unless an MPPT charge controller is incorporated into the system, the nominal voltage of the battery bank in a stand-alone system must match the nominal voltage of the array.

Early small stand-alone systems were generally configured at 12 Vdc. Today most stand-alone systems incorporate a 48 Vdc battery bank. Larger systems may use a MPPT charge controller that allows the array to be configured to as much as 600 Vdc, but the battery bank in these systems still remains at nominal 48 Vdc.

A 48-volt configuration is selected. Battery storage is measured in amp hours, so it is necessary to divide the load demand (watt-hours) by the system's nominal voltage to determine required storage capacity, in amp-hours. The daily loads are 6 kWh, so in this case, 6,000 Wh / 48 volts = 125 Ah. A 48-volt battery bank with 125 Ah of storage capacity will meet the daily load demands of the cabin if the system were 100% efficient.

Efficiency of the Inverter

The only piece of equipment between the battery bank and the load is the inverter. So the conversion efficiency of the inverter must be taken into account when sizing the battery bank.

No inverter is 100% efficient in converting the DC energy contained in the battery bank into useable AC power. While some very expensive inverters might function at a 98% efficiency, 90% efficiency might be more realistic in this example.

Assume the inverter operates at 90% efficiency, 125 Ah / .90 = 138.89 Ah storage capacity is now required for one day of load demand.

Days of Autonomy

Since it is the battery bank that drives all the loads, the designer must calculate the number of days the system will need to continue to provide power, assuming there is no sunlight available to help recharge the battery bank.

This determination is referred to as **days of autonomy**. In other words, the number of days the battery bank would need to operate independent of any outside source of power (such as the sun or a generator).

Clearly this is the worst case scenario. Even on the bleakest of days there will be some sunlight trickling in to recharge the battery bank. But there might be some mechanical problem preventing power from reaching the system. So it is theoretically possible to have zero energy reaching the battery bank over the course of a day.

*Since this is a weekend cabin, it is safe to assume it is never occupied more than two days in a row, and never more than two days out of every seven. So two days of autonomy are ample. The 138.89 Ah of battery storage capacity will provide enough power for one day. This figure will need to double to ensure two days of power. In this case, 138.89 Ah * 2 days = 278 Ah of storage capacity.*

Determine Depth of Discharge

A deep-cycle battery will last longer if it is discharged to a fairly shallow depth during its discharge cycle. However, the less of the battery's capacity used, the more batteries required. It is definitely a "pay me now or pay me later" situation. The system can get by with fewer batteries by allowing them to discharge to a greater depth, but they will likely need to be replaced sooner than if a more shallow discharge level is selected.

A 40% depth of discharge was decided upon. This means that the inverter is set to allow only 40% of the battery bank's capacity to be used before the system is shut down. This will increase the amount of storage required, since the system will not use the full amount of the battery's storage capacity. In this case, 278 Ah / .40 = 695 Ah of storage capacity now required.

Select a Battery

Again, a number of factors will go into selecting a deep cycle battery. These include: price, brand reputation, warranties, availability, storage capacity, voltage, reliability, etc. Batteries will likely be nominally rated as 2-volt, 6-volt or 12-volt (although 24-volt and 48-volt batteries are available).

It is always best to design the system with as few parallel strings as possible (and in no case more than three). Selecting a larger battery with a lower voltage and connecting more in a string may be the best option.

Battery capacity will be rated in amp-hours, with a number of discharge rates provided (for example, 5-hour, 20-hour or 100-hour). 20-hour (C20) discharge rates are normally used when comparing the capacity of one battery as opposed to another.

All batteries in the battery bank should be of a similar make, model and capacity. They should also be the same age (ideally), since the capacity of each battery will adjust itself to the overall capacity of the system.

In other words, if a brand new battery is added to an older system, it will soon only perform as well as the other batteries in the battery bank. For this reason, if a battery goes bad in an older system, the owner must often replace the entire battery bank, rather than place a good battery into an aging and poorly performing system.

A review of available options has led to the selection of a Trojan lead-acid deep cycle battery. This battery is rated at nominal 6-volts, with a capacity of 360 amp-hours at a 20-hour discharge rate (C20).

Determine how many Battery Strings are needed

It was determined how many amp-hours of storage the system requires. A battery was also selected (so it is now known how much storage capacity each battery contains).

Specific applications may influence which discharge rate is used in this calculation. If there is an application that will use energy quickly, the 5-hour rating might be used to determine the battery's capacity (for example, a well pump that runs for five hours but then sits idle for 19 hours each day).

However, in most cases, calculations will be based on the 20-hour (C20) discharge rating.

The total storage demand is 695 Ah. Each battery has a capacity of 360 Ah. Hooking them together in a string changes the voltage, but not the Ah capacity. So 695 Ah / 360 Ah = 1.93 strings. Round up to two strings.

Determine how many Batteries in each String

The voltage of the battery bank must be matched to the system voltage. If the system designed is a nominal 12-volt system, and the battery selected is a nominal 12-volt battery – no further adjustment is required. But if the battery is a nominal 6-volt, then two batteries will need to be connected together in series to match the 12-volt system rating.

The system is designed to function at 48 nominal volts. The battery selected is a nominal 6-volt battery. Eight (8) batteries connected in a string (series) are required to match the system voltage.

So, the battery bank will consist of two (2) strings of eight (8) batteries. A total of 16 batteries (2 x 8) will need to be purchased for this system to operate as designed.

TABLE 11-4: AMBIENT TEMPERATURE ADJUSTMENT MULTIPLIERS FOR LEAD-ACID BATTERIES

Capacity Multiplier	Temperature	
	Celsius	Fahrenheit
1	26.7°C	80°F
1.0	23.9°C	75°F
1.02	21.2°C	70°F
1.04	18.3°C	65°F
1.07	15.6°C	60°F
1.1	12.8°C	55°F
1.13	10°C	50°F
1.17	7.2°C	45°F
1.22	4.4°C	40°F
1.27	1.7°C	35°F
1.37	-1.1°C	30°F
1.42	-3.9°C	25°F
1.51	-6.7°C	20°F
1.63	-9.4°C	15°F
1.77	-12.2°C	10°F
1.95	-15.0°C	5°F
2.17	-17.8°C	0°F
2.21	-20.6°C	-5°F
2.25	-23.3°C	-10°F

Ambient Temperature Adjustment

Lead-acid battery capacity is affected by the ambient air temperature. Cooler ambient temperatures will reduce the system's storage capacity; however the cooler temperatures will improve the cycle life of the battery. The cycle life of a battery bank will decrease by 50% for every 10°C over 25°C (77°F).

The rated Ah capacity of lead-acid batteries are determined at 25°C (77°F). As average operating temperatures drop, a multiplier (Table 11-4) can be used to calculate the increased capacity needed to achieve the desired capacity.

If, for example, it was determined that a 400 Ah battery bank was required, but it would be located in a very cold environment with average temperatures of 10°F (-13°C), then:

400 AH battery bank capacity is required at 25°C (77°F)
Operating temp -13°C (10°F) = 1.77 temperature derate factor
1.77 X 400 = 708 Ah battery bank now required

Sizing Example with Temperature Adjustment

Assume a weather monitoring station will operate on a mountaintop with a daily load demand of 5 kWh. There is no grid available (a properly sized array is in place) and ambient air temperatures where the batteries will be stored will fall to 5°F (-15°C). The designer has decided to use lead-acid batteries.

This is a 48 Vdc stand-alone system, and because of periodic storms, it has been determined to install three days of autonomy within the system. This system will also be set to a 50% depth of discharge (DoD).

The battery bank would be sized as follows:

5 kWh (daily load) / 48 Vdc (nominal voltage) = 104 Ah per day
3 days of autonomy x 104 Ah per day = 312 Ah
312 Ah / .50 (50% DOD) = 624 Ah
624 Ah / .95 (95% inverter efficiency) = 657 Ah
657 Ah x 1.95 (temperature derate at 5°F /-15°C = 1,281 Ah battery bank required

Installing and Testing Lead Acid Battery Banks

The storage area has been designed and inspected. The battery bank has been sized and the battery selected. Now it is time to install.

Inspect and Test Each Battery

Do a physical inspection of each battery. Look for cracks or bulges in the casing. The top of the battery, posts, and connections should be clean, free of dirt, fluids, and corrosion.

If batteries are dirty:
- clean the battery top with a cloth or brush and a solution of baking soda and water. Do not allow any material to get inside the battery.
- rinse with water and dry with a clean cloth.
- clean battery terminals and the inside of cable clamps to a bright metallic shine using a post and clamp cleaner.

Check for moisture. Any fluids on or around the battery may indicate that the electrolyte solution is leaching or leaking out. Replace any damaged batteries.

Test the state of charge of each battery. This can be done in one of two ways. Through an **open-circuit voltage test**, or a **specific gravity test**.

(OPEN-CIRCUIT VOLTAGE TEST)

(SPECIFIC GRAVITY TEST)

TABLE 11-5: SPECIFIC GRAVITY AND STATE OF CHARGE VOLTAGES FOR 12-VOLT AND 48-VOLT BATTERIES

State of Charge	Specific Gravity	12 Volt	48 Volt
100 percent	1.277	12.73 volts	50.93 volts
90 percent	1.258	12.62 volts	50.47 volts
80 percent	1.238	12.50 volts	49.99 volts
70 percent	1.217	12.37 volts	49.49 volts
60 percent	1.195	12.27 volts	48.96 volts
50 percent	1.172	12.10 volts	48.41 volts
40 percent	1.148	11.89 volts	47.83 volts
30 percent	1.124	11.81 volts	47.26 volts
20 percent	1.098	11.66 volts	46.63 volts
10 percent	1.073	11.51 volts	46.03 volts

Open-Circuit Voltage Test

Prior to testing, the battery should be disconnected from the system for at least 6 hours.

- Disconnect all loads from the batteries.
- Measure the voltage using a DC multimeter.
- Check the state of charge against the anticipated open-circuit voltages in Table 11-5.
- Charge the battery if it registers less than a 70% state of charge.

If after recharging, the battery still does not register a 100% state of charge, then the battery may have been left discharged too long (severe sulfation) or it may have a bad cell. Replace the battery.

Specific Gravity Test

While testing batteries with a multimeter is a quick and convenient way to check their state of charge, a more accurate method for flooded lead-acid batteries is to perform a specific gravity test.

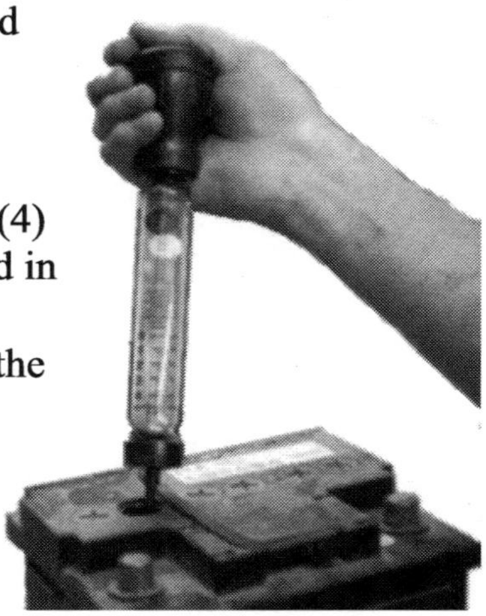

FIGURE 11-5: TESTING SPECIFIC GRAVITY OF A FLOODED LEAD-ACID BATTERY WITH A HYDROMETER

(HYDROMETER)

- Open vent caps, do not add water at this time.
- Fill and drain the **hydrometer** two (2) to four (4) times before pulling out a sample, as illustrated in Figure 11-5.
- There should be enough sample electrolyte in the hydrometer to completely support the float.
- Take a reading, record it, and return the electrolyte back to the cell.
- To check another cell, repeat the three (3) steps above.
- Check all cells in the battery.

- Replace the vent caps and wipe off any electrolyte that might have been spilled.
- Correct the readings to 80° F (26.6° C):
  Add 0.004 to readings for every 10° F (5.6° C) above 80° F (26.6° C)
- Subtract 0.004 for every 10° (5.6° C) below 80° F (26.6° C)
- Compare the readings to those listed in Table 11-5.
- Charge the battery if it registers less than a 70% state of charge.

If after recharging, the battery still does not register at 100% state of charge, then the battery may have been left discharged too long (severe sulfation) or it may have a bad cell, or the electrolyte is weak due to a spill. Replace the battery.

Wiring a Stand-Alone System

There are a number of differences when sizing wire for a stand-alone PV system as compared with a grid-tied system.

PV Source Circuit

While the array will likely be larger than a comparable grid-tied system, and the current output larger, the wiring size for the PV source circuit will be determined in a similar manner.

Take the maximum output current for the array and multiply it by 1.25 (solar variability factor) x 1.25 (NEC safety factor). For small stand-alone systems it is unusual to incorporate module level power electronics (MLPE), so the solar variability factor will have to be incorporated until the circuit reaches the charge controller.

Charge Controller Output Circuit

In a system that does not incorporate a MPPT charge controller, the ampacity rating entering the charge controller will be the same as the ampacity rating leaving the controller. So no wire size modifications are required.

However, if the voltage has been stepped down in the charge controller, then the ampacity rating of the wire in the charge controller output circuit will need to be adjusted accordingly.

For example, assume the system has a 1.5 kW array (five 300 W panels). The Isc of a single panel is 18.3 A. The Voc is 21.6 Vdc and a single string is hooked up to an MPPT charge controller that can accept up to 250 Vdc. The battery bank is rated at nominal 48 Vdc.

The PV source circuit has been sized at:

18.3 A (Isc) x 1.25 (solar variability) x 1.25 (NEC safety) = 28.6 A

So #10 AWG with a rating of 30 A has been used for this circuit.

As the circuit passes through the charge controller, the voltage is reduced from 216 Vdc (10 x 21.6 Voc - assume no temperature correction) to a nominal 48 Vdc. Nominal 48-volts rated battery banks will actually operate within a range of voltage (42 to 56 Vdc) depending on its state of charge. So the lower 42 Vdc will be used.

If the total output of the array is 1,500 W, then 1,500 W / 42 Vdc = 35.71 A. Multiply this times 1.25 (the NEC safety margin) and the current for this circuit can be a maximum of 44.64 A. So #6 AWG wire must be selected for this circuit.

Or easier still, simply take the maximum current output rating of the charge controller and multiply it times 1.25.

The solar variability derate factor does not apply in this circuit as the charge controller mitigates any variation in irradiance.

If multiple charge controllers are combined in parallel to feed the battery bank, then their ampacities are added together to size the wires connecting them to the battery bank.

If a diverted load is incorporated into this circuit, the conductor to the diverted load must be 150% of the maximum current rating of the charge controller.

The Effect of Ripple on Inverters and Batteries:

Before discussing the sizing of the wire that connects the battery bank to the inverter, it is important to understand the effect of ripple on the circuit.

Because power flows in a circuit (from the source to the load, then back to the source), the waveform of power coming from a DC source (such as a solar panel or battery) is affected by the waveform of the AC load to which it is connected, as illustrated in Figure 11-6.

FIGURE 11-6: THE DC WAVEFORM IS AFFECTED BY THE AC WAVEFORM, RESULTING IN RIPPLE.

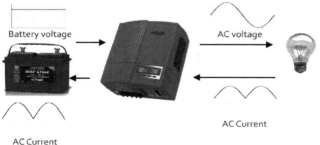

RIPPLE

While a pure DC waveform is flat, the DC waveform influenced by the system to which it is connected will fluctuate slightly, or **ripple**.

Ripple is wasted power, and has many undesirable effects in a DC circuit. It can heat components, cause noise and distortion, and may cause some digital circuits to malfunction.

Capacitors within the inverter will try to flatten out the ripple. However, the more resistance there is in the DC portion of the circuit, the greater the ripple.

Test for AC ripple on the DC portion of the system by using a multimeter set to test for AC voltage, but testing on the DC circuit. The meter will then only measure the variation in the waveform.

Excessive ripple may be the result of:
- a battery bank without adequate capacity (either too small or at too low a state of charge),
- battery cables that are too long or undersized,
- too many or low-quality switches, fuses or shunts.

Systems that experience high levels of ripple may find that:
- the lifetime of inverters decreases due to large currents in the capacitors,
- the battery lifetime is limited due to the discharge/charge effect
- the charge power is reduced due to ripple during charging,
- other connected loads will suffer from the same ripple.

For any fixed load (measured in watts), any variation in voltage will result in a corresponding increase (or decrease) in amps (as watts = amps x volts). So a drop in voltage due to ripple will result in a corresponding increase in the amps flowing through the circuit. As a result, the conductor used in the circuit must be sized to handle the increase in amps that is the result of ripple.

Inverter Input Circuit

Unlike a grid-tied system where the power flowing through the inverter is limited to the capacity of the array, a stand-alone system's inverter must handle all loads within the building.

On systems that incorporate a battery bank, the circuit that runs between the inverter and the battery bank is considered the inverter input circuit. As load demand is pulled from the batteries to the inverter, the inverter specifications will dictate the size of the wire needed for this circuit.

An inverter with a 4,000 W continuous load rating may actually deliver 8,000 W for several seconds under surge conditions. All of this power must be pulled from the batteries, through the inverter. The wire from the battery bank to the inverter must be rated to handle the maximum surge current, not the continuous operating current.

Sizing of conductors and overcurrent protection for this circuit is accomplished through the following equation:

> Maximum inverter rating (watts) / inverter efficiency / lowest battery bank voltage/ AC ripple x 1.25 (NEC safety margin)

A battery bank configured at a nominal 48-volts rating will actually operate within a range of voltage (42 to 56 V) depending on its state of charge. The lower the volts, the greater the amps required to draw the same amount of power. So select the lowest voltage the inverter will accept from the battery bank (probably around 44 volts).

For a 6,500 W (maximum surge rating), 95% efficient inverter, connected to a 48-volt battery bank, then...

$$6{,}500 \text{ VA} / .95 / 44 \text{ Vdc} = 155.5 \text{ amps}$$

The effect of current ripple must also be taken into account when sizing this circuit. While no standard exists, it is assumed that the voltage ripple of the system should not exceed 3%.

Factoring in a voltage ripple of 3%, then...

$$6{,}500 \text{ VA} / .95 / 44 \text{ Vdc} = 155.5 \text{ amps} / .97 \text{ (3\% Vac ripple)} = 160.3 \text{ amps}$$

Then adjust the amps with the NEC safety margin used in all wire sizing, and...

$$160.3 \text{ amps} \times 1.25 \text{ (NEC safety margin)} = 200.38 \text{ amps}$$

The inefficiency of the inverter must be taken into account because it affects the amount of power that must be drawn from the battery to service the load. If the inverter was 100% efficient, the 60 W of power would be needed to service a 60 W load. But if the inverter is only 90% efficient, then 66 W of power would need to be drawn from the battery to service a 60 W load.

Inverter Output Circuit

Generally the wire sizing of the inverter output circuit is accomplished by taking the maximum inverter output current (obtained from the inverter specifications) and multiply it by the 1.25 NEC safety margin.

Wiring the Battery Bank

Choosing the correct size (diameter) and length of cable is important for the overall efficiency of a battery bank. Cables that are too small or too long will result in power loss and increased resistance.

FIGURE 11-7: TYPICAL CABLES FOR BATTERY BANK

(FROM ENERDRIVE)

Short wires connecting batteries together must be equal in length, as should the longer cables connecting the battery bank to the inverter (Figure 11-7). This allows for equal resistance applied to all batteries, ensuring they charge and discharge at the same rate.

Undersizing the wires in a battery bank is a common error. Generally #4/0 AWG cables should be used if possible, but wire size at a minimum should be:
- #6 AWG for loads of up to 90 amps,
- #2 AWG for loads of up to 150 amps,
- #1/0 AWG for loads of up to 250 amps,
- #4/0 AWG for loads of up to 400 amps.

Cables used in battery bank systems are typically fine stranded, as they are quite large and this type of cable is easy to bend. The NEC requires that all fine-stranded cables must be terminated with specifically designed terminals, lugs, devices or connectors. In other words, they cannot simply be stripped and wrapped around a terminal, then tightened into place.

When making the battery connection, lightly brush and coat the battery terminal and the cable lug with protective grease. Torque the connection hardware to its recommended setting.

After the series connection has been made (for example, four 12-volt batteries to create a 48-volt system), check the voltage to ensure all connections have been properly made.

If the measured voltage is not as expected, track down the cause. It may be a loose connection or reversed polarity of one of the connections. If a battery is installed with reverse polarity, the measured voltage for the string of batteries will be reduced from the expected value by twice the open-circuit voltage of an individual battery.

If multiple series are connected in parallel, connect them together following the same process as before. When testing the system voltage, the expected voltage should be the average of all the string voltages.

It is generally recommended that no more than three (3) strings of batteries be connected in parallel. If the system requires more power than can be supplied by three (3) strings, select a larger battery.

Chapter 11 Review Questions

1. Most stand-alone inverters and battery banks sold in the US operate at:
   A) 12 Vdc, 24 Vdc or 48 Vdc nominal voltage.
   B) 100 Vdc nominal voltage.
   C) 600 Vdc nominal voltage.
   D) a voltage set by local jurisdictional requirements.

2. A charge controller must be compatible with all the following EXCEPT:
   A) the battery bank's nominal voltage.
   B) the array's nominal voltage.
   C) the inverter's power rating (in watts).
   D) the array's output rating (in watts).

3. A load diverter:
   A) reduces the system load demand when the array is not generating enough power.
   B) redirects excess power to the grid for net metering through a grid-tied inverter.
   C) reduces the load from the array so the battery bank does not overcharge.
   D) redirects excess power from the array to a load dump, such as a hot water heater.

4. A charge controller that allows a higher voltage connection from the array and converts the signal to a lower voltage that is compatible with the battery bank is typically referred to as a/an:
   A) rectifying charge controller.
   B) load dump charge controller.
   C) auto-transformer charge controller.
   D) MPPT charge controller.

5. When sizing a load dump, the:
   A) diverted load current must not exceed the maximum current rating for the charge controller.
   B) diverted load must have a voltage rating greater than that of the battery bank.
   C) diverted load power rating must be at least 150% of the maximum power output rating of the array.
   D) all of the above.

6. A stand-alone PV system has the following characteristics: a 3 kW PV array, daily load demand of 10 kWh, a maximum power draw of 2 kW at any time, a 1400 Ah battery bank, a nominal battery bank voltage of 48 Vdc and 4 hours of peak sunlight. What is the minimum power rating required for this system's inverter?
    A) 2 kW
    B) 3 kW
    C) 10 kW
    D) 12 kW

7. Which of the following is NOT an important specification to evaluate when selecting a stand-alone inverter?
    A) nominal voltage
    B) IEEE 17024 compliance
    C) peak power output rating
    D) ground fault protection

8. When sizing a stand-alone system's array, the designer must use load demand and average insolation from the month with the highest _____ rather than relying on annual averages, as would be done with a grid tied system.
    A) design ratio
    B) insolation factor
    C) peak power output
    D) inverter/load ratio

9. The difference between the amount of power that goes into a battery (in charging) and the amount that leaves (available to power the load), is known as the battery's:
    A) design ratio.
    B) roundtrip efficiency.
    C) load demand factor.
    D) inverter/load ratio.

10. The voltage of an array in a stand-alone system that incorporates a MPPT charge controller, a nominal 48-volt battery bank and an inverter with a 120 Vac nominal output is limited to:
    A) nominal 48 Vdc.
    B) nominal 120 Vdc.
    C) nominal 100 Vdc.
    D) the maximum voltage limit of the MPPT charge controller.

11. A cabin sits on a mountaintop in North Carolina. The location has a latitude of 35°N and 4.8 hours of insolation. The daily load requirement is 3 kWh. The designer has installed a 48 Vdc stand-alone system that has been derated at 75%. The inverter is 95% efficient. The temperature extremes at this location are -37°C (coldest) and 43°C (hottest).The battery bank provides two days of autonomy and uses 6-volt, 250 Ah batteries to a 60% depth of discharge. The battery will be stored in a unit that maintains a constant 25°C temperature.

    How many batteries did the designer anticipate would be required?
    A.) 1
    B) 3
    C) 8
    D) 24

12. A remote cabin sits on a mountaintop in Colorado. The owner has installed a stand-alone PV system with a 600 Ah 48-Volt (nominal) battery bank. The depth of discharge is set at 50%. The battery bank was sized based on standard test conditions with average temperatures of 25°C (77°F). But experience has shown that the average temperature where the batteries are stored is actually 10°C (50°F).

    What size battery bank should be installed on this site?
    A.) 531 Ah
    B) 600 Ah
    C) 678 Ah
    D) 1,200 Ah

13. Which of the following is **NOT** a factor in determining the size (in amp-hours) of a stand-alone system's battery bank?
    A) The daily system load requirements.
    B) The nominal voltage of the system.
    C) Peak hours of sunlight (insolation) for the location.
    D) Depth of Discharge (DoD) of the battery bank.

14. Fluctuations in the DC waveform caused by connecting the DC system to an AC system are referred to as:
    A) clipping.
    B) resistance.
    C) ripple.
    D) voltage drop.

Lab Exercises : Chapter 11

11-1. Sizing a Battery Bank

The purpose of this exercise is to practice sizing a battery bank.

A 48-volt stand-alone system with a 95% efficient inverter must service a 6 kWh daily load. The owner requires 3 days of autonomy and wants the batteries to discharge at a 50% DOD.

a. Derate for inverter efficiency.
b. Determine the daily amp-hour capacity requirement.
c. Select a battery from various online options.
d. Determine the battery bank Ah capacity.
e. Determine the number of strings required.
f. Determine the number of batteries required.
G. How would the battery bank configuration change if you allowed the batteries to discharge to 90% DOD?

Lab Exercise 11-1:

Supplies Required:
- none

Tools Required:
- Calculator
- Access to the Internet

Chapter 12

Job Site Safety

People who work with or come in contact with photovoltaic systems are exposed to a certain amount of risk. Developing a culture of safety within yourself and the company you work for will help minimize these risks.

Safety First

PV installers often work at height (on ladders and rooftops) that expose them to fall hazards. Working with electrical circuits can result in injury and/or death from electrocution and arc-flash.

There are dangers from working long hours on hot rooftops in direct sunlight. The use of power tools and heavy equipment operation poses the constant risk of injury. Many a back is injured by lifting heavy loads improperly. And then there are the respiratory hazards and other risks associated with working within confined spaces.

Developing a safe workplace culture is not only good practice, it is good business and it is the law.

From the perspective of the business owner, aside from the obvious pain, suffering and trauma associated with a workplace injury, there is also:
- lost productivity due to down-time,
- time lost by an injured employee,
- time lost by others helping the accident victim,
- cleanup and start-up of operations interrupted by an accident,
- time to hire and/or train a new worker to replace the injured worker until they return to work,
- time and cost to repair or replace materials damaged in the accident,

Chapter Objectives:
- Understand the importance of a culture of safety on the job site.
- Identify good safety policies and procedures.
- Explore and describe various personal protective equipment.
- Identify safe job site practices in organizing a task and working with tools.
- Describe safe climbing techniques and fall prevention.
- Identify proper lock-out/tag-out procedures.
- Understand the role of safety data

- cost of paying the employees wages while they are recovering,
- cost of completing paperwork generated by the accident,
- any costs resulting from lawsuits.
- increased insurance and worker's compensation costs,
- OSHA penalties,
- and perhaps the inability to bid on future jobs due to a poor safety record.

BUREAU OF WORKER'S COMPENSATION

According to the **Bureau of Worker's Compensation**, the average cost (in the US in 2020) for workplace accidents caused by slip and fall injuries was $41,757. For accidents involving electrical burns, it was $52,161. If a worker is killed on the job, the average cost was $1,340,000.

OSHA (the Occupational Safety and Health Administration) is responsible for creating the laws that govern worker safety for private-sector employees in all 50 states. It was established by the US Congress in 1970 within the Department of Labor.

OSHA establishes workplace safety requirements, enforces them (through site visits and penalties for non-compliance) and offers consulting services to companies to help improve workplace safety.

Companies risk fines from OSHA if they fail to put workplace safety policies in place, do not supervise the implementation of workplace safety practices, do not enforce workplace safety practices, or do not adequately train employees in workplace policies and procedures.

NFPA 70E

NFPA 70E is an internationally accepted standard that defines electrical safety–related work practices. It's goal is to define safety processes that use policies, procedures, and program controls to reduce the risk associated with the use of electricity to an acceptable level.

Safety Begins Before Arriving at the Site

Before a single worker sets foot on the job site, the job supervisor should have asked and answered questions such as the following:
- Are safety policies and procedures in place?
- Have all employees been trained in the company's safety practices?
- Is there a process in place to enforce safety practices?
- Are employees trained to safely perform the job with which they are tasked?
- What prep-work is required at the site prior to the job?
- What safety equipment is required at the site and is it in place and in good working order?
- Are employees working from ladders or roof tops? Is fall protection required?
- Are there any severe or unusual conditions at the site that may impact the safety of workers?

- Is someone on the job site at all times who is trained in first aid and emergency procedures?
- What medical facilities are nearby and what is the plan of action should an accident occur?
- What job tasks are required after the installation has been finished and has enough time been allocated to complete them safely?

Safety Policies and Procedures

Every installation should have safety policies and procedures in place prior to beginning the job. These practices should also be well communicated to every individual who will be on the job site.

These safety practices should include (but are not limited to):
- a personal protective equipment policy,
- procedures for using power tools and extension cords,
- maintaining an uncluttered work environment,
- proper lifting and carrying procedures,
- proper ladder use policies,
- processes to reduce the risk of heat exhaustion and de-hydration,
- policies and procedures for working with solar electric PV panels,
- ensuring proper safety equipment is in place on the job site,
- avoiding electric shock and arc-hazard risks,
- establishing lock-out tag-out procedures.

Personal Protective Equipment

Nothing can assure that an individual will not sustain an injury on the job. But the use of properly maintained **personal protective equipment (PPE)** can help reduce the risk of injury should something unexpected occur.

> PERSONAL PROTECTIVE EQUIPMENT (PPE)

Some examples of personal protective equipment include:
- foot and leg protection (safety shoes, for example),
- eye and face protection (ANSI Z87.1-1989) - as shown in Figure 12-1,
- head protection (hard hats),
- protective gloves,
- hearing protection,
- personal fall arrest systems (PFAS)

FIGURE 12-1: MUST WEAR ANSI Z87.1 RATED SAFETY GLASSES WHEN WORKING WITH ELECTRICITY

Hard Hats

Employers must ensure that their employees wear head protection if:
- objects might fall from above and strike them on the head,
- workers might bump their heads against fixed objects, such as exposed pipes or beams, or
- there is a possibility of accidental head contact with electrical hazards.

Hard hats are divided into three industrial classes:
- **Class A hard hats** provide impact and penetration resistance along with limited voltage protection (up to 2,200 volts).
- **Class B hard hats** provide the highest level of protection against electrical hazards, with high-voltage shock and burn protection (up to 20,000 volts). They also provide protection from impact and penetration hazards by flying/falling objects.
- **Class C hard hats** provide lightweight comfort and impact protection but offer no protection from electrical hazards.

Because ultra violet light degrades the plastic (makes it brittle), hard hats are dated (see Figure 12-2). Typically a hard hat will have to be replaced after every three years of use.

Fall Protection

Fall protection is required whenever there is the possibility that a worker may fall more than six feet. This protection may take the form of guardrails, cages, safety nets or **personal fall arrest systems (PFAS)**, such as a **safety harness** and **lanyard**, as illustrated in Figure 12-3.

The safety harness has a metal ring attached at the back to which a safety line is attached. The other end of the safety line is then secured to the roof. The anchorage for a fall-arrest system must support at least 5,000 pounds (2,270 kg). PFAS are required if workers have the potential to fall 10 feet (3 meters) or more.

Fall protection must be provided when employees are exposed to the following hazards where the possibility of falling 6 feet (1.8 meters) or more is present:
- holes (such as roof openings or skylights),
- wall openings,
- established floors, mezzanines, balconies, roofs and walkways with unprotected sides or edges,
- excavations.

A **guardrail system** is a common type of fall protection. It consists of a top rail, mid-rail, and intermediate vertical member. They also often incorporate toe boards that prevent materials from rolling off of the work surface.

Guardrail systems must be free of any imperfection that might cut a worker or snag a worker's clothing. Top rails and mid-rails must be at least ¼ inch thick (6 mm) to reduce the risk of cutting worker's hands. Steel or plastic banding cannot be used for top or mid-rails.

Procedures for using Power Tools

Power tools, such as drills, are common on the job site. They must be maintained in safe working condition, and be inspected by a competent person prior to use to ensure they are in safe operational condition.

Where ever possible, use battery-operated power tools to avoid the hazards associated with electrical cords on the site. Ground fault circuit interrupters should be installed on construction sites using single-phase temporary receptacles.

To prevent hazards associated with the use of power tools, installers should take the following precautions:
- Never carry a tool by the cord.
- Never pull the cord to disconnect a power tool from the receptacle.
- Keep cords and hoses away from heat, oil, and sharp edges.
- Disconnect tools when they are not in use.
- Secure work with clamps or a vise, freeing both hands to operate the tool.
- Maintain tools. Keep them sharp and clean for best performance.
- Follow all instructions in the user's manual.
- Be sure to keep good footing and maintain good balance when operating power tools.
- Wear proper apparel for the task. Loose clothing, ties, or jewelry can become caught in moving parts.
- Remove all damaged portable electric tools from use and tag them: "Do Not Use."

Maintain an Uncluttered Work Environment

A clean and orderly job site is a safe job site. Construction debris scattered about the site can present a trip and fall hazard. Extension cords, ropes and safety lines can also become unsafe obstacles.

Tools, equipment and debris left on the roof can not only present a tripping hazard, but might also fall and injure a worker standing below.

Also make sure the roof and ground areas are free of water, oils, ice or other materials that can make footing slippery or hazardous.

Proper Lifting and Carrying Procedures

Much of the equipment used on a jobsite is heavy and/or awkward. Batteries are especially heavy and solar panels are large and can weigh upwards of 50 lbs (23 kg). Without using proper lifting and carrying procedures, it is quite easy and common for an installer to injure his/her back.

When lifting an object off the ground:
- lift with the legs (not with the back),
- do not twist while lifting,
- load heavy items near the back of the bed of a truck to avoid having to push them along the truck bed,
- always use two people to lift and load large and heavy items,
- if the item is difficult to manage safely by hand when lifting it to the roof, find another method (crane or lift or scaffold),
- use carts and/or dollies when possible.

Proper Ladder use Policies

One obvious disadvantage of a roof-mounted system is that it is located up on a roof. The installation process makes it necessary to get the materials, equipment and people up there and to accomplish this in a safe manner.

Commonly, roof access is achieved with the use of ladders. But ladders are prone to damage, so inspect all ladders prior to each use.

When working with electricity (quite common when working on a PV system), only **fiberglass** ladders should be used. These are non-conductive and will help keep those working with electricity safe from the ladder accidentally coming in contact with a "live" wire and conducting this charge to the worker.

FIBERGLASS

Ladders are rated and color-coded based on how much weight they are designed to carry safely, as indicated in Table 12-1. Bear in mind that this weight rating does not just apply to the weight of the person climbing the ladder, but also includes the tools and materials that person might be carrying while on the ladder.

When placing a portable ladder against a building, the base should be about one-fourth the distance from the building as the height where the top of the ladder touches the building.

In other words, if the ladder touches the eave of the building at 16 ft (4.9 m) above ground level (as illustrated in Figure 12-4), then the base should be placed four feet from the vertical line where the ladder touches the structure.

TABLE 12-1: RATING SYSTEM FOR LADDERS

Description	Type	Load Rating	Color
Light Duty	Type III	200 pounds	Red
Medium Duty	Type II	225 pounds	Green
Heavy Duty	Type I	250 pounds	Blue
Extra Heavy Duty	Type IA	300 pounds	Orange
Special Duty	Type IAA	375 pounds	Yellow

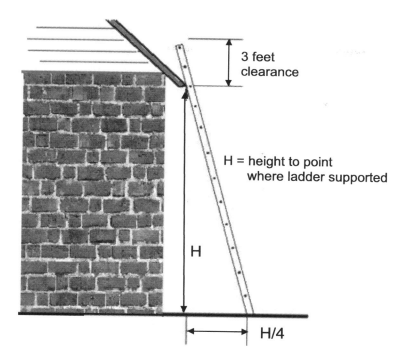

FIGURE 12-4: PROPER PLACEMENT OF A LADDER

Additionally, the top rung of the ladder should extend at least 3 ft (.91 m) above the roof line.

Three Points-of-Contact Climbing Technique
When climbing a ladder, utilize the **three points-of-contact** technique.

This technique requires that at all times the climber must face the ladder and have two hands and one foot, or two feet and one hand in contact with the **cleats** and/or **rails**. This will minimize the chance of slipping and falling from the ladder.

It is important to note that the climber must not carry any objects in either hand that can interfere with a firm grip on the ladder.

THREE POINTS-OF-CONTACT

CLEATS

RAILS

Carrying Ladders
Always carry ladders horizontally when moving them from place to place. Look up when moving ladders in order to avoid them coming in contact with overhead power lines. Use two people to carry extension ladders whenever possible.

FIGURE 12-5: BASE OF LADDER PROPERLY SECURED

(FROM DIY ADVICE)

Securing Ladders
It is safer, and often required, that a ladder placed against the building be secured at the top and at the bottom. While there is no standard way of accomplishing this, on dry level ground

FIGURE 12-6: INSTALLED POINT WHERE THE LADDER CAN BE SECURED TO THE ROOF.

FIGURE 12-7: A STRAP IS CLAMPED TO THE FASCIA BOARD TO PROVIDE LADDER STABILITY

(FROM QUALCRAFT)

the base can be secured by tying it with a rope to a stake driven into the ground or bolted to the wall (as demonstrated in Figure 12-5).

The top of the ladder should also be secured on the first trip up the ladder. There are a number of products available on the market to accomplish this task (as illustrated in Figure 12-6) or it can be secured temporarily, as demonstrated in Figure 12-7.

Other ladder climbing safety tips include:
- Always face the ladder when climbing.
- No more than one person on the ladder at one time.
- Use the 3-point climbing technique.
- Keep the area around the bottom and top of the ladder clear of debris or obstacles.
- Do not use a ladder when working in windy conditions.
- Raised ladders should never be left unattended.
- Wear slip-resistant shoes when climbing ladders.
- Keep your body centered, your belt buckle should never stray outside the rungs of the ladder.

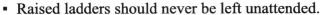

Any fixed ladder (one which is permanently mounted to the building) that is over 24 ft (7.3 m) tall must incorporate fall protection, such as a safety line or a cage.

Reduce the Risk of Heat Exhaustion and De-hydration

As the heat index rises above 103° F (40° C), there is a high risk of heat-related illnesses and injuries to workers. The heat index is a combination of heat and humidity - basically how hot it "feels" to humans. And remember, it is much hotter on a roof than the ambient air temperature.

To reduce the risk of heat exhaustion and de-hydration, workers should:
- drink water often. Workers need about four cups of cool water each hour in high temperatures.
- work in the cool of the morning and in the evening if at all possible (avoiding the hottest part of the day).
- be aware of the signs of heat-related problems. If a worker suddenly becomes confused, or uncoordinated, it may be a sign of heat stroke. Heat stroke can be fatal if not treated immediately.
- take frequent breaks and rotate workers to handle strenuous tasks.

- provide workers with personal cooling devices (for example., water-dampened clothing, cooling vests with pockets that hold cold packs, reflective clothing, or cool mist stations).
- provide shade and fans if possible.
- encourage workers to wear sunscreen, hats and sun glasses.
- set up a buddy system to watch for signs of fatigue and/or heat related distress.

Electric Shock and Arc Flash Risks

When an individual comes in contact with an exposed conductor that is carrying a current, an electric shock may occur. It doesn't take much current for the body to feel the effects of electricity.

- At 0.001 amps, skin will feel the tingle of an electric current.
- At 0.02 amps, muscles can freeze, often making it impossible to let go of the electrical source.
- At 0.1 amps, ventricular fibrillation can occur. This is an uncontrollable contraction of the heart that may result in death if medical help is not immediately available.
- At 0.5 amps, the heart can cease to function.
- At 1.5 amps, skin can start to burn and death is often the immediate result.

The damage to the body from even small amounts of electrical current can be severe and rapid.

Electric shocks are not the only risks inherent when working with electricity. An arc flash, such as illustrated in Figure 12-8, can occur when electricity jumps (short circuits) between two connectors.

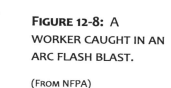

FIGURE 12-8: A WORKER CAUGHT IN AN ARC FLASH BLAST.

(FROM NFPA)

Common injuries from arc flashes include:
- burns. An arc flash contains plasma that can reach temperatures of 35,000°F (19,400°C), or four times the surface temperature of the Sun. Some studies indicate that up to 80% of the injuries received by electrical workers are the result of burns from arc flashes.
- eye damage. Unprotected eyes can result in severe damage and even blindness when subjected to the intense heat and light of an arc flash.
- blast pressure. Blast pressure waves have thrown workers across rooms and knocked them off ladders. Pressure from an arc flash can be higher than 2000 psf (96 kPa). Workers can be injured by the blast, thrown into a more dangerous place, or suffer hearing loss.

The best response to the threat of electric shocks and arc flash is prevention, through:
- knowledge. The better an installer's understanding of how electricity works, the safer the work environment. Ensure that all systems are properly grounded.
- situational awareness. Stay alert to the hazards involved in working with and near electricity.
- insulated tools and protective devices. Using properly insulated tools and wearing insulated gloves, shoes and clothing will help protect against electrical shocks.
- testing. Always assume conductors and metal near conductors are "live." Test them for voltage using a multimeter prior to working with or near them.
- mark and avoid buried cables.
- proper work procedures, such as a well communicated and enforced lock-out, tag-out procedure.

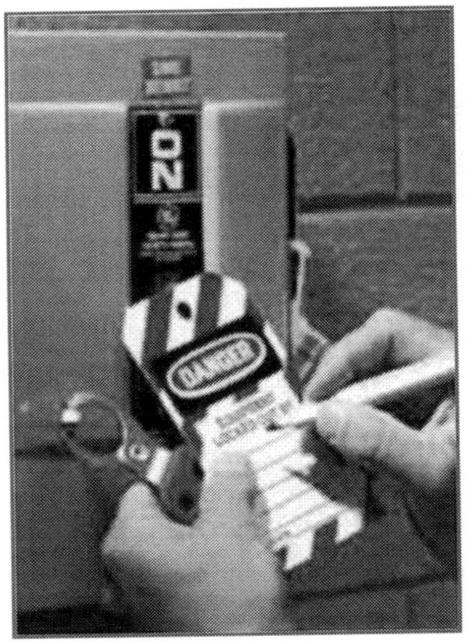

FIGURE 12-9: DEVELOP AND USE A COMPREHENSIVE LOCK-OUT, TAG-OUT PROCEDURE WHEN WORKING WITH ELECTRICITY

LOCK-OUT/TAG-OUT

Lock-Out Tag-Out Procedures

The **lock-out/tag-out (LOTO)** procedure is designed to ensure that electrical devices are properly shut off and not started up again prior to the completion of maintenance or servicing work.

Procedures may vary, but typically they include:
- turning off the device or disconnect to ensure it has been "isolated and rendered inoperative."
- locking the device in some manner so it cannot easily be turned back on by mistake.
- tagging the device with a note (as shown in Figure 12-9) to let others know the device should remain inoperative until the service or maintenance work has been completed.

MSDS or SDS Documents

Another safety concern on the job site may include working with hazardous chemicals or substances (such as is present in battery systems).

MATERIAL SAFETY DATA SHEET (MSDS)

Manufacturers of products that may contain hazardous material are to provide a **Material Safety Data Sheet (MSDS)** and these should be available on the job site and a copy provided to the customer.

Over the past several decades the Globally Harmonized System of Classification and Labelling of Chemicals also known as GHS has been adopted by most nations and the traditional MSDS name for these documents has changed to **Safety Data Sheets (SDS)**. The GHS is an internationally agreed-upon standard managed by the United Nations that was set up to replace the assortment of hazardous material classification and labeling schemes previously used around the world.

SAFETY DATA SHEET (SDS)

The SDS is a document that provides health and safety information about products, substances or chemicals that are classified as hazardous substances or dangerous goods. The SDS should provide information on:
- the manufacturer or supplier,
- the name, ingredients and properties of the product,
- how the product can affect people's health,
- precautions for using or storing it safely.

Chapter 12 Review Questions

1. In the U.S., rules regarding climbing safety are generally formed by:
    A) the International Building Code (IBC).
    B) the National Electrical Code (NEC).
    C) the American National Standards Institute (ANSI).
    D) the Occupational Safety and Health Administration (OSHA).

2. Equipment worn to minimize exposure to hazards that cause serious workplace injuries and illnesses such as safety helmets, gloves, eyeglasses, earplugs, full-body suits, vests, hard hats, safety footwear, and respiratory protective equipment (RPE) are referred to as:
    A) workplace safety equipment (WSE).
    B) personal protective equipment (PPE).
    C) OSHA compliant equipment (OCE).
    D) health and safety equipment (HSE).

3. Which of the following would **NOT** be considered personal protective equipment?
    A) a hard hat
    B) a guard rail
    C) harness and lanyard
    D) gloves

4. To avoid electrical shocks if a ladder comes in contact with a charged conductor, ladders should:
    A) be made of fiberglass.
    B) be less than 24 feet tall.
    C) be rated to carry the weight of the person climbing as well as any tools and/or material they may be carrying.
    D) be secured at the top and at the bottom of the ladder.

5. Which ladder is rated to carry 250 pounds?
    A) Light Duty, Type III, Red
    B) Medium Duty, Type II, Green
    C) Heavy Duty, Type I, Blue
    D) Extra Heavy Duty, Type IA, Orange

6. When placing a ladder against a building, the:
    A) top of the ladder should extend at least three feet above the point where it comes in contact with the structure.
    B) base of the ladder should be offset from the point where it comes in contact with the building by four feet.
    C) the ladder must incorporate fall protection if the operator is in danger of falling at least six feet from one level to another.
    D) none of the above.

7. When climbing a ladder, always:
   A) use the three-point climbing technique.
   B) secure the top and bottom with a rope when there is a potential to fall more than six feet.
   C) conduct a lock-out, tag-out process before mounting the first cleat.
   D) wear a ANSI-approved eye protection and hard hat.

8. Which of the following is **NOT** a common injury from electric shocks or arc flashes?
   A) burns
   B) dehydration
   C) eye damage
   D) blast pressure

9. Which of the following is a practice to avoid electric shocks while working on a system?
   A) Mark and avoid all buried cables.
   B) Use a three-point climbing technique.
   C) Wear sunscreen, hats and sun glasses.
   D) Always ensure that disconnects are in the closed position before working on electrical wires.

10. Which of the following is **NOT** part of a normal lock-out tag-out process?
    A) Identify and mark all buried cables.
    B) Turn off all connected devices and ensure disconnects are in the open position.
    C) Lock any breakers and/or disconnects so they cannot be closed.
    D) Notify co-workers or occupants of the building that the system has been intentionally shut down.

11. The document that provides health and safety information about products, substances or chemicals that are classified as hazardous substances or dangerous goods is referred to as the :
    A) Safety Data Sheet (SDS).
    B) Authority having Jurisdiction (AHJ).
    C) Product Safety Instructions (PSI).
    D) OSHA 10 Guidelines.

Lab Exercises: Chapter 12

12-1. Climbing Safety

The purpose of this exercise is to practice safe climbing techniques and become familiar with personal fall arrest systems (PFAS).

> **Lab Exercise 12-1:**
>
> *Tools Required:*
> - Fiberglass extension ladder
> - Hard hat

a. Place the extension ladder properly against a one-story building.
b. Secure the top and bottom of the ladder or have someone hold it steady during this exercise.
c. Put on a safety harness and attach the lanyard.
d. Climb the ladder using the three points-of-contact technique.
e. Examine the roof and determine if there are any structural concerns as well as the proper orientation of the rails of any racking system that may be placed on the roof.
f. Descend the ladder using the proper climbing technique.

12-2. Harness and Lanyard

The purpose of this exercise is to become familiar with personal fall arrest systems (PFAS).

> **Lab Exercise 12-2:**
>
> *Tools Required:*
> - Safety harness and lanyard
> - Hard hat

a. Examine the harness and lanyard and understand how these are worn.
b. Have someone assist in putting on the harness and lanyard.
c. Examine the hard hat, make sure it has not expired.
d. Adjust and wear the hard hat.

Chapter 13

Paperwork

There is no escaping the fact that every project comes with its fair share of paperwork. It is the grease that keeps the wheels of progress in motion (or stops it dead in its tracks). The installation of a photovoltaic system is no exception.

Project Documentation

Some paperwork that is typically required in the design and installation of PV systems include:
- permits
- interconnect agreements
- insurance
- manufacturer's instructions
- structural details
- commissioning forms
- design diagrams
- labels
- warning signs.

Permits

After the project has been designed, but before installation can begin, it is important to obtain all the proper **permits** required for the particular jurisdiction. Requirements will vary dramatically from location to location - so there is (unfortunately) no substitute for a bit of legwork at various governmental departments.

However, for a grid-tied system, the first stop should be to speak with the local electrical utility provider. They will likely be able to help guide the designer regarding permits and **licenses** that may be required. But they will also have a few requirements of their own.

Chapter Objectives:
- Identify the various documents required in planning and permitting a residential PV system.
- Organize a complete and comprehensive application package for permitting.
- Recognize the various labels and warnings that must be incorporated into the PV system and note where they are to be placed.
- Create an operation and maintenance documentation plan.

PERMIT

LICENSE

Utility

Most utilities have created standard forms required from homeowners who intend to install a PV system. Bear in mind that there will likely be fees associated with each of these requirements.

These include:
- **interconnection agreement.** This document sets forth all the terms and conditions under which the system can be connected to the utility grid. It includes information about the homeowner's obligation to obtain permits and insurance. It also outlines how the system must be operated and maintained.
- **net metering agreement.** This document, often called the "sale/purchase agreement", spells out the conditions under which the excess power from the system will be purchased by the utility, and the rates that will be charged to the homeowner for the power purchased from the utility.
- insurance requirements. Most utilities require the homeowner provide insurance to protect the utility's system and personnel from any problems that might be the result of connecting the photovoltaic array to the grid. Basic homeowner's insurance may already be adequate, or a rider may need to be purchased to meet this requirement.

The representative from the utility company may also be able to outline other permits that may be required within their jurisdiction.

The utility will generally request information regarding the system, such as:
- copy of the most recent utility bill,
- electric meter ID number,
- number of electric vehicle(s) charging on site,
- name of any authorized third party to act on behalf of the customer for the interconnection process (solar installation company),
- system size,
- building square footage,
- any rebate programs applied for,
- point of interconnection in relation to main breaker,
- distance of AC disconnect from main service panel,
- installation specifications (mounting method, tilt, azimuth),
- equipment make, model and quantity,
- single line diagram,
- service panel short circuit interrupting rating (SCIR),
- photographs of the site,
- copy of final building permit/inspection certificate.

- INTERCONNECTION AGREEMENT
- NET METERING AGREEMENT

Authority Having Jurisdiction

Requirements vary widely from place to place, as will the fees charged. But applying for a permit requires preparation. Typically an authority having jurisdiction (AHJ) will require:

- an **electrical permit** (assessing the system from an electrical perspective, such as wiring size, bonding and grounding, equipment ratings, etc).
- a **mechanical permit** (assessing the structural elements such as the racking system design, effects of the PV load on the structure of the home, etc. These are generally required if the array is to be mounted on a building.
- licenses and/or certifications may also be required by the local jurisdiction. It may be necessary to check with the state's licensing board to determine what licenses are required to install PV systems. For example, a licensed electrician might be required to pull the permit or to install the final connection to the utility. Or a municipality may require that the system be designed by a certified professional (certified in PV design through an organization such as ETA-I or NABCEP).

Application Package

Much of the success of the permitting process depends upon how prepared the applicant is when first approaching the permitting agency. A complete application packet will be invaluable years down the line in monitoring and maintaining the system. Don't neglect this important step of the overall process. There is no such thing as providing too much information at this stage.

The application should include:
- the site address and contact information.
- the make and model of all specified equipment (as well as copies of the manufacturer's specifications).
- system component warrantees.
- a one-line drawing of the PV design.
- the racking manufacturer's installation instructions (displaying the array support structure). Also provide an engineer-stamped drawing if available (many racking manufacturers provide this at no charge).
- any structural assessment of the building's ability to support the load (from an engineer or the rack manufacturer). For a roof mounted system, information may include: age of roof, roof type, rafter size and spacing, rafter span, weight of array.
- a **plan view** that provides a "bird's eye" view of the site, showing all the locations of the various components.
- photos of the site.
- a copy of any necessary license (if required).
- a copy of any necessary certification (if required).
- a copy of a utility bill (for grid-tied systems).

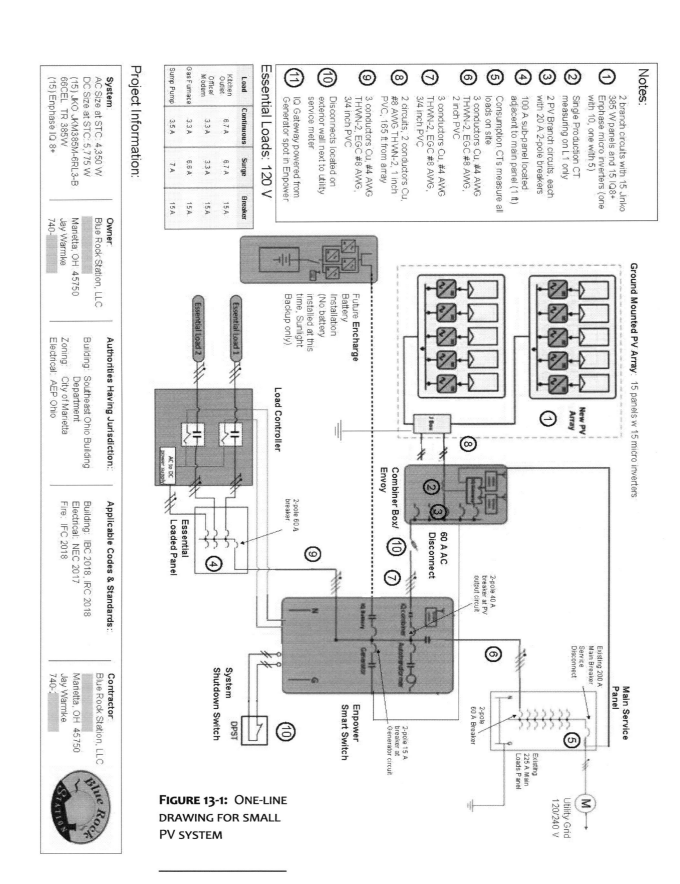

FIGURE 13-1: ONE-LINE DRAWING FOR SMALL PV SYSTEM

Notes:

1. 2 branch circuits with 15 Jinko 385 W panels and 15 IQ8+ Enphase micro inverters (one with 10, one with 5)
2. Single Production CT measuring on L1 only
3. 2 PV Branch circuits, each with 20 A 2-pole breakers
4. 100 A sub-panel located adjacent to main panel (1 ft)
5. Consumption CTs measure all loads on site
6. 3 conductors Cu, #8 AWG, THWN-2, EGC #8 AWG, 2 inch PVC
7. 3 conductors Cu, #4 AWG THWN-2, EGC #8 AWG, 3/4 inch PVC
8. 2 circuits, 2 conductors Cu, #8 AWG THWN-2, 1 inch PVC, 165 ft from array
9. 3 conductors Cu, #4 AWG THWN-2, EGC #8 AWG, 3/4 inch PVC
10. Disconnects located on exterior wall next to utility service meter
11. IQ Gateway powered from Generator spot in Empower

Essential Loads: 120 V

Load	Continuous	Surge	Breaker
Kitchen Outlet	6.7 A	6.7 A	15 A
Office/Modem	3.3 A	3.3 A	15 A
Gas Furnace	3.3 A	6.6 A	15 A
Sump Pump	3.5 A	7 A	15 A

Project Information:

System:
AC Size at STC: 4,350 W
DC Size at STC: 5,775 W
(15) JKO JKM385M-6RL3-B
66CEL TR 385W
(15) Enphase IQ 8+

Owner:
Blue Rock Station, LLC
Marietta, OH 45750
Jay Warmke
740-

Authorities Having Jurisdiction:
Building: Southeast Ohio Building Department
Zoning: City of Marietta
Electrical: AEP Ohio

Applicable Codes & Standards:
Building: IBC 2018, IRC 2018
Electrical: NEC 2017
Fire: IFC 2018

Contractor:
Blue Rock Station, LLC
Marietta, OH 45750
Jay Warmke
740-

- battery bank location and venting (for stand-alone and coupled systems).
- a schedule of warning labels that will be installed onto the system.

The One-Line Drawing
A **one-line drawing** is a simple electrical drawing that uses boxes to represent the various components of the system and lines to represent the conductors and conduits connecting these components together.

> ONE-LINE DRAWING

A very detailed one-line drawing will not only assist in the permitting process, but will be of great help during the installation of the system as well. The drawing can be made by hand or as is more common, created with the assistance of a CAD program.

The one-line drawing (as illustrated in Figure 13-1) should include the:
- site address and the installer's contact information,
- make and model of all the equipment,
- PV array system information (rating, modules, strings, STC panel specifications),
- hot and cold temperatures used for sizing strings (if required),
- wire sizes, types and run lengths,
- conduit sizes and wire counts within each raceway,
- general location of equipment (inside, outside, on roof, etc),
- specifications for the point of interconnection with the grid,
- conductor size calculations,
- backup load ratings,
- voltage drop calculations.

The Plan View
The plan view (Figure 13-2) provides a "bird's eye" view of where the system and its various components will be located on the property. Images from Google Earth can be used in creating this document.

The plan view is quite helpful in determining if all the components will fit on the site, or if any obstacles will prevent the design from being implemented in the "real world."

The permit office will often want to see where the electric meter is currently located and where the array disconnect switch will be located.

The Permitting Process
The solar permitting processes across the U.S. and within other nations are quite varied. Each state and city has its own code requirements, zoning, and targeted solar requirements. In addition to these variations, the size and type of solar array may make a difference in how permits are handled.

Complicated permitting, fees and wait times add significantly to the cost and frustration of installing a solar array. The Lawrence Berkeley National

FIGURE 13-2: PLAN VIEW FOR SMALL PV SYSTEM

Solar PV System Plan View

Address: Blue Rock Station, 1190 Virginia Ridge Rd, Philo, OH 43771

Contact Info:

Scope of Work: 7.5 kW array,
30 - 250 W panels
30 - Enphase micro inverters

Laboratory found that cities with streamlined permitting processes reduced installation costs between 4 to 12 percent.

Nations such as Germany and Australia have successfully streamlined basic design and installation processes, driving down the cost and wait-time associated with residential solar. In many cases, they have eliminated permits for residential PV systems altogether - instead requiring homeowners to simply apply for an interconnection request online.

Recently, several solar industry organizations such as the Solar Energy Industries Association (SEIA), the Solar Foundation and the National Renewable Energy Laboratory (NREL) have worked together to create permitting software, known as the **Solar Automated Permit Processing Plus (SolarAPP+)**. The goal of this program is to decrease the time and money it takes to review and approve solar permit applications.

SolarAPP+ allows installers to become accredited via a central online registration portal. It removes the need to appeal to local authorities for permission to install a system.

The main features of the plan include:
- replacing the current multi-step process with a skills training and a certification program that ensures contractors are compliant with all applicable codes, laws, and practices,
- a free, standardized online platform for local governments to "register" and automatically screen qualifying systems,
- a list of equipment standards and certified equipment,
- a standardized system design standard for qualifying solar projects,
- instantaneous permitting for systems installed by certified contractors.

SolarAPP+ was officially released in July 2021. The online tool will be provided to local governments at no charge - funded by registration fees charged to contractors.

As of December 2023, more than 160 communities in the United States have adopted SolarAPP+, using it to approve more than 32,800 projects, saving an estimated 33,000 hours of permitting staff time.

Labels and Warning Signs

The National Electrical Code (NEC) sets out a number of requirements for the labeling of PV systems.

Labels must be permanently affixed to equipment or wiring and may not be hand written. They must also be durable enough to withstand the environmental conditions to which they are exposed. A number of manufacturers make labeling packages available that comply with the various NEC updates.

Most enclosures will require some sort of label. Labeling requirements outlined in the 2023 NEC include:

FIGURE 13-3: MULTIPLE POWER SOURCE LABEL REQUIRED UNDER NEC 705.10 & NEC 710.10

On the Building
A label or plaque (Figure 13-3) should be installed at each service equipment location that indicates where it can be disconnected and contact information of off-site personnel responsible for servicing those power sources.

FIGURE 13-4: LINE AND LOAD SIDE SHOCK HAZARD LABEL REQUIRED UNDER NEC 705.20(7) AND NEC 690.13(B)

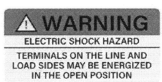

Combiner Boxes / Enclosures
PV systems have two sources of power, the array and the grid. As a result, it is often possible that both the line and load sides of a connection or disconnect can be energized when the circuit is broken. When this is the case, a warning label, such as Figure 13-4 must be placed on the enclosure.

FIGURE 13-5: TURN OFF AC DISCONNECT LABEL REQUIRED UNDER NEC 110.27(C) & OSHA 1910.145(F)(7)

A warning sign, such as in Figure 13-5, should be on all enclosures warning that the AC disconnect should be turned off prior to working on PV system wires.

DC Disconnect
The DC disconnect should be labeled indicating the shock hazard, the fact that it is a DC disconnect, as well as the maximum DC voltage under which the system might operate, as indicated in Figure 13-6.

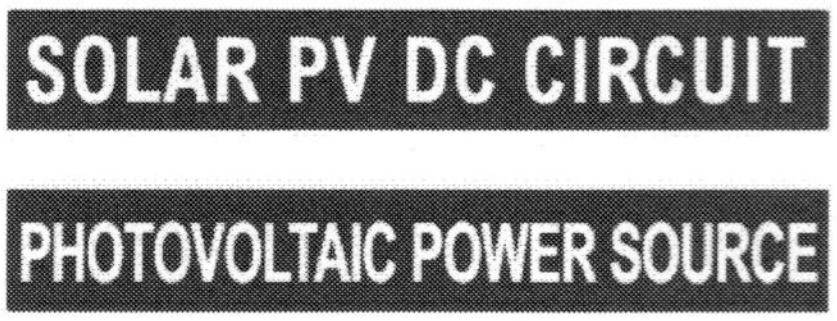

FIGURE 13-6: LABELS FOR DC DISCONNECT REQUIRED UNDER NEC 705.20(7) & NEC 690.13(B) & NEC 690.7(D)

Conduit or Raceways
NEC 690.31(D)(2) requires that labels, such as illustrated in Figure 13-7, must be placed no more than every 10 feet (3 m) apart on all conduit and/or raceways that contain PV system DC conductors.

FIGURE 13-7: LABELS FOR DC CONDUIT REQUIRED UNDER NEC 690.31(D)(2)

The labels should either say "Solar PV DC Circuit" or Photovoltaic Power Source."

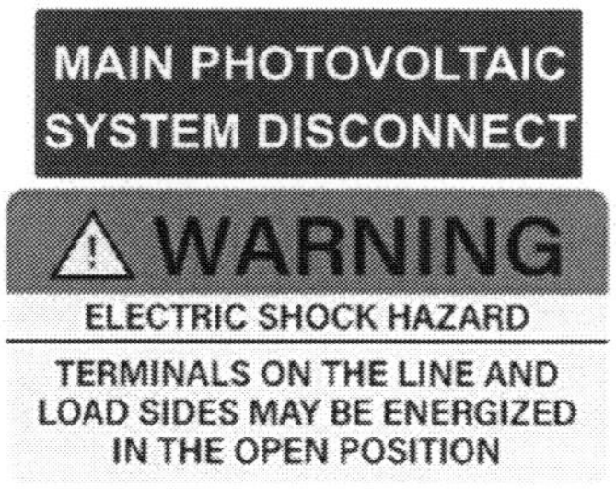

FIGURE 13-8: LABELS FOR THE AC DISCONNECT REQUIRED UNDER NEC 705.20 (7)

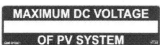

AC Disconnect
The AC disconnect should be labeled as indicated in Figure 13-8. It is indicated as the main service disconnect, and also warns that the line and the load sides may be energized when turned off.

FIGURE 13-9: Plaque indicating location of the system disconnect

When the AC disconnect cannot be located within visual sight of the service entrance (generally the utility meter), a sign or plaque should be placed at the site giving instructions as to where the disconnect is located (Figure 13-9).

Permission must be granted by the AHJ under this special situation.

The Service Panel (Load Side Connection)

When making a load side connection, there are a number of labels that are required on the utility service panel. Since the service panel can now be fed both from the utility as well as from the PV array, labels similar to those in Figure 13-4 should be affixed to the outside of the box.

The outside of the service panel box should also indicate that the system is supplied by more than one power source, with a label as shown in Figure 13-10.

FIGURE 13-10: Dual Power Source label required under NEC 705.30 (C)

This label should also be affixed to the utility meter to inform first responders, as well as the utility company, that there are multiple power sources present on the site.

Within the service panel box and next to the breaker that is connected to the PV array power source, a label, such as the one depicted in Figure 13-11, should be placed next to the breaker. This will warn that the breaker should not be moved, which might violate the 120% rule for load side connections of PV systems.

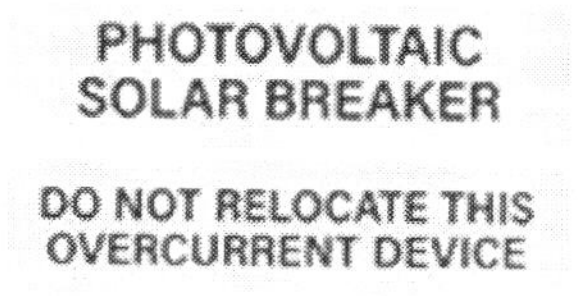

FIGURE 13-11: PV Breaker label required under NEC 705.12 (B)(2)

Battery Bank Disconnect

Systems that have an energy storage system (battery backup) must also have a way of disconnecting that power source. The disconnect should be located within 10 feet (3 m) of the ESS.

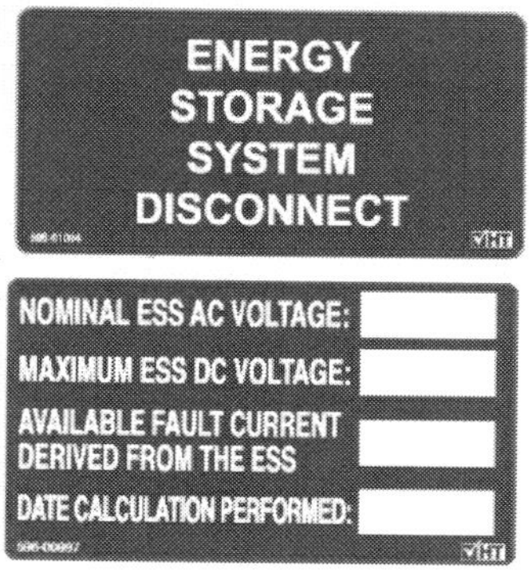

FIGURE 13-12: ESS Disconnect label required under NEC 706.15 (C)

Chapter 13: Paperwork   Page 357

It should also be clearly marked with a label similar to that in Figure 13-12.

Rapid Shutdown Labels

Systems that incorporate rapid shutdown (this should include all post-2014 rooftop installations on occupied buildings) should have a label such as shown in Figure 13-13, located on the main PV disconnect switch. In most cases this will be the AC disconnect.

FIGURE 13-13: LABEL FOR AC DISCONNECT WHEN RAPID SHUTDOWN IN PLACE REQUIRED BY NEC 690.12(D)(2)

Near the service entrance (normally the utility meter) there should also be a plaque or label that indicates the system is equipped with rapid shutdown protection.

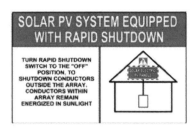

FIGURE 13-14: LABEL FOR SYSTEMS MEETING THE 2014 NEC RAPID SHUTDOWN REQUIREMENTS

Figure 13-14 is required for systems that meet the 2014 NEC requirement that conductors extending greater than 10 ft (3 m) from the array (outside) or more than 5 ft (1.5 m) inside a building are required to be de-energized to below 30 volts within 10 seconds of initiating rapid shutdown.

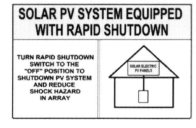

FIGURE 13-15: LABEL FOR SYSTEMS MEETING THE 2017 NEC RAPID SHUTDOWN REQUIREMENTS

Rapid shutdown systems that meet the requirements in the 2017 NEC should have a label similar to Figure 13-15. The 2017 NEC requirements that conductors WITHIN the array shut down to less than 80 volts within 30 seconds of initiating the rapid shutdown.

Older versions of the code required a specific color coding (red for 2014 and yellow for 2017). In 2023 these color codes were removed. The code now simply states that the label should be legible and contrast with the background (NEC 690.12(D)).

Commissioning Forms

The final step in the PV installation process involves testing and documenting the function and output of each component of the system. A complete record of the system's performance should be documented on a system commissioning form.

This commissioning form will provide a baseline performance record of how the system was functioning on the day it began operations.

Operation and Maintenance Documentation

Once the system is up and running, it is good practice to maintain documentation that will ensure the system functions for many years to come.

This **operation and maintenance (O&M) documentation** should include at a minimum:
- procedures for verifying the proper system operation.
- emergency shutdown procedures.
- a maintenance schedule.
- future site improvements that might affect the array.
- warranty documents for all components of the PV system.
- a troubleshooting checklist.

> OPERATION AND MAINTENANCE (O&M) DOCUMENTATION

Chapter 13 Review Questions

1. Which of the following will likely be required by the local utility before installation of the array can begin?
   A) a net metering agreement
   B) proof of insurance
   C) an interconnect agreement
   D) all of the above

2. Which of the following will likely **NOT** be required by the local building department prior to the installation of a PV system?
   A) an electrical permit
   B) a mechanical permit
   C) an interconnect agreement
   D) a license or certification

3. A simple electrical drawing that uses boxes to represent the various components of a PV system and lines to represent the conductors and conduits connecting these components together is called a(n) _____.
   A) one-line drawing
   B) plan view
   C) application package
   D) as built diagram

4. An online application software package that is increasingly being adopted by AHJ's across the US is known as:
   A) SolarAPP+
   B) PVWatts
   C) SAM
   D) Open Solar

5. When installing a PV system on a building (as opposed to a ground mount):
   A) a label must be placed at the main PV disconnect if the building is **NOT** equipped with rapid shutdown.
   B) the 2023 NEC no longer requires that rapid shutdown labels be placed on the electric meter.
   C) a label must be placed at the main PV disconnect if the building **IS** equipped with rapid shutdown.
   D) systems installed prior to 2014 must incorporate a rapid shutdown label at the main PV disconnect.

6. _____ provide a baseline performance record of how the system was functioning on the day it began operations.
   A) Operation and maintenance documents
   B) The plan view
   C) The electrical permit
   D) Commissioning forms

7. Emergency shutdown procedures are an example of information that may be included in _____.
   A) operation and maintenance documents.
   B) the plan view
   C) the electrical permit
   D) commissioning forms

Lab Exercises: Chapter 13

Lab Exercise 13-1:

Supplies Required:
- None

Tools Required:
- Internet access
- Calculator

13-1. One-Line Drawing

The purpose of this exercise is to practice designing a small residential system and creating a one-line drawing.

a. Select a site address and conduct a remote site assessment using Google Earth or Open Solar. Determine the best location for the site.
b. Estimate the monthly load demand for the site (use a utility bill if the example is a "real world" case study.
c. Calculate the array size.
d. Determine number of strings and if they will be combined or run directly to the inverter (through a junction box).
e. Calculate the ampacity requirements of all circuits.
f. Determine the conduit type and size of all circuits.
g. Perform a voltage drop calculation for the system. (You will need to estimate distances).
h. Create a one-line drawing for the system.

Chapter 14
Commissioning the System

Chapter Objectives:

- Outline what is involved in a well planned commissioning process.
- Identify the steps involved in a final installation checkout.
- Identify the steps involved in a visual inspection of the system.
- Verify compliance with all applicable codes.
- Conduct electrical verification tests.
- Conduct system function tests.
- Verify array power and energy production.
- Identify the various components involved in calculating the derate factor of the system.
- Outline what documentation is required during and after the commissioning process.

After the PV system has been installed, but before it has been turned on, there are a number of steps that should be completed. This process is known as **commissioning** the system.

This process involves visual observations, as well as tests, measurements, and documentation to verify that the system has been properly installed and will operate in a safe and efficient manner.

COMMISSIONING

Commissioning the System

A complete and thorough commissioning process serves a number of purposes.

It will:
- verify that the installation matches the plans.
- ensure compliance with all applicable code requirements.
- confirm that the system will operate safely when energized.
- test and verify that all the components are operating as specified.
- document the state of operation when the system was first employed.
- verify that the operation and maintenance documentation has been completed.

Many of the same tests conducted during the commissioning process will later be repeated as part of periodic maintenance. Later results can be compared to those obtained during commissioning to determine if there is any unexplained deterioration of performance.

A well-planned commissioning process includes:
- a final installation checkout.
- visual inspection of the system.
- verification of code compliance.
- electrical verification tests.
- system functioning tests.
- verification of array power and energy production.
- verifying AC power output.
- documenting all results.
- training the system owner.

Final Installation Checkout

The final installation checkout is conducted by the installer before initiating any tests. A **punch list** can be developed during the design phase of the project and then verified during the checkout process.

PUNCH LIST

These lists typically verify that:
- all structural components such as racking systems, conduit supports, battery cabinets, etc have been installed.
- all electrical circuits are in place and conduit/raceways installed as required.
- all wiring connections have been made.
- all equipment has been calibrated and set according to manufacturer's instructions.

Visual Inspection of the System

Once the punch list has been completed, it is time to visually inspect the system. Prior to inspecting, ensure that all disconnects are open, fuses are removed, and proper lock-out/tag-out (LOTO) procedures are in place.

It may be helpful to create a checklist of items to inspect during this phase of the commissioning process. These may include (but are certainly not limited to):

Racking Systems:
- Were footings installed at the proper locations and spaced properly?
- Was the proper hardware used and tightened to the appropriate torque?
- Is the racking system properly connected to the footings?
- Are the modules properly connected to the racking system with the appropriate hardware?

Conductors:
- Is the conductor insulation type installed suitable for the environment?
- Do wire gauge sizes match the design drawing?
- Are all conductors in place?
- Is there any visual sign of damage to any of the conductors?

- Is the color coding correct (green for ground, red for positive, etc)?
- Are all connections tight?
- Is the appropriate polarity maintained in the connection?

Conduits and Raceways:
- Have the correct conduit types been used in each environment?
- Is the conduit supported correctly?
- Have expansion fittings been installed where needed?
- Are all conductors secured "out of reach" of untrained persons that may come in contact with them?
- Are proper labels attached to the conduits and raceways?

Disconnects:
- Are the disconnects rated for the current of the circuit to which they are attached?
- Are they located in the proper places (as per the utility's requirements) and mounted properly?
- Is the correct conductor (ungrounded) being disconnected within the device?
- Is the disconnect rated for its environment (indoor, outdoor)?
- Are the disconnects properly labeled?

Grounding System:
- Have all the panels been grounded to the racking system?
- Have all the metal enclosures been properly grounded?
- Have any metal conduit and/or raceways been properly grounded?
- Have the grounding electrodes been placed in their proper location and bonded to the grounding conductor?
- Confirm that the ground fault protection fuse in the inverter is present.

Overcurrent Protection Devices (OCPD):
- Ensure the current rating of the OCPD is properly sized for the circuit.
- Ensure the voltage rating of the OCPD for the circuit.
- Do the OCPDs appear to be in good condition?

PV Modules:
- Are the correct modules installed (match the plans and specifications)?
- Are the correct number of modules installed in the array?
- Are the strings configured according to the plans?
- Are any of the panels damaged?
- Are the connectors tightly attached?
- Ensure that the proper connections are made within the junction box (negative to negative connector, etc), even if the unit came pre-wired from the manufacturer.

Inverters:
- Is there adequate clearance around the inverter for maintenance and cooling?

- Has the inverter been properly installed and configured as per the manufacturer's specifications?
- Is the inverter secure from environmental damage?
- Does the inverter's operating voltage window comply with the array's DC voltage output after it has been adjusted for temperature extremes?
- Does the inverter's output match the local utility?
- Is the inverter ANSI/UL listed?
- Has the inverter been properly labeled?

Charge Controllers:
- Has the charge controller been properly installed and configured as per the manufacturer's specifications?
- Is the charge controller secure from environmental damage?
- Does the charge controller's operating voltage window comply with the array's DC voltage output after it has been adjusted for temperature extremes?
- Does the charge controller's output voltage match the nominal voltage of the battery bank?

Battery Bank:
- Are the batteries installed in proper enclosures?
- Are the batteries properly vented to the outside?
- Are the correct batteries installed (match the plans and specifications)?
- Are the correct number of batteries installed in the bank?
- Are the strings configured to the proper nominal voltage?
- Are any of the batteries damaged or leaking?

Verification of Code Compliance

If not already done so during the design and installation portion of the project, ensure that the system is in compliance with all applicable codes.

NEC articles that apply to PV systems include (but are not limited to):
- Article 110: Requirements for Electrical Installations
- Article 230: Services
- Article 240: Overcurrent Protection
- Article 250: Grounding and Bonding
- Article 300: Wiring Methods
- Article 310: Conductors for General Wiring
- Article 314: Outlet, Device, Pull, and Junction Boxes
- Article 338: Service-Entrance Cable: Types SE and USE
- Article 344: Rigid Metal Conduit: Type RMC
- Article 356: Liquidtight Flexible Nonmetallic Conduit: Type LFNC
- Article 358: Electrical Metallic Tubing: Type EMT
- Article 400: Flexible Cords and Cables
- Article 408: Switchboards and Panelboards
- Article 445: Generators
- Article 450: Transformers

- Article 480: Storage Batteries
- Article 705: Interconnected Electric Power Production Sources
- Article 706: Energy Storage Systems

Electrical Verification Tests

The installation is complete, it has been visually inspected, all the components have been confirmed to be in compliance with the appropriate codes - now comes the time to test each of the components to ensure they are working properly.

Again, ensure that all disconnects are open and that proper lock-out/tag-out procedures have been followed prior to testing. Common electrical tests include:

- Continuity and **resistance tests** to verify the integrity of the bonding and grounding systems, conductors, connections and terminations. This can also be used to test the operation of disconnects as well as overcurrent protection devices.   `RESISTANCE TEST`

- **Polarity tests** to verify that the proper polarity has been maintained during the installation and connection process. Reversed polarity can result in damage to the modules, equipment and battery systems.   `POLARITY TEST`

    Before connecting the MC4 connectors on each panel, simply insert the positive lead from a multimeter set to read the appropriate level of DC voltage into the positive (male) connector, and the negative lead into the negative (female) connector. If the resulting reading is positive, then the polarity is correct. If the result is a negative number, then the polarity has been reversed.

    Also check to see that each panel is providing the expected voltage (they should all be about the same if exposed to similar conditions).

- Test the **voltage of each string** at the combiner box. Testing the open circuit voltage of each string can only be done on the line side (incoming from the array). Remember, the overcurrent protection should be in the open position.   `STRING VOLTAGE TEST`

- **Insulation tests** are necessary to check the integrity of the insulation around the conductors. Like a leak in a water pipe, any imperfection in the insulation can allow electricity to escape. This "leaking" is caused by conductors that may have been damaged during installation.   `INSULATION TEST`

    Insulation is subject to many external factors that can cause it to perform at a less-than-acceptable level. Excessive heat or cold, moisture, vibration, dirt, oil, and corrosive fumes can all contribute to deterioration. For this

MEGOHMMETER

FIGURE 14-1: A MEGOHMMETER, OR INSULATION TESTER

reason, routine insulation testing is necessary. Insulation testing is done with the use of a **megohmmeter** (as illustrated in Figure 14-1), or insulation tester.

To test the insulation of the PV source circuit:
- Connect the negative lead of the insulation tester to the grounding connection in the combiner box.
- Connect the positive lead of the insulation tester to the negative output connection in the combiner box.
- Apply 1000 Vdc to the negative connection and wait for the insulation resistance to achieve at least 1MΩ.
- If confirmed, document on the checklist that the insulation test has passed inspection.
- If the resistance did not pass, then test each of the negative input cable leads individually to see if there are variations in resistance. If one is low, then the location of damage needs to be found and repaired.

For stand-alone systems, additional tests may include:
- measurement of battery voltage and capacity.
- verification of charge controller set points and maximum power point tracking system.
- verification of charging currents and load control functions.

String Inverter Startup Sequence

Once the visual inspection is complete, the panels have been checked for polarity and performance, and the voltage of the strings measured, it is time to turn on the inverter.

Typically, the steps will include the following:
- verify all connections.
- verify the correct AC voltage is present at the AC disconnect (there is proper power coming from the utility).
- verify the correct DC voltage and polarity are present at the DC disconnect(s).
- close the AC disconnect.
- verify there is now the correct AC voltage at the inverter's AC terminals.
- close the DC disconnect(s).
- verify there is now the correct DC voltage and polarity at the inverter's DC terminals.
- turn on the inverter.
- wait for the inverter to step through its internal startup sequence.

System Function Tests

Once the system is operational, the next step is to determine that all components are functioning as expected. How this is accomplished will vary depending upon the components selected, and manufacturers will provide detailed steps to conduct this verification process.

- Verify that all disconnects stop the flow of electricity when in the open position.
- Verify that the inverter is receiving the anticipated voltage from the array or the battery bank.
- Verify that grid-interactive inverters shut down when disconnected to the grid, and re-energize when the connection to the grid is re-established.
- Verify that charge controllers are properly charging the battery bank.

Verify Array Power and Energy Production

The output of solar modules are rated based on standard test conditions. It would, however, be a rare day indeed if those conditions (25°C, 1000 watts per square meter of irradiance) happened to occur as the panel was being tested. So adjustments must be made to determine if the panels (and thus the array) are performing as expected under current environmental conditions.

Calculating the power output based on current conditions is accomplished in the following manner:

1. Adjust the array's maximum power voltage (Vmp) based on the current temperature.

    Measure with an **infrared thermometer** on the backside of the module (to measure actual cell temperature, as demonstrated in Figure 14-2). Find the difference (or delta Δ) between this reading and 25°C.

    INFRARED THERMOMETER

    Then multiply it by the temperature coefficient of the module (for Vmp). If the cell temperature was higher than 25°C, subtract the result from 100% and multiply the result by the Vmp (at STC) to find the temperature adjusted Vmp. If the cell temperature was lower than 25°C, add the result to 100% and multiply the result by the Vmp (at STC) to find the temperature adjusted Vmp.

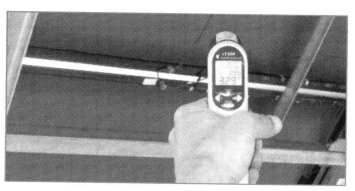

FIGURE 14-2: MEASURING BACK OF PANEL WITH INFRARED THERMOMETER

For example, if the Vmp (at STC) of selected panel is 30.3 Vdc with a temperature coefficient of 0.344%/°C and the measured cell temperature was 37° C, then:

$$37° C - 25° C = \Delta\ 12° C \times 0.344\%/°C = 4.13\ \%$$
(less efficient than at STC)
$$100\% - 4.13\ \% = 95.87\ \%\ /100 = .9587 \times 30.3\ Vdc = 29.05\ Vdc$$

So the temperature adjusted Vmp for this panel when the cell temperature is 37°C will equal 29.05 Vdc (slightly less than its rating under standard test conditions).

Conversely, if the measured cell temperature were COLDER than standard test conditions, say 7°C, then:

$$25° C - 7° C = \Delta\ 18° C \times 0.344\%/°C = 6.19\ \%$$
(more efficient than at STC)
$$100\% + 6.19\ \% = 106.19\ \%\ /100 = 1.0619 \times 30.3\ Vdc = 32.18\ Vdc$$

So the temperature adjusted Vmp for this panel when the cell temperature is 7° C will equal 32.18 Vdc (slightly higher than at STC).

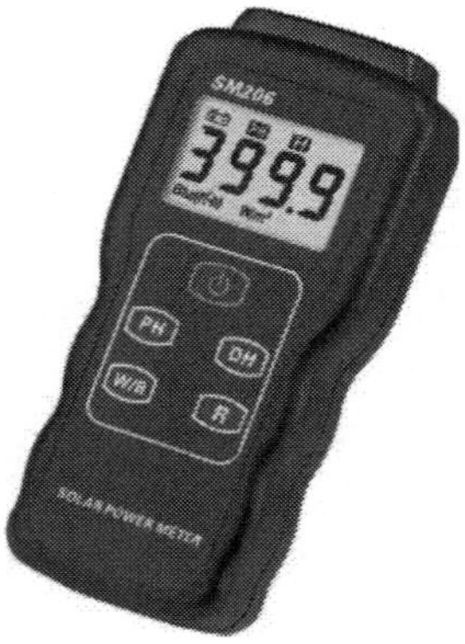

FIGURE 14-3: A PYRANOMETER, OR A SOLAR IRRADIANCE POWER METER

PYRANOMETER

2. Adjust the array's maximum power current (Imp) based on the measured irradiance.

Again, it will be highly unlikely that the irradiance at the time of commissioning the system will be exactly 1,000 watts per square meter. So the maximum power current will need to be adjusted based on the available sunlight.

Using a **pyranometer** or sun meter (such as shown in Figure 14-3) held at the base of the panel and at the angle of the array facing the sun, measure the irradiance at the front of the panel. Take several readings and average them - since the readings will tend to vary from moment to moment.

Divide the reading(s) by 1,000 W/m². Then multiply the result by the maximum power current (Imp) as rated at STC.

For example, if the averaged reading was 778 W/m², and the Imp for a specific panel was 8.27 A, then...

$$(778\ W/m^2) / (1000\ W/m^2) = .778 \times 8.27\ A = 6.434\ A$$
of current given the measured irradiance.

3. Multiply the temperature-adjusted voltage determined in step 1 by the irradiance-adjusted current determined in step 2 to obtain the expected power output of the array under the current measured conditions.

   In this example:
   $$29.05 \text{ Vdc} \times 6.434 \text{ A} = 186.9 \text{ watts}$$

   So even though the panel was rated at 250 watts under STC, given the current conditions, each panel will only produce about 187 watts.

4. Next, multiply the power production of each panel by the total number of panels to calculate the DC output from the array.

   $$186.9 \text{ W (per panel)} \times 16 \text{ panels} = 2,992 \text{ W}$$

5. Multiply the DC output from the array by the total system efficiency after accounting for losses within the system (derate the system).

   So, if it is determined that the system has a derate factor of .91, then the array DC output should be multiplied by this factor.

   $$2,992 \text{ W (array production)} \times .91 \text{ derate} = 2,723 \text{ W}$$

   In this example, the inverter should be generating 2,723 W of AC power at the time of commissioning.

6. Check the value calculated in step 5 with the display on the inverter.

   If they match, the system is operating as anticipated.

7. Document the Results
   Good documentation is critical to the commissioning process. The installer should:
   - complete a commissioning form, such as the example depicted in Figure 14-4.
   - record the results of the visual inspections.
   - record all the test results.
   - interpret and/or summarize the test results.
   - note any performance anomalies.
   - note any special maintenance needs or corrective actions.
   - detail any changes to the system design that have been incorporated into the installed system.
   - date the results and obtain any signatures from the responsible persons.
   - photograph the system and all its components to provide a visual record of the system at the end of the commissioning process. If possible, photograph the display results from the inverter and other equipment that demonstrate it is working as designed.

FIGURE 14-4: SOLAR COMMISSIONING FORM

SOLAR COMMISSIONING FORM

A: GENERAL SITE INFORMATION

SITE CONTACT NAME: _____
SITE ADDRESS: _____
LEAD TECHNICIAN'S NAME: _____
DATE: _____ TIME: _____ WEATHER: _____
CURRENTLY PRODUCING WATTS (TOTAL FROM ALL INVERTER DISPLAYS): _____
TOTAL SYSTEM WATTS (AT STC): _____
MAXIMUM SYSTEM VOLTAGE: _____ SITE HIGH TEMP: _____ LOW TEMP: _____
UTILITY METER NUMBER: _____
ACTUAL GRID VOLTAGE (AT POINT OF INTERCONNECTION): L1-L2: _____ L1-N: _____ L2-N: _____
BACK OF MODULE TEMP: _____ SOLAR IRRADIANCE: _____

B: ARRAY INFORMATION

	Array A	Array B	Totals:
Array true azimuth (degrees)			n/a
Array tilt (degrees)			n/a
Shading (Pathfinder at 4 corners of array)	%	%	% avg
Inverter (Manufacturer / Model)			n/a
Number of Inverters			
Module (Manufacturer/ Model)			n/a
Number of Modules			

C: INVERTER INFORMATION

Label	Array (A,B) / String (1,2,3,4)	Serial Number	Output (W)
Inverter 1			
Inverter 2			
Inverter 3			
Inverter 4			

D: STRING INFORMATION

String	# of Modules	Voc	Imp
String 1			
String 2			
String 3			
String 4			

Notes:

Chapter 14 Review Questions

1. A checklist of all the steps involved in installing a PV system is typically referred to as a _____.
   A) commissioning form
   B) compliance form
   C) inspection list
   D) punch list

2. Which of the following is **NOT** an electrical test normally conducted during the commissioning process?
   A) polarity testing
   B) resistance testing
   C) reactance testing
   D) performance testing

3. Which of the following steps in the commissioning process will identify that the charge controller is properly charging the battery bank?
   A) Visual inspection of the system.
   B) Verification of code compliance.
   C) System functioning tests.
   D) Verification of array power and energy production.

4. A 5 kW array has been installed on the site using 250 watt panels with the following specifications:
   Voc = 38.3 Vdc   Vmp = 29.8 Vdc   Isc = 8.9 A   Imp = 8.4 A
   Temp Coefficient = 0.485 %/ °C   Series Fuse Rating = 15 A

   You have calculated the system derate factor at .87. On the day you are commissioning the array, the following measurements are taken:
   Cell Temp = 42 °C   Irradiance = 934 W/m²   Relative Humidity = 72%

   What should be the energy production reading noted on the inverter read?
   A) 186.63 W
   B) 292.12 W
   C) 3.73 kW
   D) 4.06 kW

Lab Exercises : Chapter 14

14-1. Verify Array Energy Production

The purpose of this exercise is to compare the actual energy production of an array with the rated energy production of that array as measured at STC.

Lab Exercise 14-1:

Supplies Required:
- Solar panel

Tools Required:
- Infrared thermometer
- Pyranometer
- Calculator

a. Place a solar panel in direct sunlight (this single panel will represent an entire array).
b. Using an infrared thermometer, take a temperature reading at the back of the panel.
c. Determine the voltage (Vmp) the panel should generate when adjusted to current temperature.
d. Using a pyrometer, measure the current irradiance striking the panel in watts per square meter.
e. Determine the amps (Imp) the panel should generate when adjusted to the current irradiance.
f. Calculate the output in watts that the panel should generate when adjusted for current installed conditions.

14-2. Connecting Array to Combiner Box

The purpose of this exercise is to become familiar with configuring a commercial combiner box.

Lab Exercise 14-2:

Supplies Required:
- residential combiner box (with minimum two ports)
- two solar panels (min) with connector conductors
- Wire, conduit

Tools Required:
- multimeter
- Screwdrivers
- Wire Strippers
- MC4 crimping set
- Fish Tape
- other tools as required due to the combiner box configuration

a. If the combiner box is not already configured, follow the manufacturer's instructions to prepare the box for installation.
b. Create MC4 jumpers from the string leads to connect to the combiner box.
c. Connect the positive lead from one panel (or string) to the overcurrent protection in the combiner box. Then connect another panel or string to another overcurrent protection device.
d. Connect the negative leads to the negative busbar.
e. Connect the equipment ground to the ground busbar.
f. Using a multimeter, test each PV source circuit for voltage (with the breakers turned off). Explain why each string generates a different voltage (assuming they do).
g. Turn on the breakers and test again. Why have the voltages merged?
h. Calculate the proper breaker or fuse rating required for this configuration.
i. Determine the proper size wire required for the PV output circuit (assuming no heat or conduit fill adjustments are required).
j. Attach proper PV output circuit wires.
k. Pull the PV output circuit wires to the DC disconnect.

14-3. Connecting the DC Disconnect

The purpose of this exercise is to become familiar with configuring a DC disconnect in a string inverter configuration.

a. Determine the line and the load side connection points.
b. Connect the positive and negative wires to the line side connections of the disconnect.
c. Connect the equipment ground to the ground busbar.
d. Using a multimeter, test the PV output circuit for voltage (with the breakers turned on in the combiner box).
e. Turn off the breakers and test again. What is the result?
f. Determine the proper size wire required for the Inverter Input circuit (assuming no heat or conduit fill adjustments are required).
g. Attach proper Inverter Input circuit wires.
h. Pull the Inverter Input circuit wires to the String Inverter.

Lab Exercise 14-3:

Supplies Required:
- String Inverter
- DC disconnect with min of two poles
- Wire, conduit

Tools Required:
- multimeter
- screwdrivers
- Wire strippers
- Fish Tape

14-4. Connecting the String Inverter

The purpose of this exercise is to become familiar with configuring a string inverter.

a. Attach the positive and negative leads from the array in the proper location on the inverter.
b. Connect the line 1, line 2 and neutral wires to the AC output of the inverter.
c. Connect the equipment ground to the ground busbar.
d. Determine the proper size wire required for the Inverter Output (AC) circuit (assuming no heat or conduit fill adjustments are required).
e. Pull the Inverter Output circuit wires to the AC disconnect.

Lab Exercise 14-4:

Supplies Required:
- String Inverter
- AC disconnect with min of two poles
- Wire, conduit

Tools Required:
- multimeter
- screwdrivers
- Wire strippers
- Fish Tape

14-5. Connecting the AC Disconnect

The purpose of this exercise is to become familiar with configuring an AC disconnect.

a. Determine the line and the load side connection points.
b. Connect the hots (line 1 and line 2) from the inverter to the load side connections of the disconnect.
c. Connect the neutral wire to the neutral busbar.
d. Connect the equipment ground to the ground busbar.
e. Determine the proper size wire required for the Inverter Output (AC) circuit from the AC disconnect to the service panel.
f. Connect the line 1 and line 2 wires that will continue to the service panel to the line side of the disconnect.
g. Attach the neutral and equipment ground to the appropriate busbars.
h. Pull the Inverter Output circuit wires to the Service Panel.

Lab Exercise 14-5:

Supplies Required:
- AC disconnect with min of two poles
- Service panel
- Wire, conduit

Tools Required:
- multimeter
- screwdrivers
- Wire strippers
- Fish Tape

Lab Exercise 14-6:

Supplies Required:
- AC disconnect with min of two poles
- Service panel
- Wire, conduit
- Appropriately sized double-pole breaker

Tools Required:
- multimeter
- screwdrivers
- Wire strippers
- Fish Tape

14-6. Connecting the Array to the Service Panel

The purpose of this exercise is to become familiar with configuring a load side connection to a service panel..

a. Assume the inverter used is a 3 kW unit. What is the appropriate circuit breaker size to connect to in the service panel? Where will it be located?

a. Connect the hots (line 1 and line 2) from the AC disconnect to the PV array circuit breaker.

b. Connect the neutral wire to the neutral busbar.

c. Connect the equipment ground to the ground busbar.

d. Perform a continuity test between all the pieces of equipment in the system to ensure proper bonding has been installed.

Chapter 15

System Maintenance & Troubleshooting

One of the advantages of a PV system is that it requires very little maintenance to function properly. With no moving parts, the bulk of the system simply sits there generating energy, unless acted upon by an outside force.

But, as with any system devised by humans, things happen. Parts age, break, wear out, are hit by falling objects, chewed on by squirrels or submerged in flood waters. If it can be imagined - it has probably happened.

So a bit of troubleshooting will be required when the system is not functioning up to expectations - and periodic maintenance is required to ensure that it continues to operate as designed.

Monitoring Performance

Most systems come with monitoring software or a display of some sort. Some are Internet-based, some can be accessed by smart phone, and some require a periodic trip down into the basement to check the displays to make sure everything is behaving itself.

A routine monitoring schedule should be established. Most new owners are excited enough that they monitor their system quite often. But over time, it may require more effort.

With routine monitoring, any problems can be identified and addressed quickly.

Chapter Objectives:
- Understand how system monitoring can assist in the maintenance and troubleshooting of a PV system.
- Identify common design and installation mistakes that can lead to future trouble.
- Note common problems and ways these issues can be corrected.
- Create a proper maintenance routine to keep the PV system operating efficiently.

Troubleshooting

Many problems experienced over the life of a PV system likely have their roots in a poor initial design and/or installation.

Common mistakes that will impact performance may include:
- not accounting for future shading. Remember, trees grow and neighbors build things. While the array may have been free of shading issues during the early days of its life, shading may become a factor as the system ages.
- poor site survey. The orientation of the array may not have been optimized. The roof structure may not have been up to the task. Supports for ground mounted systems may not have been placed in a way that avoids shifting and sinking. Wiring and systems may be affected by heat issues that were not anticipated.
- the balance of systems (BOS) may be undersized or oversized. Most electronic equipment such as charge controllers or inverters, operate best within a limited voltage range. If the power is too great or not enough, performance will suffer.
- undersized wire and/or conduit. Loss due to voltage drop or current capacity affected by heat is a common problem when the conductors and conduit have not been sized properly.
- poor installation. A badly designed and installed grounding system, poor mounting, lack of space for cooling (panels on rooftop, or equipment near buildings), poorly connected conductors - all can combine to cause a host of problems as the system ages.

Typical System Problems

Some common problems that may occur within a PV system include:
- lower than expected power output from the array,
- sudden system failure,
- battery failure.

It should be noted that all troubleshooting (and maintenance for that matter) should begin with a thorough visual inspection of the system. Nine times out of ten, this process will identify the problem (the wind blew a panel off the roof, a squirrel chewed through a wire, a circuit breaker was tripped, etc).

Low System Output

When the system was first installed, a series of steps referred to as commissioning were taken and the results of this testing was documented. If current system performance does not match the initial system performance, something is wrong.

One option is to repeat all the steps of the commissioning process. This is probably the most complete and reliable method of discovering where the problem lies. But short of undertaking that rather time consuming process, a few common problems that can affect the output of the array might be explored.

These include:
- dirt and/or debris on the modules. The fix is pretty easy. Simply clean the panels and clear away any debris.
- shading issues. Visually inspect to see if a tree or shrub has grown enough to shade a portion of the array. Remove the shading element if possible.
- problems with a module. Physically inspect the module for cracks, burn marks, loose connections, etc. If possible, test each module with a multimeter. If not possible, shade a significant portion of each module (with a tarp or piece of cardboard) and note if the array output declines. If shading a module does not affect output, there may be a problem with that module.
- **heat fade.** Solar modules perform less well as temperatures increase. The amount of heat each module is exposed to should have been anticipated and accounted for in the design process. However, a weak module may react even more adversely to heat than anticipated. Test this by cooling the deck of the roof below the panels by spraying water from a hose. Do not spray the water directly on the panels, as it might cause the glass to crack if the shock between the excessive heat and the cold water is too abrupt. If production returns to anticipated levels, then heat may be the cause of the lower output.

  > HEAT FADE

- the array may be incorrectly wired. Make sure all strings have the same number of modules. Check all string voltages in the combiner box, making sure the system is turned off and all breakers are in the open position. This should have been discovered during the initial commissioning process.
- burnt terminals and/or diodes. A visual inspection should show if there has been a catastrophic failure. If the array has been struck by lightning, for example, these connections may fail. The damage should be fairly obvious.
- loose connections. Check the wiring system to ensure all connections are secure. Again, make sure the system is turned off (lock-out/tag-out) before checking these connections.
- overcurrent protection. Check all fuses and breakers to make sure they have not been blown or tripped. There may have been a power surge and the overcurrent protection has done its job and shut down part of the system.
- check conduit fittings. Wind, physical disruption, or simply age may have caused fittings to loosen, allowing water to enter conduits.
- inspect the inverter. Make sure the inverter is functioning properly and the cooling system (fan) is operating. Check for clogged filters if present.

Sudden System Failure

There are also times when the system shuts down completely. Reasons for this may include:
- the grid is down. On grid-tied systems, the array is designed to shut down when the power from the grid is interrupted. There may be no problem and the system is simply working as it should.
- a ground fault. If the inverter detects a ground fault on the DC side of the system, the ground fault interruption device should shut down the system and indicate that there is a ground fault that must be addressed.
- a bad inverter. While the PV array has no moving parts, most string inverters do - and they will eventually fail. Follow the troubleshooting instructions provided by the inverter manufacturer to determine if the problem is with the inverter.
- something is turned off. Check to see that the inverter, all disconnects, breakers and other system equipment are in the on (closed) position.
- a fuse has blown. Some sudden surge (a lightning strike, for example) has caused a fuse to blow or a breaker to open.

Battery Failure

The inclusion of a battery bank complicates matters a bit. Problems with the battery bank may simply be an indication of problems elsewhere in the system. Remember, when working with batteries, always wear approved eye, face and hand protection to avoid injury.

Most battery management systems incorporated with ESS units are not designed to be repaired in the field. Consult the manual for specific troubleshooting procedures.

Smaller systems and systems that incorporate lead-acid batteries will require periodic maintenance and do allow for some troubleshooting in the field.

Problems may include:
- poor connections. Using a multimeter, test the voltage of all batteries connected in series. The readings should be within a few tenths of a volt of each other. Amp readings for all batteries connected in parallel should be the same and within 5% under charging and discharging conditions.
- bad battery. Batteries do fail. Disconnect and test each battery. All voltage readings should be within a tenth of a volt or so.
- excessive loads. If the array has been undersized, or the loads have increased, the battery bank may not be able to fully recover from load demand. Many charge controllers and/or inverters maintain data regarding the performance of battery banks. Daily records of amp-hours delivered to and extracted from the battery bank will indicate if the battery bank is able to routinely fully recover from daily load demands.

- too many parallel strings. The battery bank design should not call for more than three parallel battery strings. The resulting bank will tend to lose its equalization, resulting in accelerated failure of any weak cells.
- poor temperature control. When the temperature falls below 0° C (32°F), batteries will lose about 25% of their available charge. Avoid heaters that will keep some batteries warmer than others.

System Maintenance

Maintaining a PV system generally does not require a great deal of time.

A good annual maintenance checklist includes:
- periodic debris removal. Remove leaves, trash and other clutter from around the PV system. This not only avoids shading issues, but removes a fire hazard, habitat for insects and vermin, and possible drainage problems.
- shading control. Trim trees, shrubs and other vegetation that may grow to shade the array.
- cleaning the array. Generally the rain will take care of this, but in areas that experience infrequent rain, it might be necessary to wash the panels periodically.
- weather sealing. The flashing and weather sealing of the footings for roof mounted systems should be inspected annually and resealed as required.
- mechanical inspection. Inspect all mounting systems to ensure they are still tightly fitted and torqued to specification.
- wiring inspection. Inspect all conductors in all circuits. Make sure the insulation is intact, and all connections are tight. This is especially a concern for the PV source circuit where conductors are more generally exposed to the elements than in other circuits.

Battery Bank Maintenance

Maintenance for battery banks incorporated into stand-alone systems, where they are cycled (charged and discharged) regularly, should be repeated every month or so. For grid-interactive systems with a battery bank that is cycled less often, six months or so can pass between maintenance tasks.

These tasks include:
- check battery connections. Battery terminals are made of soft metal (generally lead). They can easily become loose over time. Check them and tighten as needed.
- check water levels. Unsealed lead acid batteries will vent hydrogen over time, depleting the water within them. Add distilled water to the fill line of all batteries as needed.
- equalize flooded batteries. Most charge controllers allow for the periodic equalization of batteries. This can be done either manually or automatically (as determined by the settings on the charge controller). It

should be done every three months or so. Follow the battery's manufacturer's recommendations. Ensure that the battery is well vented prior to equalization. Also, ensure that it has not been overfilled with water. Hydrogen gas will be emitted and water will boil off during the equalization process. Check water levels after equalization and fill if required.
- inspect and clean. Inspect the racks, terminals, cases and trays to ensure they remain intact. Clean any corrosion or sulfation from surfaces.
- measure voltages. Test each battery, either with a multimeter or a hydrometer to ensure voltages remain in specified ranges..

Chapter 15 Review Questions

1. A proper maintenance and troubleshooting regime should always begin with:
   A) periodic commissioning.
   B) periodic compliance.
   C) periodic monitoring.
   D) periodic equalization.

2. All of the following are common problems caused by poor design, **EXCEPT**:
   A) oversizing of wire and conduit.
   B) a poor site survey.
   C) failure to account for future shading.
   D) balance of systems undersized.

3. A grid-tied PV system is generating less power than expected. Which of the following is most likely the cause of such a situation?
   A) heat fade
   B) reversed polarity
   C) the grid is down
   D) a ground fault

4. A grid-tied PV system has suddenly stopped working completely. Which of the following is most likely the cause of such a situation?
   A) heat fade
   B) shading
   C) a ground fault
   D) dirt on the modules

5. The batteries in a stand-alone system seem to fail prematurely. Which of the following is likely NOT a cause of such a situation?
   A) excessive loads
   B) poor temperature control
   C) too many batteries in series
   D) too many batteries in parallel

6. A weak module reacting with excessive voltage loss due to increased cell temperature is known as:
   A) heat loss.
   B) heat resistance.
   C) heat drop.
   D) heat fade.

Formulas

Common formulas used in photovoltaics:

Power Equation:

Watts = Amps x Volts

(W = I x V, or P = I x E)

Energy is essentially power over time. So energy is measured as:

Watts (power) x hours (time) = **Watt-hour (Wh)**

Ohm's Law:

Volts = Amps x Resistance

(V = I x R)

Voltage Drop of a Circuit:

Voltage Drop = Amps at Maximum Power Point x Resistance (per 1000 feet) x twice the distance of the circuit (in feet)

Single Phase: $V_d = I_{mp} \times R \times 2d / 1{,}000 \text{ ft}$

Three Phase: $V_d = I_{mp} \times R \times 1.73d / 1{,}000 \text{ ft}$

Voltage between phases in 3-phase electric:

1.73 (the square root of 3) larger than the voltage of each phases: 1.73 x 120 volts = 208 Volts

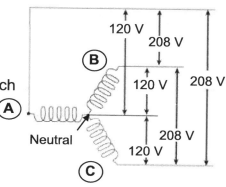

Three Phase Wye 120 / 208 V

Calculation of Weighted Daily Load:
watts used x hours x days per week operated / days per week

Peak Sun Hours
Peak Sun Hours = Total Daily Energy from the Sun (watts) / 1000 (watts)/hr

Angle of Incidence Biases:
summer bias = latitude - 15°
winter bias = latitude + 15°

Total Solar Resource Fraction (TSRF):
TSRF = Tilt & Orientation Factor (TOF) x Shading Factor (SF)

10-Foot Tap Rule
10% x [Feeder Breaker + (Inverter Output x 1.25)]

25-Foot Tap Rule
33.33% (1/3) x [Feeder Breaker + (Inverter Output x 1.25)]

Total Area Required for Rooftop Array
Total Area Required = Array Power Output / (1000 W/m^2 x Panel Efficiency)

Sizing an Array:
(average monthly load) / 30.5 days per month = kWh per day / (average solar insolation) = 100% efficient system size / (system derate)

Sizing Battery Backup:
load demand / system voltage / depth of discharge / inverter efficiency

Acronyms

a-Si	amorphous silicon
AC	alternating current
ACELA	American Clean Energy Leadership Act
ACES	American Clean Energy and Security Act
ACWP	actual cost of work performed
AESO	Alberta Electric System Operator
AFCI	arc-fault circuit interrupter
AGM	absorbent glass mats
Ah	amp-hour
AHJ	authorities having jurisdiction
AIC	ampere interrupting capacity
AIR	ampere interrupt rating
ANSI	American National Standards Institute
ARRA	American Recovery and Reinvestment Act
ASCE	American Society of Civil Engineers
ASES	American Solar Energy Society
ASHRAE	American Society of Heating, Refrigerating and Air-Conditioning Engineers
AS/NZS -	Australian/New Zealand Standard
AWG	American Wire Gauge
BIPV	building integrated photovoltaic
BOS	balance of systems
BS	British Standard
C	Celsius
CAD	computer-aided design
CAES	compressed air energy storage
CAISO	California Independent Systems Operator
CCOHS	Canadian Centre for Occupational Health and Safety
CdTe	cadmium telluride
CEC	Canadian Electrical Code
CENELEC	Central European Normalisation Electrotechnique
CFR	Code of Federal Regulations
CHAdeMO	DC fast charging protocol
CHP	combined heat and power
CIS/CIGS	copper indium gallium selenide
CPV	concentrator photovoltaic modules
CRES	competitive retail electric service
CSA	Canadian Standard Association
Cts	current transformers
CV	cost variance
d	distance
DC	direct current
DCD	DC disconnect
DER	distributed energy resources
DG	distributed generation
DISCO	distribution company or distribution utility
DOD	depth of discharge
DSIRE	Database of State Incentives for Renewables & Efficiency
DSM	demand side management
E	electromotive force
EC	European Commission

EDU	electric distribution utilities
EERS	energy efficiency resource standard
Eg	band gap energy
EGC	equipment grounding conductor
EIS	environmental impact statement
EMT	electrical metallic tubing
EPS	electrical power systems
ERCOT	Electric Reliability Council of Texas
ESS	energy storage systems
ETA-I	Electronic Technicians Association International
ETL	Electrical Testing Labs/Intertek
EU	European Union
EU-OSHA	European Agency for Safety and Health at Work
EV	earned value
EV	electric vehicle
eV	electron volt
EVSE	electric vehicle supply equipment
EWG	exempt wholesale generator
f	Fahrenheit
FERC	Federal Energy Regulatory Commission
FIP	feed-in premium
FIT	feed-in tariff
FLA	flooded lead acid (battery)
FMT	flexible metallic tubing
GATS	Generation Attribute Tracking System
GCR	ground cover ratio
GE	grounding electrode
GEC	grounding electrode conductor
GFCI	ground-fault circuit interrupter
GFP	ground fault protection
GFPD	ground fault protection device
GFPE	ground-fault protection of equipment
GW	gigawatt
HOA	homeowners associations
HV	high voltage
HVDC	high-voltage direct current
Hz	Hertz
I	ampere (amp)
IBC	International Building Code
ICC	International Code Council
IEC	International Electrotechnical Commission
IECC	International Energy Conservation Code
IEEE	Institute of Electrical and Electronics Engineers
IESO	Ontario Independent Electricity System Operator
IFC	International Fire Code
IGCC	International Green Construction Code
ILR	inverter load ratio
Imp	maximum power current
in-lb	inch-pound
IOU	investor-owned utility
IRC	International Residential Code
IREC	Interstate Renewable Energy Council
Isc	short circuit current
ISO	Independent Systems Operator

ITC	investment tax credit
IV curve	amps/voltage curve
KPI	key project indicator
kPa	kilopascal
kVA	kilo volt-amp
kW	kilowatt
kWh	kilowatt-hour
LAZ	limited access zone
LCO	Lithium cobalt oxide
LCOE	levelized cost of energy
LED	light emitting diode
LFNC	liquid-tight flexible nonmetallic
LFP	Lithium iron phosphate
Li-ion	Lithium-ion
LMO	Lithium manganese oxide
LOTO	lock-out/tag-out
LSE	load-serving entities
LTO	Lithium titanate oxide
m^2	square meter
mm^2	square millimeter
MACRS	modified accelerated cost-recovery system
MC	metal-clad cable
MC3	multi contact unit with a 3mm^2 contact assembly pin
MC4	multi contact unit with a 4mm^2 contact assembly pin
MISO	Midcontinent Independent System Operator, Inc.
MLPE	module-level power electronics
Mono-SI	monocrystalline
MPP	maximum power point
MPPT	maximum power point tracking (or transfer)
MSDS	material safety data sheet
MWh	megawatt hour
NABCEP	North American Board of Certified Electrical Practitioners
NBPSO	New Brunswick Power System Operator
NCA	(Lithium) nickel cobalt aluminum oxide
NEC	National Electrical Code
NEM	net energy metering
NEMA	National Electrical Manufacturers Association
NERC	North American Electric Reliability Corporation
NFPA	National Fire Protection Association
NiFe	nickel iron battery
NJATC	National Joint Apprenticeship and Training Committee
Nm	Newton-meter
NMC	(Lithium) nickel manganese cobalt oxide
NOCT	nominal operating cell temperature
NOMT	nominal module operating temperature
NREL	National Renewable Energy Laboratory
NRTL	national regional testing laboratory
NYISO	New York Independent System Operator
O&M	operation and maintenance
OCPD	overcurrent protection device
ohm-cm	ohm-centimeters
OJT	on-the-job
OSHA	Occupational Safety and Health Administration
PACE	property assessed clean energy

PBI	performance (or production) based incentives
PCC	point of common coupling
PERC	passivated emitter and rear cell
PFAS	personal fall arrest system
PPE	personal protective equipment
PHEV	plug-in hybrid electric vehicle
PHV	plug-in hybrid vehicle
PJM	Pennsylvania-Jersey-Maryland Interconnection
PMAX	maximum power point
PN junction	positive/negative junction
POC	point of connection
Poly-SI	polycrystalline
PPA	power purchase agreement
PPE	personal protective equipment
PSC	Public Service Commission
psf	per square foot
PUC	Public Utility Commission
PV	photovoltaics
PV	planned value
PVC	rigid polyvinyl chloride
PVES	photovoltaic energy system
PVRSE	rapid shutdown equipment
PVRSS	rapid shutdown system
PWM	pulse width modulation
QC	quality control
R	resistance
REC	renewable energy certificate
REP	retail electric provider
RFP	request for proposal
RMC	rigid metal conduit
RMS	root mean square
ROI	return on investment
RPS	renewable portfolio standards
RSI	rapid shutdown initiator
RTO	regional transmission organization
RTP	real-time pricing
SAM	System Advisor Model
SCIR	short-circuit interrupting rating
SEP	State Energy Program
SEPA	Solar Electric Power Association
SF	shading factor
SoC	state of charge
SPI	Solar Power International
SPP	Southwest Power Pool
SREC	solar renewable energy certificate
STC	standard test conditions
TC	temperature coefficient
TDSP	transmission & distribution service provider
TDU	transmission and distribution utility
TOF	tilt and orientation factor
TOU	time of use rates
TSRF	total solar resource factor
UL	Underwriter's Laboratory
USE-2	underground service entrance (high heat)

UPS	uninterruptible power source
USP	utility scale power
UV	ultra violet
V	volt
V2G	vehicle to grid
V2H	vehicle to home
VA	volt-amperes
VAR	volt amps reactive
Vmp	maximum power voltage
VNM	virtual net metering
Voc	open circuit voltage
VRLA	vented and recombinant or valve regulated lead-acid
W	watt
WEEB	washer electrical equipment bond
Ω	ohm

Renewable Energy Podcasts Available

You are invited to join us each week for a discussion of what is happening in the world of solar PV, upcoming solar events, and a deep dive into a topic of interest within the industry.

There are several ways to access this information.

Solar News Weekly is a weekly podcast created in conjunction with ASES (American Solar Energy Society). You can access this 6-minute or so news roundup either on our Youtube channel at:

https://www.youtube.com/@bluerockstation/

or on the audio podcast at: https://www.podbean.com/pu/pbblog-ybvhp-116755c

Also - please join us each week on Tuesdays at Noon (Eastern) for an hour-long get together with your fellow solar enthusiasts for Solar Noon Tuesday.

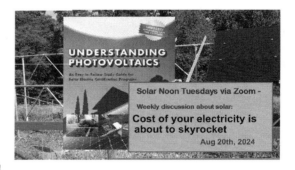

Videos from these sessions have been awarded 30 minutes continuing education for certifications from ETA International and past sessions can be found at:

https://solarpvtraining.com/solar-noon-tuesday/

Weekly Zoom access to the live show is:

https://zoom.us/j/91067004366?pwd=b1d...

Meeting ID: 910 6700 4366

Password: 625623

Figures

Chapter 1

Figure 1-1: Growth of Solar Electric (2010-2017)...1
Figure 1-2: Solar PV Capacity by Country (as of 2022) 2
Figure 1-3: Decline in the wholesale price of PV Modules (2008 - 2023)..............................3
Figure 1-4: Cost breakdown of Installing Residential PV System..4
Figure 1-5: Traditional electric utility demand curve...9
Figure 1-6: Duck shaped electric utility demand curve as rooftop solar added....................10
Figure 1-7: Antoine-César Becquerel (1788- 1878)..11
Figure 1-8: Albert Einstein (1879- 1955)..12
Figure 1-9: North American Board of Certified Energy Practitioners..................................13
Figure 1-10: Electronics Technicians Association..14
Figure 1-11: National Electrical Code..14
Figure 1-12: State-by-state adoption of NEC..14
Figure 1-13: UL Service Marks...18
Figure 1-14: ETL Service Mark...19
Figure 1-15: CSA Service Mark...19
Figure 1-16: CE Service Mark...20
Figure 1-17: Occupational Safety and Health Administration (OSHA)................................20

Chapter 2

Figure 2-1: Cross section of a PV cell...28
Figure 2-2: Monocrystalline PV module...29
Figure 2-3: Polycrystalline PV module...29
Figure 2-4: Thin-film PV module..30
Figure 2-5: Split-Cell PV module..31
Figure 2-6: Thin Film Perovskite cells and Silicon-Perovskite Tandem cells......................33
Figure 2-7: Bifacial PV module...34
Figure 2-8: Busbar and Feeder conductors on a silicon solar cell.......................................34
Figure 2-9: Relationship between solar cell, solar module, solar panel and solar array.....36
Figure 2-10: Typical IV Curve for PV panel at STC..39
Figure 2-11: Typical IV Curve for PV panel at various levels of irradiation........................40
Figure 2-12: Temperature adjusted IV curve..41
Figure 2-13: Six PV cells connected in series..42
Figure 2-14: Six PV cells connected in series with final cell shaded..................................42
Figure 2-15: Six PV cells connected in series with bypass diode in place..........................42
Figure 2-16: Strings of cells bypassed within a panel if shading becomes an issue...........43
Figure 2-17: Blocking diodes incorporated into a PV system..44
Figure 2-18: Rating information sticker on back of a specific PV panel.............................44
Figure 2-19: Solar Panel Data Sheet..45

Chapter 3

Figure 3-1: Diagram of a stand-alone PV system...56
Figure 3-2: Diagram of a grid-tied PV system..56
Figure 3-3: Average installed cost of Residential Solar Grid-Tied System in US..............................57
Figure 3-4: Time-of-Use Pricing structure 58
Figure 3-5: Load Shifting 59
Figure 3-6: Peak shaving affect on peak load demand by incorporating battery bank.....................59
Figure 3-7: Diagram of a DC-Coupled Multimode system with whole home backup......................61
Figure 3-8: Diagram of an AC-Coupled Multimode PV system...62
Figure 3-9: Diagram of an Enphase AC Coupled System..63
Figure 3-10: Kill-a-Watt meter hooked up to a toaster 66
Figure 3-11: Examples of weighted loads...70

Chapter 4

Figure 4-1: A typical electric utility distribution sub-station..81
Figure 4-2: Transformer drum at customer's site..81
Figure 4-3: Standard US Single-Phase Wall Outlet..82
Figure 4-4: Single-phase waveform..83
Figure 4-5: Split phase waveform...83
Figure 4-6: Typical Single-Phase Split Phase Service Panel with double-pole breaker feeding a load............84
Figure 4-7: US Outlet wiring for 120 volt outlet (left) and 240 volt outlet (right).............................84
Figure 4-8: Three-Phase Power sine wave form...85
Figure 4-9: Utility provided Single-Phase Power diagram..85
Figure 4-10: Diagram of a simple electrical circuit..86
Figure 4-11: Open and closed electrical circuits...87
Figure 4-12: Four 12-volt, 90 Ah batteries connected together in series..87
Figure 4-13: Four 12-volt, 90 Ah batteries connected together in parallel.......................................88
Figure 4-14: Six 12-volt solar panels connected together in series...88
Figure 4-15: Six 12-volt solar panels connected together in parallel..88
Figure 4-16: Six 12-volt solar panels connected together in series and parallel..............................89
Figure 4-17: DC power WaveForm..89
Figure 4-18: AC power signal..90
Figure 4-19: Ohm's Law Triangle..91
Figure 4-20: A continuity test of a switch using a standard multimeter..95
Figure 4-21: Equipment grounding conductor bonding the metal frames of two PV panels together........96
Figure 4-22: Bonding and grounding system..97

Chapter 5

Figure 5-1: Solar panel (also called a solar module)...105
Figure 5-2: Testing a solar panel..106
Figure 5-3: Example of solar panel specifications 107
Figure 5-4: Junction box with Bypass diodes..107
Figure 5-5: MC3 (also referred to as a Solarline 1) connector..108
Figure 5-6: MC4 (also referred to as a Solarline 2) connector..108
Figure 5-7: Combiner box with overcurrent protection..109
Figure 5-8: Pre-wired combiner box..109
Figure 5-9: Soladeck junction box installed on rooftop system..110
Figure 5-10: Surge protection attached to combiner box..110

Figure 5-11: Cable mount offset from IronRidge...113
Figure 5-12: Labels on conductors...118
Figure 5-13: Typical labeled PV array DC disconnect...118
Figure 5-14: Common back-fed single-pole, two-pole and three-pole circuit breakers......120
Figure 5-15: Ground-Fault Circuit Interrupter (GFCI) circuit breaker................................121
Figure 5-16: Standard US GFCI wall outlet...121
Figure 5-17: GFPE fault powered circuit breaker...121
Figure 5-18: Typical small charge controller..122
Figure 5-19: Morningstar TriStar MPPT 60-amp charge controller....................................123
Figure 5-20: Stand-alone PV system with load diverter and load dump incorporated......123
Figure 5-21: Exploded View of Enphase IQ Battery 3T ESS Unit...124
Figure 5-22: Fronius Symo Advanced Grid-Tie Inverter...126
Figure 5-23: Stand-alone inverter specifications sheet...126
Figure 5-24: Waveforms utilized within various inverters..128
Figure 5-25: Example of grid-tied inverter specifications...129
Figure 5-26: Remote Battery Rapid Shutdown Initiator..130
Figure 5-27: SOL-ARK 12K Bimodal inverter..133
Figure 5-28: Tigo Module Level Power Electronic device..134
Figure 5-29: Micro inverter..134
Figure 5-30: PV system incorporating micro inverters..134
Figure 5-31: Power Optimizer system...135
Figure 5-32: Online monitoring system for Enphase system..136
Figure 5-33: A typical AC-rated disconnect..137
Figure 5-34: Installed Current Transformer (CT)..138
Figure 5-35: A typical US service panel with PV system connected...................................138
Figure 5-36: A Span smart service panel..139
Figure 5-37: Emergency Utility Disconnect built into the meter base...............................139
Figure 5-38: Typical low-cost digital multimeter...140
Figure 5-39: Zero-center ammeter display (Left) and current clamp measuring tool (Right)......140
Figure 5-40: Assortment of various crimp connectors..141
Figure 5-42: Flat pry bar or roofing bar (also called a ripping bar)....................................141

Chapter 6

Figure 6-1: Azimuth measured in relation to true south..153
Figure 6-2: Magnetic declination map of the United States...153
Figure 6-3: Adjusting for solar North using a compass..154
Figure 6-4: Loss in Array output when oriented off Solar South.......................................154
Figure 6-5: The ideal angle of a solar panel will vary depending on the month of the year.........155
Figure 6-6: Amount of energy in sunlight striking a square meter of surface area..........155
Figure 6-7: Winter and Summer bias for solar panel orientation......................................157
Figure 6-8: Energy Generating Chart for Columbus Ohio..158
Figure 6-9: Measuring the angle of the roof using a smart phone.....................................159
Figure 6-10: Sun chart for 48°N showing the altitude and azimuth over the course of a year.........160
Figure 6-11: Example of a properly mounted roof mounted PV array...............................160
Figure 6-12: A ballasted array mounted on a flat roof...161
Figure 6-13: A relatively small ground-mounted array..162
Figure 6-14: Avoid inter row shading by offsetting the back row......................................163

Figure 6-15: Landscape and portrait orientation of PV panels..163
Figure 6-16: Pole-Mounted PV array..164
Figure 6-17: Flexible solar shingles integrated into the roof..164
Figure 6-18: Solar panels utilized to provide covered parking..164
Figure 6-19: Shadow cast from one array to the next reduces the output of a panel by 90%.................165
Figure 6-20: Solar Pathfinder with stand and carrying case..165
Figure 6-21: SunEye 210 Shade Tool..166
Figure 6-22: Trace the shading line on the sun chart or take a photograph..166
Figure 6-23: Remote assessment design tool 167
Figure 6-24: A number of system losses can reduce the total solar resource..169
Figure 6-25: Properly installed service panel with plenty of room for expansion...................................174
Figure 6-26: Connecting a grid-tied PV system with a generator..177
Figure 6-27: Cost breakdown per watt of an installed residential PV system..179

Chapter 7

Figure 7-1: Derate factors for solar array..194
Figure 7-2: Panel specifications Hyperion 400 W bi-facial Solar Panel...198
Figure 7-3: One Line Drawing of an array consisting of three strings of 10 panels.................................203
Figure 7-4: 4-in-1 micro inverter...204
Figure 7-5: Dual Input Power Optimizer...206
Figure 7-6: String Calculation Chart for Power Optimizers..206
Figure 7-7: Partial specifications for Sunny Boy String Inverters..207
Figure 7-8: Partial specifications for Enphase IQ8 Series micro inverters..208
Figure 7-9: Clipping diagram 208
Figure 7-10: Diagram of an AC-Coupled Multimode PV system 210
Figure 7-11: Connections within a OutBack Radian inverter 211
Figure 7-12: Electrical diagram of a grounded inverter...213

Chapter 8

Figure 8-1: Shipment of panels, connectorized with MC4 pigtails..222
Figure 8-2: MC4 field connection kit..222
Figure 8-3: Crimping the barrel contact for an MC4 connector..222
Figure 8-4: Assembling and tightening the MC4 connector...223
Figure 8-5: The Midnight Solar Combiner Box specifications...225
Figure 8-6: Circuits in a system that incorporate power optimizers..226
Figure 8-7: Circuits in a system that incorporate micro inverters...227
Figure 8-8: Cable ring with three cable sections..229
Figure 8-9: Metal-clad (MC) type cable..230
Figure 8-10: Field threading RMC conduit..231
Figure 8-11: Conduit expansion fittings, PVC (left) and EMT (right)...233
Figure 8-12: Maintain 10 inch minimum clear distance between roof deck and conduit 234
Figure 8-13: Manual one-shot EMT conduit bender...235
Figure 8-14: Bending EMT conduit...235
Figure 8-15: Typical Conduit body...236
Figure 8-16: Using a fish tape to pull wire through conduit...236
Figure 8-17: Cable tugger..237
Figure 8-18: Metallic Warning tape for buried conductors..238
Figure 8-19: DC disconnect built into a SolarEdge inverter...238

Figure 8-20: Enphase micro inverter Q cable connection..240
Figure 8-21: Enphase combiner box with communication (IQ Gateway) integrated.....................241
Figure 8-22: Small fused AC disconnect indicating line and load side connections......................242
Figure 8-23: Single-Phase inverter Load side connected to Standard Residential Service panel.............243
Figure 8-24: Standard Residential Service panel using 120% Rule..243
Figure 8-25: Main lug load center connected to the main service panel through feed-thru lugs.................244
Figure 8-26: Supply side grid connection...245
Figure 8-27: Insulated piercing connector..246
Figure 8-28: Line Tap supply side connection above the main breaker in a residential service panel..........246
Figure 8-29: Electric junction box, allowing PV system to tap into the service from the utility...................246
Figure 8-30: A bonding jumper making an electrical connection between two rail sections.......................248
Figure 8-31: Bonding rails with a bare copper #6 AWG wire using stainless steel lug............................249
Figure 8-32: The Equipment Grounding System..250
Figure 8-33: Grounding system for solidly grounded inverter...251

Chapter 9

Figure 9-1: Constant Current/ Constant Voltage charging process..266
Figure 9-2: Secure well lit, well ventilated battery storage room 270
Figure 9-3: Vented Battery storage unit for smaller PV system..270
Figure 9-4: Tesla Powerwall 2..271
Figure 9-5: LG Chem battery storage unit matched with a SMA inverter 272
Figure 9-6: Enphase AC Battery system..273
Figure 9-7: Level 1 EV charging connector...274
Figure 9-8: Level 2 EV charging Station..274
Figure 9-9: DC Fast-Charging EV charging Station...275

Chapter 10

Figure 10-1: Flush-to-Roof mounted footing with flashing incorporated into the design................282
Figure 10-2: Flush-to-Roof mounting layout on roof..283
Figure 10-3: Placing 20 panels on a 35 ft x 21 ft roof area...285
Figure 10-4: Wind loading zones on a roof..288
Figure 10-5: Calculating Tributary Area of a Single Footer..289
Figure 10-6: Structural cross section of a typical shingle roof..290
Figure 10-7: Reducing load by staggering footers on rafters..291
Figure 10-8: Commercial rooftop system designed to meet 2012 international fire code..............292
Figure 10-9: Top down mounting bracket...293
Figure 10-10: Proper orientation of panel on rail..294
Figure 10-11: Digital Torque wrench..294
Figure 10-12: Torqued lugs marked with QC gel..295
Figure 10-13: Rail-less solar panel mounting system..295
Figure 10-14: Shared Rail solar panel mounting system...296
Figure 10-15: Ballasted mounting system for flat roof..296
Figure 10-16: East/West flat roof mounting system..296
Figure 10-17: Squirrel guard installed on rooftop solar array...297
Figure 10-18: A top-down mounting clamp with bonding..297
Figure 10-19: Bonding lug..297
Figure 10-20: Helical anchor support for ground mounted PV system..299
Figure 10-21: Single Post Foundation Support...299

Figure 10-22: Double Post Foundation Support..300
Figure 10-23: Placing a vertical support for a ground mounted PV array...300
Figure 10-24: Example of online racking design tool...303

Chapter 11

Figure 11-1: Block diagram of a PV stand-alone system...308
Figure 11-2: AC-to-DC adapter for common DC native loads..309
FIGURE 11-3: Specifications for a TriStar MPPT Solar Charge Controller...310
Figure 11-4: Specifications of a stand-alone inverter..312
Figure 11-5: Testing specific gravity of a flooded lead-acid battery with a hydrometer......................................322
Figure 11-6: DC waveform affected by the AC waveform, resulting in Ripple 324
Figure 11-7: Typical cables for battery bank..326

Chapter 12

Figure 12-1: Must wear ANSI Z87.1 rated safety glasses when working with electricity......................................335
Figure 12-2: Stamp in hard hat indicates it was manufactured in September, 2004. 336
Figure 12-3: Harness and lanyard..336
Figure 12-4: Proper placement of a ladder...339
Figure 12-5: Base of ladder properly secured...339
Figure 12-6: Installed point where the ladder can be secured to the roof...340
Figure 12-7: A strap is clamped to the fascia board to provide ladder stability..340
Figure 12-8: A worker caught in an arc flash blast..341
Figure 12-9: Lock-out, tag-out procedure when working with electricity..342

Chapter 13

Figure 13-1: One-line drawing for small PV system..352
Figure 13-2: Plan view for small PV system...354
Figure 13-3: Multiple Power Source Label Required under NEC 705.10 & NEC 710.10..356
Figure 13-4: Line and Load Side Shock Hazard Label Required under NEC 705.20(7) and NEC 690.13(B).........356
Figure 13-5: Turn off AC Disconnect Label Required under NEC 110.27(C) & OSHA 1910.145(f)(7).....................356
Figure 13-6: Labels for DC Disconnect Required under NEC 705.20(7) & NEC 690.13(B) & NEC 690.7(D) 356
Figure 13-7: Labels for DC Conduit Required under NEC 690.31(D)(2)...356
Figure 13-8: Labels for the AC disconnect Required under NEC 705.20 (7)...356
Figure 13-9: Plaque indicating location of the system disconnect 357
Figure 13-10: Dual Power Source label Required under NEC 705.30 (C)..357
Figure 13-11: PV Breaker label Required under NEC 705.12 (B)(2)...357
Figure 13-12: ESS Disconnect label Required under NEC 706.15 (C)..357
Figure 13-13: Label for AC Disconnect when Rapid Shutdown in place required by NEC 690.12(D)(2).............358
Figure 13-14: Label for systems meeting the 2014 NEC Rapid shutdown requirements....................................358
Figure 13-15: Label for systems meeting the 2017 NEC Rapid shutdown requirements....................................358

Chapter 14

Figure 14-1: A megohmmeter, or insulation tester...368
Figure 14-2: Measuring back of panel with infrared thermometer..369
Figure 14-3: A pyranometer, or a solar irradiance power meter..370
Figure 14-4: Solar Commissioning form..372

Tables

Chapter 2
Table 2-1: Hail resistance ratings as measured by UL 2218 47

Chapter 3
Table 3-1: Load analysis for small one-bedroom cabin 67
Table 3-2: Load analysis for small one-bedroom cabin after making energy efficiency adjustments 68
Table 3-3: Maximum power draw analysis for small one-bedroom cabin 69

Chapter 4
Table 4-1: Copper conductor resistance characteristics measured in ohms (Ω) per 1,000 feet 92
Table 4-2: Copper conductor resistance characteristics measured in ohms (Ω) per 1,000 meters 93

Chapter 5
Table 5-1: Allowable ampacities for insulated conductors 111
Table 5-2: Temperature correction factors for insulated conductors at 30°C 112
Table 5-3: Ampacity adjustments for conduit fill 114
Table 5-4: Common types of conductors used in PV systems 115
Table 5-5: Insulation color-coding of AC and DC wires 116
Table 5-6: Color coding conventions for DC, AC and Three-Phase AC current carrying conductors 117
Table 5-7: Typical charge characteristics of a Lithium Ion battery cell 125

Chapter 6
Table 6-1: Peak hours of Solar Radiation for Columbus Ohio at various degrees of altitude 158
Table 6-2: Array angle at various roof pitches 159

Chapter 7
Table 7-1: Ambient air temperature correction factors 201

Chapter 8
Table 8-1: Number of conductors permitted in conduit 232
Table 8-2: Support distances for non-metallic conduit 233
Table 8-3: Expansion Characteristics of PVC Rigid Nonmetallic Conduit 233

Chapter 9
Table 9-1: An example of a deep-cycle battery's life expectancy at various DOD settings 262
Table 9-2: How C-rating affects the capacity of a Surrette Deep Cycle Solar Series 5000 battery 264

Chapter 10
Table 10-1: Pull out capacity of common lag bolts 290
Table 10-2: Maximum anchor spacing of supports on rooftop PV system 291

Chapter 11
Table 11-1: Sample design ratio calculations at 28 degrees altitude 314
Table 11-2: Sample design ratio calculations at 58 degrees altitude 315
Table 11-3: Battery bank sizing worksheet 317
Table 11-4: Ambient temperature adjustment multipliers for lead-acid batteries 320
Table 11-5: Specific gravity and state of charge voltages 322

Chapter 12
Table 12-1: Rating system for ladders..338

Index

1
10-Foot Tap Rule 247
120% rule 175, 243-245, 357

2
25-Foot Tap Rule 247

4
4-in-1 Micro Inverter 204

6
60-cell 31

7
72-cell 31

A
Absorbed Glass Mat (AGM) 260
Absorption charge 265
AC combiner box 226, 240, 273
AC disconnect 63, 119, 129-130, 134, 135, 137-138, 171, 177,
    221, 226, 240-242, 246-247, 255-256, 309, 350, 354, 356-358, 368
AC-Coupled Multimode systems 62
Adjustable frame 157-158
Age Derate 195
Agonic line 153, 154
Albedo 34
Alternating current (AC) 89
Altitude 37, 155-161, 167-169, 314-316
Ambient air temperature 38, 94, 113, 201-202, 229, 320-321, 340
American Society of Civil Engineers 286
American Society of Heating, Refrigerating and Air-Conditioning Engineers (ASHRAE) 202
American wire gauge (AWG) 47, 91-93, 97, 102-104, 111, 225, 228-229, 232, 239
Ammeter 140
Amorphous 30, 32, 35
Ampacity 113-114, 137, 205, 228-230, 239, 247, 274
Ampere Interrupt Rating 120
Amp-hours (Ah) 87-88, 263-264, 267, 317-319, 380
Amps 39, 66, 78-80, 88-90, 93, 106, 109, 122, 137, 209, 221, 224, 228-230, 239, 263, 317, 326
ANSI/ UL1703 36
Anti-islanding 63, 128-129, 131-132, 135
Apparent power 210
Arc fault 117
Arc fault protection device 247
Arc fault protection 252
Arc Flash 341
Array boundary 129, 205
Article 250 .... 15, 366
Article 690 15
Article 705 .... 15, 367

AS/NZS 3000:2007 Wiring Rules... 16
ASCE Standard #7... 286
ASHRAE.................see American Society of Heating, Refrigerating and air-conditioning engineers
Authority having jurisdiction (AHJ)... 351
Automatic transfer switch (ATS)... 61
Avoided Cost.. 58
Azimuth.. 152-154, 156, 158-160, 168-169, 350, 372

B

Back-Fed Breaker... 121
Backfeed.. 44
Back-up load panel.. 62-63
Balance of Systems (BOS)... 4, 160, 170-171, 378
Ballasted... 299
Ballasted mounting system... 162, 296
Band gap.. 28-29
Basis.. 173
Battery bank... 44, 55-62, 87, 118, 122, 125, 127, 139, 170-171,
    196, 262, 265-266, 268-269, 272, 307-312, 315-321, 324-327, 353, 366, 369, 380-381
Battery management system (BMS).. 124
Battery-based inverter.. 127, 130, 211-212, 272-273
Behind-the-meter.. 5-6
Bell Labs... 12
Bi-Directional Meter... 56, 139
Bifacial Modules... 34
Bifacial PV.. 31
Bi-Facial Solar Cells... 35
Bimodal inverter... 131, 133
Blocking diode... 43-44, 90
Bonded.. 96, 248, 282, 365
Bonding... 15, 96-97, 110, 248, 251, 351, 366-367
Bonding and grounding.. 96-97, 351, 367
Bonding jumper... 248, 296, 302
Branch circuit... 205, 239-241
Brownfields... 299
BS 7671.. 16, 18
Buck/boost charge controller... 123
Building Integrated Photovoltaic... 164
Bulk charge... 265
Bureau of Worker's Compensation... 334
Busbar.. 34, 138, 174-175, 242-244, 246
Busbars... 28, 31, 34-35, 41, 83
Bypass diode... 42-43, 107

C

Cable Lube.. 237
Cable Tugger... 237
California Energy Commission's (CEC)... 38
California Solar mandate... 192
Canadian Electrical Code.. 16, 225
CE Mark... 20
CENELEC... 17
Centre for Occupational Health and Safety... 21
Charge controller.. 43-44, 105-106, 118-125, 170-171, 212, 229,
    265-266, 271, 308-311, 316-317, 319, 323-324, 366, 368, 369, 378, 380-381
Charging rate.. 267
Circuit.. 28-29, 39, 41, 44, 80, 86-87, 89-91, 93-94, 96, 97, 106,
    119-120, 140, 195, 200, 202, 209, 223-225, 227, 229-230, 239, 251, 310-311, 317, 325, 365
Circuit breaker... 65, 83, 110, 120-121, 138, 171, 174-175, 228-229, 378, 380

Term	Pages
Clearance	171, 365
Cleats	339
Clipped	208
Closed circuit	87, 94, 380
Code	269, 303
Cold Cranking Amp	260
Color coding	83, 116-118, 358, 365
Combiner box	56, 63, 65, 105, 109-111, 120, 130, 134, 138, 180-181, 199, 203, 221, 223-227, 229-230, 234, 240-241, 247, 249, 273, 298, 302, 308, 316, 356, 367-368, 379
Commissioning	13, 17, 172, 349, 358, 363-365, 367, 369-370, 375, 378-379
Conductor	11, 27-28, 42, 78, 87, 89-93, 96-97, 116-117, 119, 141, 236, 239, 251, 353, 364-365
Conductors	31, 34, 85, 112, 117, 121, 138, 173, 176, 229, 231-234, 237-238, 242, 245-246, 268, 301-302, 325
Conduit	113-114, 170-171, 176, 221, 229-238, 246, 302, 353, 354, 364-366, 378-379
Conduit body	236
Conduit run	233-236
Connectors	107-108, 141, 221, 227, 365
Constant voltage	89, 124-125, 265-267
Continuity test	94-95, 367
Continuous power	272
Continuous power rating	68, 313
Continuous rating	69, 127
Conversion efficiency	35, 284
Conversion losses	271
C-rate	264
Crimping	140, 222, 235
Critical design month	314-316
Critical load	177
Current limit	125, 240
Current Transformer	137-138
Cycle	69, 83, 85, 90, 125, 196, 230, 259-263, 265, 267, 318, 320
Cycles per second	83, 89-90

D

Term	Pages
Daily load	55, 66-70, 193, 316-318, 320-321, 380
Data monitoring	109
Days of autonomy	317-318, 321
DC disconnect	61-62, 105, 109-111, 118-119, 137, 171, 210, 214, 225-226, 229, 238-239, 308, 356, 368
DC Fast-Charging	275
DC to DC voltage converter	123
DC-Coupled Multimode system	61
DC-to-AC ratio	207
DC-to-DC Converter Circuit	226
DC-to-DC converters	135
DC-to-DC voltage converters	123
Dead load	281, 283, 285
Dead weight	161, 285
Declination map	153
Deep cycle	260, 263-264, 319
Deep cycle battery	260, 263-264, 319
Delta	200, 202, 369
Demand charge	60, 274, 275
Demand ramping	10-11
Demarcation point	175, 245
Depth of Discharge (DOD)	262, 272, 310, 321
Derate factor	191-192, 194-195, 212, 315, 321, 324, 363, 371
Design Ratio	314-315
Desulfate	266
Dielectric barrier	213
Diode	28, 42-43, 54, 107, 195

DIPS	132
DIRECT BURIED CABLE	237
DIRECT CURRENT (DC)	89, 116
DISCONNECTING COMBINER BOX	250
DISTRIBUTED ENERGY RESOURCES (DER)	18, 131
DISTRIBUTED ENERGY SYSTEM	8, 11, 18, 57, 63, 131
DISTRIBUTION	81, 85-86, 173, 176, 244
DIVERTED LOAD	311, 324
DOPING	27
DRAG FORCE	161
DRILLED AND GROUTED	298-299
DRIVEN PILE	298-299
DUAL INPUT	206
DUAL-AXIS TRACKING SYSTEM	156, 195
DUCK CURVE	10-11
DYNAMIC REACTIVE POWER COMPENSATION	132

E

EARTH SCREW	298, 299
EASEMENTS	301
EAST/WEST MOUNTING SYSTEM	296
EFFICIENCY	29-35, 46, 60-61, 94, 107, 122, 135-136, 161, 178, 195-196, 205, 209, 250, 261, 272-273, 284, 315-318, 321, 326
EIA	183
ELECTRIC VEHICLE (EV)	164, 173, 273-274
ELECTRICAL BOS	171
ELECTRICAL CHARGE	78, 86
ELECTRICAL CURRENT	12, 78
ELECTRICAL GRID	80
ELECTRICAL METALLIC TUBING (EMT)	231-232, 234-235, 366
ELECTRICAL PERMIT	351
ELECTRICAL POTENTIAL	78, 86
ELECTRICAL SERVICE PANEL	90, 138, 171, 250
ELECTRICAL SERVICE PANEL	138
ELECTRICAL TESTING LABS/INTERTEK (ETL)	19
ELECTRICITY	5, 8, 11-12, 27-29, 42-44, 57, 65, 67, 77-79, 87, 89-91, 96, 97, 107, 113, 126, 128, 136, 139, 179, 182-183, 192, 268, 311, 313, 338, 369
ELECTROLYTE	260, 321-323
ELECTROMAGNETISM	78
ELECTROMOTIVE FORCE	78, 80
ELECTRON HOLE	28
ELECTRONIC TECHNICIANS ASSOCIATION (ETA)	13-14, 351
ELECTRONS	12, 27-29, 87, 89
EMERGENCY UTILITY DISCONNECT	139, 149
ENERGY COMMUNITY	6
ENERGY EFFICIENCY	3, 29, 41, 46, 65, 67-68, 94, 107, 127, 133-135, 157, 195, 317-318, 325, 331, 371
ENERGY STORAGE SYSTEM (ESS)	15, 62-64, 124, 172, 181, 259, 261, 263, 265, 267, 269, 271, 273-275, 309, 357, 367
ENGINEERED DRAWING	282
EQUALIZATION CHARGE	266
EQUIPMENT GROUNDING	96-97, 116, 249-251
EQUIPMENT GROUNDING CONDUCTORS	249
EUROPEAN AGENCY FOR SAFETY AND HEALTH AT WORK (EU-OSHA)	21

F

FALL PROTECTION	178, 334, 336, 340
FEEDER CONDUCTORS	34
FEEDER TAP CONDUCTOR	247
FEED-IN PREMIUM (FIP)	5-6
FEED-IN TARIFF	5

FEED-THRU LUGS..244-245
FIBERGLASS...260, 338
FISH TAPE..260, 338
FIXED POWER FACTOR...236
FIXED-MOUNT..132
Flexible Metallic Tubing (FMT)..156-157
FLOATING CHARGE..231
FLOODED...265
FLUSH-TO-ROOF..260, 267-268, 322, 381
FOOTINGS...282-283, 303
FREQUENCY...282, 297, 300, 364, 381
FRONT-OF-THE-METER..69, 83, 89, 90, 115, 127, 131-133, 176, 209
Frost Line..5
FUNCTIONALLY GROUNDED..300
FUSE.............. 44, 46, 79, 94-95, 109-110, 117, 119-120, 140, 209, 224, 227-229, 242, 325, 364-365, 379-380
Future Proof..170

G

GALVANIC CORROSION..249, 294-295
GELLED..260
GENERATION METER...260
GENERATOR...136, 139
GRID.................................5-6, 56, 78, 81, 95, 119, 128, 131, 173, 176-177, 210, 308-309, 312, 318, 366
GRID... 1, 8-9, 56-58, 61-62, 65, 68, 121, 126, 128, 139, 151, 174, 209,
        241, 245, 307-309, 312-313, 350, 369, 372, 380-381
GRID FALLBACK PV SYSTEM...62, 121
GRID INTERACTIVE PV SYSTEM..58, 369, 381
GRID PARITY..1, 5
GRID-TIED INVERTER..60, 128-129, 131-133, 199, 209-210, 312-314, 316
GRID-TIED SYSTEM..60, 130, 200, 229, 239, 315, 325
GROUND COVER RATIO..167
GROUND ELECTRODE CONDUCTOR..250
GROUND FAULT....................................... 117, 121, 131, 213, 245, 251-252, 256, 312, 314, 337, 365, 380
GROUND FAULT PROTECTION DEVICE...251
GROUND MOUNTED SYSTEM...162, 281, 298, 300-302, 378
GROUND ROD...96, 176, 250-251, 302
GROUND SCREW..298
GROUNDED..96, 117, 251, 365
GROUNDED SYSTEMS..213
GROUND-FAULT CIRCUIT INTERRUPTER..121
GROUND-FAULT PROTECTION OF EQUIPMENT...121
GROUNDING ELECTRODE..96-97, 173, 176, 250-251, 302, 365
GROUNDING ELECTRODE CONDUCTOR...176, 251
GUARDRAIL SYSTEM...336

H

Half Cut Cells..31
HARD HATS...79, 335-336
HARMONICS..79, 335-336
HAZARD ASSESSMENT..132
HEAT FADE...170, 177
HEIGHT REQUIREMENTS..379
Helical anchor...171
HERTZ (HZ)..298-299
HETEROJUNCTION CELLS...66, 83, 89-90, 127, 133
HIGH VOLTAGE BATTERY SYSTEM..32
Hot Spots..271
HYBRID INVERTERS..31
HYBRID SYSTEMS...133, 211
HYDROMETER..63
....322, 382

I

IEC 60068-2-68	37
IEC 60364	16-17
IEC 61701	37
IEC 61730	36, 196
IEC 62109	131
IEC 62716	37
IEEE	18, 38, 131
IEEE Standard 1547	18
Impedance	11, 214
Inaccessible	302
Incidence angle	155
Independent power producers	86
Inflation Reduction Act (IRA)	6
Infrared thermometer	369, 379
Input voltage	123, 127, 199, 310, 312
Insolation	74, 152, 157, 169, 193, 314-316, 329-330
Insulation	96, 108, 115-118, 141, 213, 223, 297, 364, 367-368, 381
Insulation tests	367
Insulator	27-28
Inter row shading	162-163
Interactive system	57, 262, 273
Interconnection	1, 4, 8, 18, 131, 174, 178, 209, 350, 353, 355, 372
International Building Code	286
International Electrotechnical Commission (IEC)	17
International Energy Conservation Code (IECC)	16
International Fire Code (IFC)	15, 172, 270, 284, 291, 293
Inter-row shading	284
Inverter	4, 11, 43, 56-57, 60-64, 67-71, 85, 95, 97, 105, 110, 117-119, 125-137, 170-171, 175-181, 191, 193, 195-225-226, 238-244, 246-247, 249-252, 269, 271-273, 308-309, 312-314, 316-318, 321, 324-326, 365-366, 368-369, 371-372, 379-380
Inverter input circuit	225, 239, 325
Inverter load ratio	207-208
Inverter output circuit	225, 239, 242
Inverter/Charger	62-63, 210, 308, 312
Investment tax credit (ITC)	6
Investor-owned utilities	8, 86, 139
Irradiance	37, 40-41, 151-152, 283, 324, 369-372
Islanding	57, 61, 63, 128-129, 131-132, 135, 273
Isolated	55, 118, 124, 134, 137, 213-214, 307, 342
IV curve	39-41

J

Joule	80
Junction box	4, 31, 34, 43, 56, 105, 107, 110, 118, 177, 221-222, 225-227, 229, 232, 240, 246, 249, 365-366
Jurisdiction	170, 196, 285, 349-351

K

Kilowatt hour (kWh)	65
kVA	191, 210-212

L

Labels	20, 118, 349, 353, 355-358, 365
Landscape	162-163, 284, 293, 299
Lanyard	336
Latitude	156-157, 160, 162, 165, 193
Lead-acid batteries	259-263, 315, 320-321
Lead-acid deep cycle	SEE LEAD-ACID BATTERIES
Level 1 charging	274
Level 2 charging	274

LFNC (LIQUIDTIGHT FLEXIBLE NONMETALLIC).. 232, 366
LICENSES.. 349, 351
LIDAR.. 349, 351
LIFT FORCE.. 167, 168, 187
LIGHT INDUCED DEGRADATION (LID)... 161
LINE SIDE CONNECTION... 195
LITHIUM-ION.. 175, 242, 245
LIVE LOAD... 172, 259- 267, 269, 271, 315
LIVE WEIGHT.. 283, 286, 294
LOAD ANALYSIS.. 161, 285
LOAD ASSESSMENT... 55, 66-68, 127, 191, 193, 313, 316
LOAD DIVERTER... 60
LOAD DUMP... 123
LOAD SHIFTING... 123-124, 311
LOAD SIDE CONNECTION... 59-60, 137
LOCKOUT/TAGOUT.. 174, 242-243, 357
                                                                                                                        342, 364, 367

M

MAGNETIC SOUTH.. 153-154
MAIN LUG LOAD CENTER... 244-245
MAINS CURRENT... 90
MATERIAL SAFETY DATA SHEET (MSDS).. 342
MAXIMUM BATTERY CHARGE.. 212
MAXIMUM INPUT VOLTAGE.. 199, 209, 310, 316
MAXIMUM OPERATING INPUT CURRENT.. 209
MAXIMUM OUTPUT CURRENT... 209, 230, 323
MAXIMUM POWER (PMAX)... SEE MAXIMUM POWER POINT
MAXIMUM POWER CURRENT (IMP)... 40, 44, 88, 94, 107, 370, 372
MAXIMUM POWER DRAW.. 55, 67-71, 127, 212, 313
MAXIMUM POWER POINT... 39, 44, 122, 202, 204, 368
MAXIMUM POWER POINT TRACKING (MPPT). 44, 122-123, 133, 135, 204-205, 209, 225, 310-311, 316-317, 323
MAXIMUM POWER VOLTAGE (VMP).. 39, 44, 106, 202-203, 369-370
MC CONNECTORS... 108
MC4 CONNECTOR.. 108-109, 222-223, 302, 367
MECHANICAL PERMIT.. 351
MEGOHMMETER... 368
METAL-CLAD (MC) CABLE.. 231
MICRO INVERTER... 62-63, 130, 134-136, 148, 178, 180-181, 199, 204-205,
    207-208, 215, 217-218, 226-227, 239-240, 247, 255, 272-273, 354
MINIMUM INPUT VOLTAGE.. 199
MODIFIED SINE WAVE... 127-128, 313
MODULE MISMATCH.. 195
MODULE NAMEPLATE DC RATING... 195
MODULE-LEVEL POWER ELECTRONICS (MLPE)... 130, 206
MONOCRYSTALLINE.. 29, 36, 46, 105, 107
MONOFACIAL.. 34
MPPT VOLTAGE INPUT RANGE... 204
MULTIMETER... 94-95, 106, 140, 322, 325, 342, 367, 379, 380, 382
MUNICIPAL SYSTEMS... 8, 86, 139

N

NABCEP (NORTH AMERICAN BOARD OF CERTIFIED ELECTRICAL PRACTITIONERS).............. 13, 351
NAMEPLATE CAPACITY.. 265
NASA... 12, 15, 110
NATIONAL ELECTRICAL CODE (NEC)... 92-93, 96, 108, 111-114, 119, 132, 174, 199, 224,
    228, 231-232, 236, 239, 248-249, 302, 309, 325, 355, 366
NATIONAL ELECTRICAL MANUFACTURERS ASSOCIATION (NEMA)....................................... 137, 172-173
NATIONAL FIRE PROTECTION ASSOCIATION (NFPA).. 14, 16, 172
NATIONAL RECOGNIZED TESTING LABORATORY... 19
NATIONAL RENEWABLE ENERGY LABS (NREL).. 4, 157, 179, 194

NEC Table 310-16 113
Net Meter 5, 102, 139, 192, 196, 350
Net-Metering 7-9, 58, 193
Neutral 82
NFPA 1 .... 15-16, 172, 269-270
NFPA 70 .... 14-15, 334
NFPA 70E 334
NFPA 855 .... 172-173, 269-270
NMOT 39
Nominal Voltage 44, 82, 88, 94-95, 106, 123, 263-264, 309-310, 312, 316-318, 321, 366
Non-isolated Inverters 213
Normal Operating Cell Temperatures (NOCT) 38
NREL .... see National Renewable Energy Labs (NREL)
N-Type 28

O

Occupancy categories 281, 287
Occupational Safety and Health Administration (OSHA) 13-14, 20, 79, 270, 334
Off-grid PV system 55, 127
Ohm 90, 140
Ohm's Law 77, 90-92
Ohm's Law Triangle 91
One-Line Drawing 203, 351-353
One-Shot Bender 235
Open Circuit 39, 44, 87, 94, 106, 171, 200, 202, 236, 241, 251, 268, 310, 364, 367, 369, 379, 380
Open Circuit Voltage (Voc) 39
Open-Circuit Voltage Test 321-322
Operation and Maintenance (O&M) documentation 359
OSHA .... see Occupational Safety and Health Administration (OSHA)
OSHA 10 Certification 21
OSHA 30 Certification 21
Overcurrent protection 17, 105, 109-110, 120, 137, 209, 214, 221, 224,
    227-228, 271, 307, 311, 313, 317, 325, 365-367, 379
Overcurrent Protection Device (OCPD) 120, 365

P

Parallel 42, 44, 46, 87-89, 105, 109, 160, 199, 205-206, 229-230, 282, 319, 324, 327, 380-381
Passive Cooling 161
Payback period 5, 183
Peak load demand 11, 59, 273
Peak shaving 59-60
Peak sun hours 152
Peak watts 191, 197
PERC 33-34, 79-181
Performance Test 367-368
Permits 178, 281, 349-351, 353, 355
Perovskite 32
Personal Fall Arrest Systems (PFAS) 335-336
Personal Protective Equipment (PPE) 335
Photon 12, 28
Photovoltaic cell 3, 12, 41
Photovoltaic effect 11
Photovoltaic modules 3
Piers 298-300
Pigtail 121, 221
Plan view 351, 360-361
P-N Junction 28
Point of Common Coupling 11
Point of connection 243
Polarity Test 367

Pole Mounted	163, 281
Poles	137, 202, 298, 300
Polycrystalline	29-30, 36, 46, 105, 107
Polyphase Systems	85
Polysilicon	3
Portrait	162-163, 284-285, 293
Power Conversion System (PCS)	124
Power Equation	79-80, 210
Power Factor	132, 210
Power Optimizers	130, 134-136, 199, 206, 226, 230
Power Rating	39, 55, 68, 127, 132, 272, 311, 313
Power Tolerance	198
Powerwall	See Tesla Powerwall
PTC	38, 51
P-Type	28
Public Service Commission	86
Public Utility Commission	86
Pull-out Loads	286-288, 290
Pull-up Force	287, 289
Pulse Width Modulation (PWM)	122, 265-266
Punch List	364
Pure Sine Wave	127-128, 313
PV + Storage	259
PV Hazard Control System (PVHCS)	130
PV Inverter	62, 210
PV Jumpers	222
PV Modules	3-4, 36-37, 88, 228, 365
PV Output Circuit	225-226, 229-232, 239
PV Source Circuit	225-227, 229, 231, 238-239, 302, 323, 368, 381
PV String Circuit	112, 226-227, 231
PV System Disconnect	245
PV USA Test Conditions (PTC)	See PTC
PV Watts	157, 167-169, 193
PV Wire	115, 228
PVC (Rigid Polyvinyl Chloride)	231-232, 234
Pyrometer	370, 375

R

Raceway	113-114, 118, 221, 226-227, 229, 231, 234, 238, 249, 302, 353, 356, 364-365
Rail-less Systems	295
Rails	114, 227, 248-249, 283, 289, 293, 295, 297, 336-337, 339
Ramp Rate	132
Rapid Shutdown	128, 130, 205
Rapid Shutdown Initiator	148-149
Rapid Shutdown System	358
Rate of Charge	265
Rate of Discharge	264-265
Rated Capacity	209, 228, 264, 310
Rated Current	107
Rated Power	39, 46
Rated Voltage	106, 223
Reactive Power	132, 210
Real Power	210
Rectifier	275
Remote Site Assessment	151, 166, 187, 361
Remote Site Assessment	166
Renewable Energy Certificate (REC)	5-7, 9. 136
Renewable Portfolio Standard (RPS)	7
Reserve Capacity	260

Resistance	35, 37, 43, 46-47, 77, 90-95, 111, 140, 173, 195, 232, 238, 247, 250, 264, 298-299, 324, 326, 336, 367-368
Revenue-Grade Meter	136
Rigid Polyvinyl Chloride (PVC)	231
Ripple	307, 324-326
Rise and Run	158
Roof Mounted	160, 281-282, 351, 381
Roof Zone	281, 287-288
Root Mean Square (RMS)	90
Roundtrip Efficiency	261, 272-273, 315
Rule 21	131
Rural Electric Cooperatives	86

S

Safety Data Sheets (SDS)	343
Safety Harness	336
Schedule 40	232
Schedule 80	232, 237
Sealed	260, 268
Secondary Load Center	245
Seismic Loads	285-286
Selenium	11-12
Self Consumption	59
Self Discharge	267
Self-Consumption Grid-Tied	60
Self-contained energy storage systems (ESS)	271
Semiconductor	27, 30, 32
Series	42-43, 46, 87-89, 105, 109, 122, 199, 205-206, 221, 223-224, 264, 316-317, 320, 327, 378, 380
Series Charge Controller	122
Service Entrance	115, 140, 171, 173, 357-358
Service Panel	61-63, 90, 210
Service Point	175-176
Set Point Voltage	124-125
Shading Factor	168-169
Shared Rail System	295
Shear Loads	286
Shockley–Queisser limit	32
Short Circuit Current (Isc)	39
Short-Circuit Current Rating	174
Short-Circuit Interrupting Rating	174
Silicon	7, 12, 27-30, 32-34, 41, 195
Sine Wave	84-85, 89-90, 127-128, 313
Single Phase	90, 140
Single-Axis Tracking System	156
Single-Phase	82, 84-85, 92, 94-95, 134, 137, 209, 239, 243-244, 274, 337
Site Survey	151, 153-159, 161, 163, 165, 167, 169, 171, 173, 175-177, 179, 181, 183, 185, 187, 189, 378
Skin Effect	115
Smart Electrical Panels	138
Smart Inverter	131-132, 147, 207
Smart Switch	61, 63, 176, 273
Smart Switch	61, 63, 181
Snow Loads	285-286
Soft Costs	4
Soft Start Reconnection	132
Soiling	195
Solar Array	67-69, 88, 118, 122, 123, 125-127, 132, 139, 153, 156, 170, 199, 265, 285, 312
Solar Cells	7, 27, 29-35, 37, 39, 41-43, 45, 47, 105
Solar Constant	37
Solar Elevation	155

Term	Pages
SOLAR IRRADIANCE	151-152, 370, 372
SOLAR MODULE	3-4, 27, 29, 31, 33, 35-36, 39-41, 43, 45, 47, 105-106, 155, 191, 196, 248, 369, 379
SOLAR NOON	152-153, 166, 208
SOLAR PATHFINDER	165-166, 168
SOLAR RENEWABLE ENERGY CERTIFICATE (SREC)	7
SOLAR SET-ASIDE	7
SOLAR SHINGLES	30, 46, 164
SOLAR VARIABILITY CORRECTION	224, 226
SolarAPP+	186, 355
SOLID CONDUCTORS	115
SOLID GROUND	117
SOLIDLY GROUNDED	116-117, 119, 213-214, 250-251
SPECIFIC GRAVITY TEST	321-322
SPLIT CELL	31
SPLIT MODULES	31
SPLIT PHASE	83-84, 90
SQUARE WAVE	127, 146, 313
SQUARE WAVE	127-128
STAND-ALONE INVERTER	68, 126-127, 133, 312-314
STAND-ALONE PV SYSTEM	55-57, 106, 123, 126, 307, 309, 316, 323
STANDARD	172, 269
STANDARD TEST CONDITIONS (STC)	37-38, 41, 44, 200, 202, 206, 228, 283, 353, 369-372
STANDOFF MOUNTING	282
STATE OF CHARGE (SoC)	29, 87, 262, 265, 267, 321-325
STRANDED CONDUCTORS	115
STRING INVERTER	207
STRING VOLTAGE TEST	367
STRINGS	44, 109, 199, 201, 203, 221, 223, 229-230, 317, 320, 331, 353, 365-366, 379, 381
STRUCTURAL BOS	171
STUB HEIGHT	235
SULFATION	262, 322-323, 382
SUMMER BIAS	156-157, 161
SUN CHART	159-160, 165-166
SUN TRACKING SYSTEM	156
SunEye	168
SUPPLY SIDE CONNECTION	175, 242, 245-247
SURFACE ROUGHNESS	287
SURGE PROTECTION	110-111
SURGE PROTECTION	110
SURGE RATING	69, 127, 211, 313, 326
SYSTEM AVAILABILITY	60, 195
SYSTEM GROUND	97, 102, 176, 214, 251, 297
SYSTEM GROUNDING	176
SYSTEM LOAD	65, 244, 313

T

Term	Pages
TAKE-UP	235
TANDEM CELLS	32-33
TAPS	95, 246
TARIFF	7
TEMPERATURE COEFFICIENT	41, 46, 200, 202, 369-370
TESLA POWERWALL	64, 271-272
THERMAL RUNAWAY	269
THHN	115, 232
THIN-FILM	29-30, 36, 283
THREE POINTS-OF-CONTACT	339
THREE-PHASE	84-85, 90, 94, 117, 132, 137, 205, 207, 244, 274
TIER 1	179-181, 197
TIER 2	197

Tier 3 197
Tilt and orientation factor 168-169
Time of use pricing 59
Top-down mounting clamp 293, 297
Torque 294, 327, 364
Total solar resource 168-169
Trailing pullstring 236
Transfer switch 177, 272-273
Transformerless inverters 213-214
Transformers 10, 81, 95, 138, 366, 387
Transient 79, 96
Transmission 81, 85-86
Tributary Area 287-289
True south 152-155, 165
True-up period 139

U

UL 1741 .... 131, 133, 209, 251, 269
UL 1973 269
UL 2218 46-47
UL 61215 36
UL 61730 .... 36, 196
UL 9540 .... 269, 271
UL Classification Mark 19
UL Listing Mark 19
UL Standard 3741 130
UL1741-SA 132
UL1741-SB 132
Under load 106-107, 228
Underwriters Labs 19, 46, 120
Ungrounded systems 213
Uninterruptible Power Source (UPS) 61
Unsettled Soil 300
Usable capacity 265, 272-273
USE-2 .... 112, 228
Utility inter-tie device 133, 209
UV (ultra-violet light) resistant 108, 221

V

VA 326
Vehicle-to-grid (V2G) 273
Vehicle-to-Grid (V2G) 273
Vehicle-to-home (V2H) 273
Vibrational loosening 294
Voltage drop 77, 90-92, 94-95, 111, 134, 193, 195, 200, 222, 229, 266, 308-309, 353, 378
Voltage potential 97
Voltage rise 11
Voltage sag 95-96
Voltage swell 95-96
Voltage Violation 10
Volts 28, 39, 66, 78-80, 87-90, 93-94, 106, 122, 137, 201, 221, 229, 317-320
VRLA 268

W

Watt-hour (Wh) 79-80, 385
Waveform 82-85, 89, 94, 105, 127, 176, 230, 239, 324-325
Weighted load 70
Wind exposure categories 287
Wind load 160, 282, 284-286, 296, 303
Wind uplift force 287
Winter bias 157, 161, 315-316

Wire Identification... 240
Wire Strippers.. 141
Wire-Pulling Compound... 237

Z

zenith.. 155

For Schools & Instructors:

A few ways Blue Rock Station can help support your program:

- **Classroom PowerPoint Slides:** complete instructor tool kit that compliments and tracks with this PV study guide for in-person lectures.
- **Remote learning online course:** over 40-hours of narrated slides and videos tailored to this text
- **Weekly Zoom Sessions:** weekly online sessions with industry experts to answer instructor and/or student questions
- **Chapter Exams:** to check on student's retention
- **Hands-On Lab Guide:** that not only reinforce the text but also comply with the hands-on requirements for the ETA PV Level 1 Certification Examination
- **Solar Generator Labs:** construct a stand-alone PV system in your classroom
- **Solar Podcasts:** a variety of 30 minute podcasts focused on solar topics
- **Curriculum Support:** help in designing a complete program or simply guidance in how to integrate this text into existing programs
- **Instructor Support:** from free periodic consultation, online Zoom sessions, to a full 5-day train-the-trainer workshop
- **Lab Design Support:** assistance in identifying components of a complete lab to help train the curriculum (tools, products and consumable materials)
- **Free Desk Copies:** contact us for a free copy if you are considering incorporating this text in your training program
- **Exam Proctoring Services:** we can come and proctor the ETA certification exam at your facility
- **Classroom Field Trips:** for those within range, bring your class to Blue Rock Station's 40-acre sustainability demonstration farm, for a tour, a workshop, or a weekend adventure
- **Guest Lectures:** we can come and speak to your group on any number of topics
- **Internships/Fellowships:** each summer Blue Rock Station accepts a limited number of resident interns

Other Texts from Blue Rock Station:

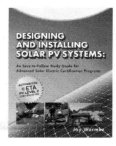 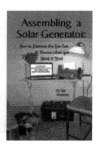 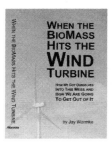

For more information about the services, products and events of Blue Rock Station, visit our website at www.solarPVtraining.com

Or give us a call at 740-674-4300, or email us at annie@bluerockstation.com